I0038591

Diseases of
OILSEED CROPS
and their Management

The Editors

Dr Hemant Kumar Singh is Associate Professor, Plant Pathology in Narendra Deva University of Agriculture and Technology, Kumarganj, Faizabad, (U.P.) India, obtained Master degree in Mycology and Plant Pathology standing first class from Banaras Hindu University, Varanasi. He received Ph.D. degree from N. D. University of Agriculture and Technology, Kumarganj, Faizabad. Having experience of more than 14 years as a teacher and scientist, he worked on different diseases of crops especially professional experience in the field of Rapeseed-Mustard Pathology and taught various courses both at undergraduate and post-graduate level. He has published more than 95 research papers, review articles in National and International Journals, 5 books, 14 book chapters and many popular articles in reputed magazines. He has contributed/assisted significantly in developing three varieties of oilseed. Dr. Singh received several awards *viz.* SURE Distinguished Scientist Award 2014 by the Society for Upliftment of Rural Economy, Varanasi (U.P.); Fellow of Indian Phytopathological Society (FPSI) by Indian Phytopathological Society, New Delhi and so many other accolades.

Dr Raj Bahadur Singh, Ex. Professor, Plant Pathology at N.D. University of Agriculture and Technology, Faizabad (U.P.) obtained M.Sc. and Ph.D. degrees from DDU University, Gorakhpur in Botany specialized with Plant Pathology. He was awarded post Doctorate Fellowship (PDF) by CSIR, New Delhi for the period of two years. He has experience of more than 35 years as teacher and scientist and worked on different diseases of crops and taught various courses at undergraduate and post-graduate levels. He has published more than 100 research paper in reputed journals. Besides, he has published many popular articles, book chapters and bulletins. He was awarded as Fellow of Pathological Society of India (FPSI), Fellow of International Society for Environmental Protection (FISEP) and Fellow of Indian Society of Oilseed Research (FISOR) from different internationally recognised scientific organisations and societies. He was elected as zonal president of Indian Phytopathological Society of Mid-eastern zone in 2002 and organized several symposium, seminar and workshop during his stay at University.

Diseases of
OILSEED CROPS
and their Management

– Editors –

Dr. H.K. Silngh

Professor R.B. Singh

Department of Plant Pathology
N.D. University of Agriculture & Technology
Kumarganj, Faizabad, U.P.

2019

Daya Publishing House®

A Division of

Astral International Pvt. Ltd.

New Delhi – 110 002

© 2019 EDITORS

ISBN: 9789388173612 (Int. Edn)

Publisher's Note:

Every possible effort has been made to ensure that the information contained in this book is accurate at the time of going to press, and the publisher and author cannot accept responsibility for any errors or omissions, however caused. No responsibility for loss or damage occasioned to any person acting, or refraining from action, as a result of the material in this publication can be accepted by the editor, the publisher or the author. The Publisher is not associated with any product or vendor mentioned in the book. The contents of this work are intended to further general scientific research, understanding and discussion only. Readers should consult with a specialist where appropriate.

Every effort has been made to trace the owners of copyright material used in this book, if any. The author and the publisher will be grateful for any omission brought to their notice for acknowledgement in the future editions of the book.

All Rights reserved under International Copyright Conventions. No part of this publication may be reproduced, stored in a retrieval system, or transmitted in any form or by any means, electronic, mechanical, photocopying, recording or otherwise without the prior written consent of the publisher and the copyright owner.

Published by   : **Daya Publishing House®**
                        A Division of
                        Astral International Pvt. Ltd.
                        – ISO 9001:2015 Certified Company –
                        4736/23, Ansari Road, Darya Ganj
                        New Delhi-110 002
                        Ph. 011-43549197, 23278134
                        E-mail: info@astralint.com
                        Website: www.astralint.com

Digitally Printed at : **Replika Press Pvt. Ltd.**

प्रोफेसर जे. एस. संधू
कुलपति
पूर्व कृषि आयुक्त, भारत सरकार
Professor J. S. Sandhu
Vice Chancellor
Ex Commissioner (Agri) GOI

नरेन्द्र देव कृषि एवं प्रौद्योगिक विश्वविद्यालय
कुमारगंज, फैजाबाद – 224 229 (उ.प्र.), भारत
Narendra Deva University of Agriculture & Technology
Kumarganj, Faizabad - 224 229 (U.P.) India

Foreword

Oilseed crops generally are one of the most important crops in the world. Their role in human diet and industrial application cannot be under estimated. India is the major oilseed producer in the world. The major oilseed crops include soyabean, coconut, oil palm, sesame, rapeseed, sunflower, safflower, olive seed, etc. The by products (hull, meal and oil) of oilseed crops had been integrated into human and animal diets due to its nutrient compositions. Majority of the oilseed meal consists of proteins and high contents of essential amino acid which are beneficial to human health and wellbeing. Likewise, the fat yields of oilseed crops are generally high, though varied from crop to crop, and methods of extraction; high polyunsaturated fatty acids contents also prevent against coronary heart disease. Apart from the food value of oilseeds, several industrial products such as biodiesel, fertilizer, medicine, cosmetics, animal feeds, fibers, paint, button etc. have also been reported. Oilseed crops are promising crop with high potentials to improve human diets, prevent malnutrition and food insecurity and to provide employment through income generation in the society.

The threat of climate change, coupled with declining land, water and agricultural labor force, besides rising input costs, have necessitated the development of efficient cost effective and sustainable disease management approaches to minimize crop losses and to produce quality and pesticide residue-free produce. Accurate identification of the diseases affecting oilseed crops is a key factor for biosecurity preparedness and adopting eco-friendly and effective strategies.

Diseases caused by several agents namely, fungi, bacteria, nematodes and virioids are primary causes which restrict quality and quantity of the produce. The present book entitled "Diseases of Oilseed Crops and their Management" includes 12 chapters contributed by eminent researchers in their respective fields. This book focuses primarily on diseases of groundnut, linseed, castor, safflower, sesame, soybean, sunflower, rapeseed mustard and niger. The symptoms, causal organism, etiology, disease cycle, epidemiology and management strategies of diseases is presented in this book. Besides these, the breeding for disease resistance in Indian mustard, biotechnology for genetic improvement of oilseed crops and the latest diagnostics tools and management strategies of the diseases of oilseed crops are also included.

It is hoped that the book is cater the needs of the research workers, teachers, students, growers not only in the discipline of plant pathology but also in the area of agriculture. I congratulate the editors of this book Drs. H K Singh and R B Singh for their sincere efforts in bringing out this useful publication.

(J S Sandhu)

Preface

Oilseed crops have pivotal role in agricultural economy. India is one of the largest producers of oilseeds in the world and occupies an important position in the Indian agricultural economy. There are nine important oilseed crops grown in India out of which seven are of edible oils i.e. soybean, groundnut, safflower and niger and two are of non-edible oils i.e. caster and linseed. In terms of acreage, production and economic value, oilseeds are second only to food grains. India rank first in the production of ground nut, third in rapeseed-mustard and fifth in soybean. Indian vegetable oil economy in the fourth largest in the world. Oilcrops role in human diet and industrial application cannot be underestimated. The by-product chull, meal and oil of oilseed crops had been integrated into human and animal diets due to its nutrient compositions. Majority of the oilseed meal consist of proteins and high contents of essential amino acid which are beneficial to human health and wellbeing. Likewise, the fat yields of oilseed crops are generally high through varied from crop to crop, and methods of extraction, high poly-unsaturated fatty acids contents also prevent against coronary heart disease. Apart from the food value of oilseeds, several industrial products such as biodiesel, fertilizer, medicine, cosmetics, animal feeds, fibers, paint etc. have been reported. Oilseed crops are promising crops with high potentials to improve human diets, prevent malnutrition and food insecurity and to provide employment through income generation in the society.

To meet the edible oil demands of ever increasing population agricultural production is being augmented through the use of new crop varieties and changed economic practices. These practices have enormously increased the incidence of several pests and diseases. Plant diseases cause serious threats to the successful cultivation of oilseed crops resulting in huge losses in their yields. The destructive potential of oilseed crop diseases in modern day agriculture has increases due to the use of cultivars having narrow genetic base over large areas. Correct disease diagnosis is the prime requirement for recommending preventive measures for effective disease management. All the available must be used in an integrated manner and holistic approach needs to be developed for the management of major diseases of oilseed crops.

The present book entitled "Diseases of Oilseed Crops and their Management" includes 12 chapters contributed by eminent researchers in their respective fields. This book focuses primarily on diseases of groundnut, linseed, castor, safflower, sesame, soybean, sunflower, rapeseed mustard and niger. The symptoms, casual organism, etiology, disease cycle, epidemiology and management strategies of diseases is presented in this book. Besides these, the breeding for disease resistance in Indian mustard, biotechnology for genetic improvement of oilseed crops and the latest diagnostics tools and management strategies of the diseases of oilseed crops are included.

We are grateful to all the contributors for their cooperation, support and timely submission of their manuscript for bringing out this publication. We are extremely grateful to Prof. J.S. Sandhu, Vice Chancellor, N.D. University of Agriculture & Technology, Kumarganj, Faizabad for agreeing to write the Foreword of the book. Inspite of the best effort, it is possible that some errors may have occurred into the compilation and editing of the book. Further queries, suggestions and criticism for the improvement of the book are always welcome and shall be thankfully acknowledged. It is a pleasure for us to extend our sincere thanks to Dr. B.B. Singh, Editor-in-chief, Astral International (P) Ltd., New Delhi and his team members for his keen interest in preparation and publishing of this book so efficiently and promptly.

Dr. H. K. Singh

Prof. R.B. Singh

Contents

List of Contributors

Name	Scientist	Address
Dr. R. C. Shakywar	Assistant Professor	Department of Plant Protection, College of Horticulture and Forestry, CAU, Pasighat - 791 102
		E-mail: rcshakywar@gmail.com
Dr. M. Pathak	Senior Scientist and Head	KVK East Siang, College of Horticulture and Forestry, CAU, Pasighat - 791 102
		E-mail: maheshpathak@rediffmail.com
Dr. Debashish Sen	Associate Professor	Department of NRM, College of Horticulture and Forestry, CAU, Pasighat - 791 102
		E-mail: dr.d.sen@gmail.com
Dr. R. B. Singh	Ex-Professor/ Oilseed Pathologist	Department of Plant Pathology, Narendra Deva University of Agriculture and Technology, Kumarganj, Faizabad-224 229 (U.P.) India
		E-mail: rbspath.2010@gmail.com
Dr. H. K. Singh	Assistant Professor/ Oilseed Pathologist	Department of Plant Pathology, Narendra Deva University of Agriculture and Technology, Kumarganj, Faizabad-224 229 (U.P.) India
		E-mail: hksndu@gmail.com
Dr. Mahesh Singh	Assistant Professor	School of Agriculture, ITM University Gwalior-475001 (M.P.), India
		E-mail: msmaurya96@gmail.com

Name	*Scientist*	*Address*
Dr. Shiwangi	Research Scholar	Department of Plant Pathology, Narendra Deva University of Agriculture and Technology, Kumarganj, Faizabad-224 229 (U.P.) India
Dr. V.S. Verma	Professor (Retd.)	Division of Plant Pathology, SKUAST-J, Chatha Jammu-180009
		E-mail: drvsverma@gmail.com
Dr. Vishal Gupta	Asstt. Professor	Division of Plant Pathology, Faculty of Agriculture, Sher-e-Kashmir University of Agricultural Sciences and Technology of Jammu, Chatha, Jammu-180009 (J&K)
		E-mail: vishal94gupta@rediffmail.com
Dr. Rishu Sharma	Research Associate	Division of Plant Pathology, Faculty of Agriculture, Sher-e-Kashmir University of Agricultural Sciences and Technology of Jammu, Chatha, Jammu-180009 (J&K)
		E-mail: rrishu.sharma@rediffmail.com
Ms Kavaljeet Kaur	Research Associate	Division of Plant Pathology, Faculty of Agriculture, Sher-e-Kashmir University of Agricultural Sciences and Technology of Jammu, Chatha, Jammu-180009 (J&K)
		E-mail: kaurr.kaval@gmail.com
Dr. P. Kishore Varma	Senior Scientist	Department of Plant Pathology, Regional Agricultural Research Station, Anakapalle 531001, Visakhapatnam District, Andhra Pradesh
		E-mail: penumatsakishore@gmail.com
Dr. Mohan V. Totawar	Professor and Head	Department of Plant Pathology, College of Agriculture, PDKV, Akola - 444 104, Maharashtra
		E-mail: mohantotawar@gmail.com
Dr. Dinesh Rai	Assistant Professor	Department of Plant Pathology, Dr. Rajendra Prasad Central Agricultural University, Pusa-848125 (Samastipur) Bihar
		E-mail: drai1975@gmail.com
Dr. Sunil Kumar	Jr. Scientist	School of Agricultural Sciences and Rural Development, Nagaland University, Medziphema - 797 106, Nagaland.
		E-mail: drsunilk81@gmail.com

Name	Scientist	Address
Dr. Shailesh Godika	Professor	Department of Plant Pathology, SKNCOA, Jobner (Jaipur)
		E-mail: shaileshgodika@gmail.com
Dr. R. P. Ghasolia	Asstt. Professor	Department of Plant Pathology, SKNCOA, Jobner (Jaipur)
		E-mail: rpghasolia.ppath@sknau.ac.in
Dr. S. K. Goyal	Asstt. Professor	Division of Plant Pathology, RARI (SKNAU, Jobner), Durgapura (Jaipur)
		E-mail: shashikant69@gmail.com
Dr. Jitendra Sharma	ARO	Centre of Excellence, Mandarin, Jhalawar
		E-mail: jitendrasharmarca@gmail.com
Dr. Ramesh Singh	Associate Professor	Head, Department of Plant Pathology, TDPG College, Jaunpur-222002 (U.P.) India
		E-mail: ramesh.ramesh.singh37@gmail.com
Dr. V. V. Singh	Principal Scientist (Plant Breeding)	ICAR-Directorate of Rapeseed-Mustard Research, Sewar, Bharatpur-321303, Rajasthan.
		E-mail: singhvijayveer71@gmail.com
Dr. H. S. Meena,	Senior Scientist (Plant breeding)	ICAR-Directorate of Rapeseed-Mustard Research, Sewar, Bharatpur-321303, Rajasthan.
		email: singh_hari2006@yahoo.co.in
Dr. P. D. Meena	Principal Scientist	ICAR-Directorate of Rapeseed-Mustard Research, Sewar, Bharatpur-321303, Rajasthan.
		E-mail: pdmeena@gmail.com
Dr. B. L. Meena	Scientist Sr Scale (Plant breeding)	ICAR-Directorate of Rapeseed-Mustard Research, Sewar, Bharatpur-321303, Rajasthan.
		email: blmeena.icar@gmail.com
Dr. Dhiraj Singh	Ex-Director (Plant Breeder)	ICAR-Directorate of Rapeseed-Mustard Research, Sewar, Bharatpur-321303 Rajasthan India
		email: dhirajmustard@gmail.com

Name	Scientist	Address
Dr. P.K. Rai	Director (Acting), Plant Pathology	ICAR-Directorate of Rapeseed-Mustard Research, Sewar, Bharatpur-321303 Rajasthan India email: rai_68@rediffmail.com
Dr. D.K. Dwivedi	Professor (Plant Breeding)	Department of Biotechnology, N.D. University of Agriculture and Technology, Kumarganj, Faizabad - 224229, (UP), India E-mail: ddwivedi2000@gmail.com
Dr. Archana Devi	Research Scholar	Department of Genetics and Plant Breeding, N.D. University of Agriculture and Technology, Kumarganj, Faizabad - 224229, (UP), India
Dr. Preeti Kumari	Research Scholar	Department of Genetics and Plant Breeding, N.D. University of Agriculture and Technology, Kumarganj, Faizabad - 224229, (UP), India
Dr. Ibandalin Mawlong	Scientist	ICAR-Directorate of Rapeseed-Mustard Research, Sewar, Bharatpur-321303 Rajasthan, India E-mail: iban02@gmail.com
Mr. Ashish Sheera	Research Scholar	Sam Higginbottom Institute of Agriculture, Technology and Sciences, Allahabad-211007, Uttar Pradesh, India E-mail: sheeraashish@gmail.com
Mr. J. K. Yadav	Research Scholar	Department of Plant Pathology, N.D. University of Agriculture and Technology, Kumarganj, Faizabad - 224229, (UP), India E-mail: jaykumaryadav4556@gmail.com
Dr. K. N. Maurya	Assistant Professor	Department of Genetics and Plant Breeding, N.D. University of Agriculture and Technology, Kumarganj, Faizabad - 224229, (UP), India
Mr. M. K. Maurya	Research Scholar	Department of Plant Pathology, N.D. University of Agriculture and Technology, Kumarganj, Faizabad - 224229, (UP), India E-mail: manishmaurya8712@gmail.com

1 Diseases of Groundnut and their Management

R.C. Shakywar, M. Pathak and Debashish Sen

Introduction

Groundnut or peanut is commonly called the poor man's nut and belongs to family Leguminoseae. It is known to western world, ranked 6th among oil seed crop and 13th among the food crops of the world. In addition to providing high quality edible oil (48-50 per cent), easily digestible protein (26-28 per cent), nearly half of the 13 essential vitamins and 7 of the 20 essential minerals necessary for normal human growth, it produces high quality fodder for livestock. Consequently plays significant role in the livelihoods of marginal-farmers through income and nutritional security. This plant is native to South America and has never been found uncultivated. The botanical name of groundnut is *Arachis hypogaea* Linn. It is derived from two Greek words, *Arachis* meaning a legume and *hypogaea* meaning below ground, referring to the formation of pods in the soil (Nautiyal *et al.*, 2001). It is a self pollinated, allotetraploid (2n=4x=40) with a genome size of 2891 Mbp and only nut that grows below the earth. The groundnut plant is a variable annual herb, which grows upto 50 cm in height. The flowers of the plant develop a stalk which enters into the soil forms a pod containing generally two seeds. They become mature in about two months, when the leaves of the plant turn yellow. The plant is then removed from the earth and allowed to dry. After three to six weeks they are separated from the plant. The groundnut is particularly valued for its protein contents, which is of high biological value and it contain more protein than meat-about two and a half times more than eggs and far more than any other vegetable food except soybean and yeast. Groundnut is grown in nearly 100 countries. India is one of the major exporting countries of groundnuts after China (Rama Rao *et al.*, 2000).

Groundnut is grown on nearly 23.95 million ha worldwide with the total production of 36.45 million tonn and an average yield of 1520 kg/ha in 2009

(FAOSTAT, 2011). China, India, Nigeria, USA and Myanmar are the major groundnut growing countries. Developing countries in Asia, Africa and South America account for over 97 per cent of world groundnut area and 95 per cent of total production. Production is concentrated in Asia (50 per cent of global area and 64 per cent of global production) and Africa (46 per cent of global area and 28 per cent of global production), where the crop is grown mostly by small holder farmers under rainfed conditions with limited inputs (Pal *et al.*, 2000).

Groundnut crop is prone attack by numerous diseases to a much larger extent than any other crops. One of the most important factors contributing to low yield is disease attack. More than 55 pathogens including viruses have been reported to affect groundnut. Some diseases are widely distributed and cause economic crop losses while others are restricted in distribution and are not considered to be economically important at present. The diseases which are of minor magnitude today may become major in a while. Among fungal foliar diseases, only a few are economically important in India such as leaf spots (early and late) and rust which are widely distributed can cause losses in susceptible genotypes to the extent of 70-80 per cent when both of them occur together. After the event, *Alternaria alternata* leaf spot is becoming increasingly important on *Rabi*/summer crop and also on *Kharif* groundnut crop. Other fungal foliar diseases like anthracnose, leaf scorch, *Phomosis* leaf spot, *Phyllosticta* leaf spot, *Pestalotiopsis* leaf spot, *Phoma* leaf diseases, *Drechslera* leaf blight and *Cylindrocladium* leaf spot are not economically important at present time. Similarly, seed and soil-borne diseases collar rot; stem rot and dry root rot have been realized as major limitations in crop production. These diseases cause severe seedling mortality resulting in irregular crop stands in sandy loam soils and reduce the yields by 30-40 per cent. Several economically important virus diseases like bud necrosis, peanut (groundnut) mottle, resettle and peanut clump disease have also started assuming importance in the recent years in India. Bud necrosis and peanut mottle virus (PMV) is a serious disease of groundnut, wide spread with wide host range and can cause severe yield losses ranging up to 60 per cent. Reports on diseases caused by nematodes in groundnut are very few in India. The root knot nematodes have been reported to cause damage in various parts of the country. Understanding the diseases and their behaviour is basic to successful and economic cultivation of groundnut or peanut. In this chapter, attempt has been made to present the current knowledge about the groundnut diseases including bacterial diseases, fungal disease, viral diseases, phytoplasmal diseases, nematodes diseases, phanerogamic plants, non parasitic diseases and physiological disorders and their management practices.

Bacterial Disease

Bacterial Wilt

Symptoms

It can cause serious losses, if a crop is infected early. Infected plants show water stress symptoms and may wilt suddenly without yellowing of the foliage, particularly, when temperatures are high.

Causal Organism

Ralstonia solanacearum

Etiology

The bacterium is a gram-negative, motile and rod shaped. The organism grows aerobically and does not form endospores. Cells are 0.5-0.7 X 1.5-2.0 µm and non-encapsulated. It is catalase positive, oxidase positive and reduces nitrates. It does not hydrolyze starch and readily degrade gelatin (Gitaitis and Hansons, 1984).

Disease Cycle and Epidemiology

The bacterium usually infects groundnut plants through the roots (through wounds or at the points of emergence of lateral roots). Soil

Figure 1.1: Bacterial Wilt Symptoms on Groundnut (Photo courtesy jnkvv.ac.in).

borne organisms can cause injury to plant roots and favour penetration of the bacterium. Plant infection can also occur through root injuries caused by cultural practices or insect damage. In some cases, plant to plant spread can occur when bacteria move from roots of infected plants to roots of nearby healthy plants, often via irrigation practices. Spread of bacteria by aerial means and subsequent plant contamination through foliage is not known to occur, thus making *R. solanacearum* a non airborne pathogen. High temperatures (29-35⁰C) play a major role in pathogen growth and disease development. Several other factors that may affect pathogen survival in soil and water may also favour disease development, including soil type and structure, soil moisture content, organic matter in soil, water pH and salt content and the presence of antagonist microorganisms **(Saddler, 2005)**. The bacterium also has an "exterior" phase (epiphyte) in which it can reside on the outside of the plant. It is of minor importance in epidemiology of the pathogen since bacteria do not survive epiphytically for long periods of time when exposed to hot conditions or when relative humidity is below 95 per cent. Under unfavourable conditions, groundnut plants infected with *R. solanacearum* may not show any disease symptoms. In this case, latently infected plants can play a major role in spread of the bacterium. *R. solanacearum* can survive for days to years in infected plant material in soils, infested surface irrigation water and infected weeds. From these sources of inoculums, bacteria can spread from infested to healthy fields by soil transfer on machinery and surface runoff water after irrigation or rainfall. *R. solanacearum* can also be propagated in infested ponds or rivers and disseminated to non-infested fields through waterways. Infected semi-aquatic weeds may also play a major role in disseminating the pathogen by releasing bacteria from roots into irrigation waters (Denny, 2006).

Integrated Disease Management

Before Plantation

1. Consider an effective weed management in and around groundnut fields and aquatic weed control around irrigation ponds.

2. Apply 3-4 years rotation and cover crops for infested fields to reduce *R. solanacearum*, weeds and nematodes.

3. Do not irrigate rotation and cover crops with *R. solanacearum* contaminated pond or surface water, avoid reinfestation.

4. Use well drained and leveled fields and do not use low-lying areas of the field.

5. Raise soil pH to 7.5-7.6 and increase available calcium (liming).

During Production

1. Exclude the pathogen by applying strict sanitation practices like as pathogen free irrigation water, transplants, stakes, machinery, *etc.* (**Coutinho, 2005**).

2. Chlorinate your irrigation water continuously if you are using surface water or *R. solanacearum* infested pond water.

3. Continue an effective weed control in and around groundnut fields and irrigation ponds.

4. Irrigate based on water need, avoid over irrigation.

5. Apply plant resistance inducer, such as Actigard if you are using moderately resistant cultivars. Actigard enhances resistance against this disease if it is used in combination with moderately resistant cultivars (**Boshou, 2005**).

6. Application of stable bleaching powder @ 12 kg/ha mixed with fertilizer in furrows while planting reduces wilt incidence by 80 per cent.

After Harvesting

1. Plough under crop residue immediately.

2. Start with suitable rotation and cover crops (*i.e.*, rye for winter, sudan-sorghum for summer) to avoid weeds that support *R. solanacearum* populations.

Fungal Diseases

Anthracnose

Symptoms

Small water soaked yellowish spots appears on the lower leaves which later turn into circular brown lesion with yellow margin 1-3 mm in diameter. In some cases lesions enlarge rapidly become irregular and cover the entire leaflets and extents to stipules and stems.

Causal Organism

Colletotrichum capsici and *Colletotrichum dematium*

Figure 1.2: Anthracnose Symptoms and Conidia of Fungus (Photo courtesy icar. tripura.center).

Systemic Position (Ainsworth *et al.*, 1973)

Kingdom: Mycota

Division: Eumycota

Subdivision: Deuteromycotina

Class: Coelomycetes

Order: Melanconiales

Family: Melanconiaceae

Genus: *Colletotrichum*

Species: *capsici, dematium*

Etiology

The fungus produce acervuli developed on affected parts which are dark, setose, erumpent and contain falcate unicellular spores.

Mode of Spread and Survival

The pathogen is seed, soil, air borne and wind also carried water droplets.

Favorable Conditions

The fungus over winters in old lesions on leaves. Warm spring conditions promote the development and release of spores from the acervuli in old lesions.

Integrated Disease Management

1. Removal and destruction of crop debris.
2. Deep summer ploughing.
3. Use of healthy/certified seeds.

4. Seed treatment with copper oxychloride or Mancozeb @ 3g/kg or Bavistin @ 2g/kg of seed is effective against anthracnose disease.

5. Foliar application of contact fungicides @ 3g/liter of water is always effective.

Crown Rot

Symptoms

Seed may be killed in pre emergence rotting. Post emergence infection cause death and rapid decay of seedlings. Young plants collapse and die soon after emergence due to rotting of elongating hypocotyls. Collar region become dark brown and shredded. In mature plants large lesion develop on stem just below the soil surface and spread upward along the branching causing wilting and death. The fungus sporulates on the surface of mature pod resulting in the path and black sooty spores.

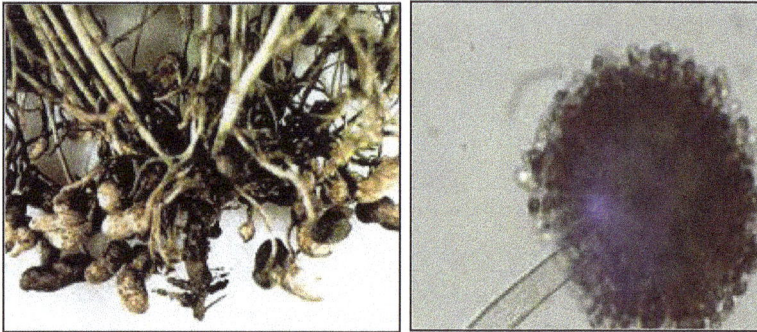

Figure 1.3: Crown Rot of Groundnut Plants (Photo courtesy plantwise.org).

Causal Organism

Aspergillus niger

Systemic Position (Ainsworth *et al.*, 1973)

Kingdom: Mycota

Division: Eumycota

Subdivision: Ascomycotina

Class: Eurotiomycetes

Order: Eurotiales

Family: Trichocomaceae

Genus: *Aspergillus*

Species: *niger*

Etiology

The pathogen is a saprotroph widespread in nature, typically found in soil and decaying organic matter such as compost heaps, where it plays an essential role in carbon and nitrogen recycling. Colonies of the fungus produce from conidiophores thousands of minute grey-green conidia (2-3 μm) that readily become airborne.

Mode of Spread and Survival

The fungus is wind borne in nature.

Favourable Conditions

Low soil moisture and high air temp between30-35⁰C favour the disease development.

Management

1. Crop rotation and field sanitation.
2. Remove and destroy previous season infested crop debris in the field.
3. Seed treatment with *Trichoderma viride* or *Trichoderma harzianum* @ 4g/kg of seed and soil application of *Trichoderma viride* or *Trichoderma harzianum* @ 25-75 kg/ha, preferably in the conjunction with organic amendments such as castor or neem or mahua cakes @ 500-700 kg/ha.
4. Seed treatment with thiram @ 3g/kg of seed is also recommended.

Damping-off

Symptoms

Damping-off diseases affect the initial establishment of a crop. Their main features include poor emergence and death of seedlings leading to poor stands in seedbeds and fields. Seeds may rot before germination. Affected seedlings that have emerged from the soil show water-soaking, browning and shriveling of the stem at the soil level. They eventually fall over and die. Damping-off diseases are favoured by excessive wetness of the soil and low soil temperatures.

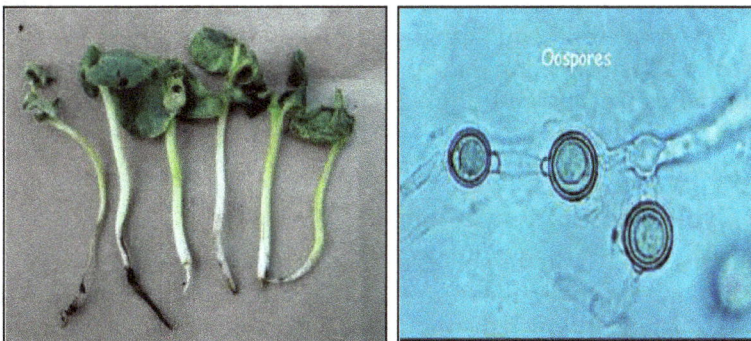

Figure 1.4: Damping Off of Groundnut (Photo courtesy infonet.biovision.org).

Causal Organism

Pythium spp., *Rhizoctonia solani*

Systemic Position of *Pythium* spp. (Ainsworth *et al.*, 1973)

Kingdom: Mycota

Division: Eumycota

Subdivision: Mastigomycotina

Class: oomycetes

Order: Perenosporales

Family: Pythiaceae

Genus: *Pythium*

Systemic Position of *Rhizoctonia solani* (Ainsworth *et al.*, 1973)

Kingdom: Mycota

Division: Eumycota

Subdivision: Duteromycotina

Class: Hyphomycetes

Order: Agonomycetales

Family: Agonomycetaceae

Genus: *Rhizoctonia*

Species: *solani*

Etiology

The *Pythium* produced sporangia on (both) above and below ground plant parts. Zoospores are motile which allow the fungus to spread in saturated soils or standing water. Each zoospore can cause a new infection. Oospores are thick-walled spores which allow the fungus to survive on equipment or in soils for long periods of time.

The *Rhizoctonia solani* reproduces asexually and exists primarily as vegetative mycelium and/or sclerotia. Unlike many basidiomycete fungi the basidiospores are not enclosed in a fleshy fruiting body or mushroom. The sexual fruiting structures and basidiospores (*i.e.* teleomorph) was first observed in detail by Prillieux and Delacroiz in 1891. The sexual stage of *R. solani* has undergone several name changes since 1891, but is now known as *Thanatephorus cucumeris*.

Disease Cycle and Epidemiology

The fungus survives in the soil as oospores, hyphae or sporangia. The fungus can cause disease in cool temperatures 12-17°C but ideal conditions are between 30-35°C, a characteristic which distinguishes it from other *Pythium* species.

Management

1. Use certified disease-free seeds
2. Avoid over-irrigation and excessive fertilization with nitrogen fertilizers.
3. Avoid fields previously planted with cotton or other related crops.
4. Seed treatment with bio-control agent's *viz., Trichoderma viride or Trichoderma harzianum* @ 4g/kg of seed reduce the inoculums of the pathogen.
5. Seed treatment with systemic fungicides like as Ridomil MZ 72 WP or Metalaxyl @ 2g/kg of seed reduce the inoculums of the damping of pathogen.

Leaf Spot

Symptoms

Lesions produced by *Alternaria arachidis* are brown in color and irregular in shape surrounded by yellow halos. Symptoms produced by *A. tenuissima* are characterized by blighting of apical portion of leaflets which turn light to dark brown in color. In the later stage of infection, blighted leaves curl inwards and become brittles. The lesions produced by *A. alternata* are small, chlorotic; water soaked that spread over the surface of the leaves. The lesions become necrotic, brown are round to irregular shape. Veins and veinlets nearby to the lesion become necrotic. Lesions increase in the area and their central portion become pale, rapidly dry out and disintegrate. Affected leaves show chlorotic and in severe attack become prematurely senescent. Lesions are coalease give the leaf a ragged and blighted appearance.

Figure 1.5: Leaf Spot on Groundnut and Conidia of *Alternaria* spp. (Photo courtesy jnkvv.nic.in)

Causal Organism

Alternaria arachidis, A. tenuissima and *A. alternata.*

Systemic Position (Ainsworth *et al.*, 1973)

Kingdom: Mycota

   Division: Eumycota

      Subdivision: Deuteromycotina

         Class: Hyphomycetes

            Order: Hyphomycetales/Moniliales

            Family: Dematiaceae

               Genus: *Alternaria*

                  Species: *arachidis, tenuissima* and *alternata*

Etiology

The fungus produce conidia (muriform) and conidiophores are singly or in groups, straight or flexuous, cylindrical, septate, pale to olivaceous brown; conidia were straight, obclavate, pale olivaceous brown, smooth, with up to 15 transverse and rarely 1 or 2 longitudinal or oblique septa and measured 50-115 × 5-10 µm (Subrahmanyam *et al.*, 1980).

Disease Cycle and Epidemiology

It can survive in own plant debris. Spores are dispersed by air and splashing water. For the disease development hot moist weather, temperature 26° C and dew or rain for 9 hours is essential for infection. The disease is also favoured by rainy weather and/or overhead irrigation.

Management

1. Avoid overhead irrigation and continuous cropping in field with same field seed.
2. Always use crop rotation with non host crops (2-3 years).
3. Use disease-free propagating material.
4. Remove infected plant parts from the main field and bury or dispose of offsite.
5. Provide good air circulation and adjust environmental controls to avoid condensation forming on plants.
6. Hot water treatment of seeds at 50°C for 30 minutes.
7. Foliar application of Mancozeb @ 0.3 per cent copper oxychloride @ 0.3 per cent or Cardendazim @ 0.2g per cent, at first appearance of leaf spot symptoms and at fortnightly intervals, thereafter.

Rust

The rust of groundnut is distributed in Central and South America, China, India, West Indies and USSR. The disease is found in Andhra Pradesh, Punjab,

Tamil Nadu, West Bengal and Uttar Pradesh. The rust of groundnut is an economic disease and causes 14-32 per cent yield loss (Subrahmanyam *et al.*, 1984).

Symptoms

The disease is found on six weeks or more old plants. The small orange colored uredial pustules appear on lower surface of the leaves. At later stages, these pustules may appear on upper leaf surface and other aerial parts of the plant except flower. The ruptured epidermis exposes a powdery mass of uredospores. The infected leaves are showed small, brown and necrotic lesions on the upper leaf surface. The severely infected leaves wither and drop prematurely. The seeds formed on infected plants are small and shriveled.

Figure 1.6: Rust Disease of Groundnut and Urediospores of *Puccinia arachidis* (Photo courtesy icar.tripura.center).

Causal Organism

The rust of groundnut is caused by *Puccinia arachidis*.

Systemic Position (Ainsworth *et al.*, 1973)

Kingdom: Mycota

Division: Eumycota

Subdivision: Basidiomycotina

Class: Teliomycetes

Order: Uredinales

Family: Pucciniaceae

Genus: *Puccinia*

Species: *arachidis*

Etiology

The uredial and telial stages of the pathogen are known till now. The uredospores are one celled, sub globose, ovoid to round, light brown, thin walled, 2-3 germ pores and measuring 24 x 21 µm with short and hyaline pedicels.

Disease Cycle

The uredospores are short lived in infected plant debris. The continuous cultivation of the crop in India without any significant break may perpetuate the disease. The uredospores found in southern India may act as potential source of disease in northern India blown by wind during monsoon season.

Management

1. Crop rotation and field sanitation.
2. Strict plant quarantine regulations should be enforced to avoid the spread of the rust on pod or seeds of disease free area.
3. Deep burying of crop residues in the soil and removal of volunteer groundnut plants are importance measure to reduce the source of primary inoculums.
4. Foliar application of aqueous neem leaf extracts @ 2-5 per cent or neem seed kernel extracts @ 5 per cent at two weeks interval, three times starting from fourth week after planting is good.
5. The application of a mixture of Carbendazim (0.05 per cent) and Mancozeb (0.25 per cent) at 2-3 weeks interval on 4-5 weeks old plants effectively controlled the disease.
6. Three to four spray of Tridemorph @ 0.2 per cent at 15-20 days interval gives good management of rust.
7. Spray application of Chlorothalonil @ 0.2 per cent or Wettable sulphur @ 0.2 per cent g on 35 and 50 days after sowing.

Stem Rot

Symptoms

Development of white cottony out growth over affected plant tissues particularly on stem were seen. Base of the plants turn yellow and then wilt down. Sheath of white mycelium develop around the affected area of the stem near the soil, due to this stem becomes shredded. White sclerotia of mustard seed size are produced on the infected tissues which later turn to brown color. Seeds in the infected pods show a characteristics bluish grey discoloration (Holt and Reddy, 1984).

Causal Organism

Sclerotium rolfsii

Systemic Position (Ainsworth *et al.*, 1973)

Kingdom: Mycota

Division: Eumycota

Subdivision: Duteromycotina

Class: Hyphomycetes

Order: Agonomycetales

Family: Agonomycetaceae

Genus: *Sclerotium*

Species: *rolfsii*

Figure 1.7: Stem Rot of Groundnut (Photo courtesy oca.testhead.blogspot.com).

Etiology

The pathogen does not produce spores. Coarse, straight, large cells (2-9 μm x 150-250 μm) have two clamp connections at each septation but may exhibit branching in place of one of the clamps. Branching is common in the slender hyphae (1.5-2.5 μm in diameter) which tend to grow irregularly and lack clamp connections. Slender hyphae are often observed penetrating the substrate. Sclerotia (0.5-2.0 mm diameter) and begin to develop after 4-7 days of mycelial growth (Reddy and McDonald, 1983).

Disease Cycle and Epidemiology

Pathogen survives in infested soil and infected plant material and act as source of primary inoculum. Use of contaminated equipment and machinery may spread sclerotia to uninfested fields. The disease is favoured with alternate dry and wet periods. High soil moisture, dense planting and frequent irrigation promote the infection.

Management

1. Removal and destruction of infected crop debris.
2. Deep ploughing and cultivation of groundnut in flat and slightly raised beds.
3. Seed treatment with *Trichoderma viride* or *Trichoderma harzianum* @ 4g/kg of seed and soil application of *Trichoderma viride* or *Trichoderma harzianum* @ 25-75 kg/ha, preferably in the conjunction with organic amendments such as castor or neem or mahua cakes @ 500-700 kg/ha.
4. Seed treatment with thiram @ 3g/kg or Bavistin @ 2g/kg of seed is also recommended.

Tikka Disease or Leaf Spot

Tikka disease is reported from all groundnut growing countries of the world such as Africa, Australia, China, India, Indonesia, Malaysia, Philippines, Sri Lanka and USA. The yield loss from tikka disease has been reported from 20-50 per cent but may be increased with association other diseases. The all groundnut varieties grown in India are susceptible to tikka disease. The disease symptoms are categorized in two types- (i) early leaf spot and (ii) late leaf spot (Smith, 1984).

Symptoms

Early Leaf Spot

Early leaf spot symptoms are starts after one month of sowing. Small chlorotic spots appear on the leaflets, with time they enlarge and turn black to brown and assume sub circular shape on upper surface of leaves. On lower surface of leaves light brown coloration is seen. Lesions also appear on petioles, stems and stipules. In severe cases several lesions coalesce and result in pre mature senescence.

Figure 1.8: Early Leaf Spot Disease of Groundnut and Conidia of *Cercospora arachidicola* (Photo courtesy icar.tripura.center).

Late Leaf Spot

Late leaf spot are start around after 55-60 days after sowing in *Kharif* and 42-46 days after sowing in *Rabi*. Black and nearly circular spots appear on the lower surface of leaflets. Lesions are rough in appearance and number is more. In extreme cases many lesions coalesce and resulting in pre matures senescence and shedding of the leaflets. Late leaf spot is more dangers as compare to early leaf spot.

Causal Organism

The causal organism of (early leaf spot) tikka disease is *Cercospora arachidicola* Hori (perfect stage of the pathogen: *Mycosphaerella arachidicola* W. A. Jenkins).

Figure 1.9: Late Leaf Spot Disease of Groundnut and Conidia of *Phaeoisariopsis personata* (Photo courtesy icar.tripura.center).

Systemic Position (Ainsworth *et al.*, 1973)

Kingdom: Mycota

    Division: Eumycota

        Subdivision: Duteromycotina

            Class: Hyphomycetes

                Order: Hyphomycetales/Moniliales

                    Family: Dematiaceae

                        Genus: *Cercospora*

                          Species: *arachidicola*

Etiology

The mycelium of *C. arachidicola* is inter and intracellular, brown, septate, branched and without haustoria. The conidiophores are 22-45 x 3-5 µm, yellowish brown, septate and conidia are hyaline or pale yellow, obclavate, 4-12 septate measuring 38-108 x 3-6 µm (Subrahmanyam and Ravindranath, 1988).

Causal Organism

The causal organism of (late leaf spot) tikka disease is *Cercosporidium personatum* or *Phaeoisariopsis personata* (Berk and Curt) Deighton (perfect stage of the pathogen: *Mycosphaerella berkeleyii* W. A. Jenkins).

Systemic Position (Ainsworth *et al.*, 1973)

Kingdom: Mycota

    Division: Eumycota

        Subdivision: Duteromycotina

Class: Hyphomycetes

Order: Hyphomycetales/Moniliales

Family: Dematiaceae

Genus: *Cercospora*

Species: *personatum*

Etiology

The mycelium of *C. personatum* is intercellular, brown, septate, branched and slender with haustoria. The conidia are hyaline, 18-60 x 6-11 µm, 2-7 septate and borne singly on short, 26-54 x 5-8 µm conidiophores. The conidiophores are produced in bunches from the hymenial layer of sub-epidermal region.

Disease Cycle and Dpidemiology

The tikka disease of groundnut is soil borne. The pathogen *C. arachidicola* and *C. personatum* disseminated by wind which is blown from leaf to leaf. The primary infection of disease is caused by conidia found on the plant debris in the soil. The spores remain viable in the soil for a long time and infect the succeeding crop under favourable environmental conditions. High humidity and relatively low temperature is essential for initiating the fungal infection. It is observed that the high nitrogen fertilizer increases disease intensity.

Management

1. The disease can be controlled by long crop rotation and sanitation practices.
2. The intercropping with pearl millet, sorghum and use of phosphatic fertilizers also reduced the disease incidence.
3. The early sown cultivars reduce the disease.
4. Deep burying of crop residues in the soil and removal of volunteer groundnut plants are important measure to reduce the source of primary inoculums.
5. Foliar application of aqueous neem leaf extracts @ 2-5 per cent or neem seed kernel extracts at two weeks interval, three times starting from fourth week after planting is good.
6. The use of Dithane Z-78 @ 0.2 per cent, Dithane M-45 @ 0.2 per cent, Cosan, Breston @ 0.1 per cent and copper sulphate mixture @ 15-25 kg/ha effectively managed the disease. Some other effective systemic fungicides are Benomyl, Bavistin, Brestanol and Cercobin.

Nematode Disease

Kalahasti Malady

Symptoms

Infected plants appears in patches in the fields are stunted greener than normal foliage. Small brownish lesion appears on pegs and on young developing pods. Peg length reduced and in advance stage of disease the entire pods become blackened. Discolorations are also seen on roots (Boswell, 1984).

Figure 1.10: Kalahasthi Malady on Groundnut Pod (Photo courtesy agropedia.iitk.ac.in).

Causal Organism

Tylenchorhynchus brevilineatus

Management

1. Always grow resistant/tolerant varieties such as Tirupathi-2 and 3(TPT 2, 3) (Dropkin, 1980).
2. The disease incidence is less in Groundnut fields sown after rice.
3. Apply Carbofuran 4 kg *a.i.* (133kg/ha) 25-30 days after sowing along with irrigation water (Minton, 1984).

Root-knot Nematodes

The root knot nematode *Meloidogyne arenaria, M. hapla, M. javanica* is a minute round worm which causes typical gall on the roots of groundnut.

Symptoms

Groundnut plants infected with the groundnut root knot nematode commonly develop enlarged roots and pegs which widen into galls of various sizes. It damages the plants by devitalizing root tip and causing formation of swelling in roots. The above ground symptoms are reduced growth and small, pale green or yellowish leaves, pretending to wilt and generally appears as clearly defined patches in the field. Symptoms also associated with the infection are premature leaf fall, wilting, decline in production and loss in field. Pods also become infected and develop knobs, protuberances, or small warts. Plants infected with root knot nematodes may show various degrees of stunting and chlorosis and usually linger throughout the growing season, seldom killing the plant prematurely (Rodrigues, 1984[a]).

Management

1. Deep ploughing during hot weather, soil solarization and flooding can significantly decrease levels of infestation of root-knot nematodes in soils.

Figure 1.11: Root-knot Nematodes on Groundnut Plant (Photo courtesy peanut.nscu. edu).

2. Rotation of groundnut with non host crop like pearl millet or sorghum, wheat, corn *etc* provide satisfactory level of control.

3. Non fumigant systemic nematicides like Carbofuran 3G @ 3g/m, aldicarbs and phenamiphos (Nemacur) are most effective when applied in furrows during sowing @ 2-3 kg of active ingredient per hectare (Rodrigues, 1984[b]).

Viral Disease

Peanut Stem and Bud Necrosis or Bud Rot or Bud Blight

Symptoms

Chlorotic spots appear on young leaflets and necrotic rings and streaks are developed. Terminal bud necrosis are occurs when temperature is relatively high. As the plant matures it becomes stunted with short internodes and proliferation of auxillary shoots (Gibbs and Harrison, 1976).

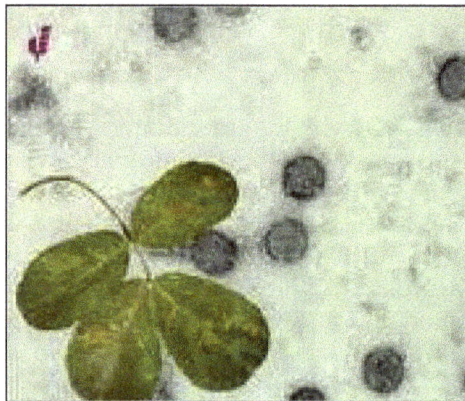

Figure 1.12: Peanut Bud Necrosis Virus on Groundnut Leaf (Photo courtesy printasia. com).

Mode of Spread

The virus is mainly transmitted by thrips.

Causal Organism

Pea nut bud necrosis virus

Survival and Favorable Conditions

The virus survives in the host or the thrips and act as a source of inoculums for the vector. The thrips are carried by winds. The population of vector rapidly increases in the month of January-March in *rabi* and August-September in *kharif* and hence the crops suffer a heavy loss in both the season (Reddy, 1984a).

Management

1. Early sown crops during *Rabi* and *Kharif* summer season is less affected.
2. Always grow resistant/tolerant varieties like ICGS 11, ICGS 44 and R8808.
3. Removal and destruction of alternate weed host plants.
4. Increase plant density, do early sowing, mixed the crop with pearl millets to restrict the vector movements.
5. Seed treatment with Chloropyriphos @ 6ml/kg seed followed by Mancozeb @ 3g or Carbendazim @ 2g/kg seed.
6. Foliar spray with Monocrotophos @1.5ml/liter of water or Dimethoate @ 2ml/liter of water (Reddy *et al.*, 1991).

Groundnut Rosette Disease

It consists of three types of symptoms namely- groundnut chlorotic rosette, groundnut green rosette and groundnut mosaic. The disease is caused by a complex of different strains of groundnut rosette umbravirus (Reddy, 1984[b]).

Symptoms

Symptoms vary depending on strain(s) present. They include yellowing, mottling and mosaic symptoms on leaves and stunting and distortion of the shoots. Older leaves are dark green, reduced in size, and show downward rolling of leaflet margins. If the plants are infected at young stage, they may not produce nuts (Reddy, 1987).

Causal Organism

Groundnut mosaic virus

Mode of Spread and Survival

The virus is transmitted by aphids (*Aphis craccivora* and *A. gossypii*), which feed on the undersides of the leaves (Kimmins *et al.*, 2002).

Management

1. Early sown and high density planting should be used in rainy season.

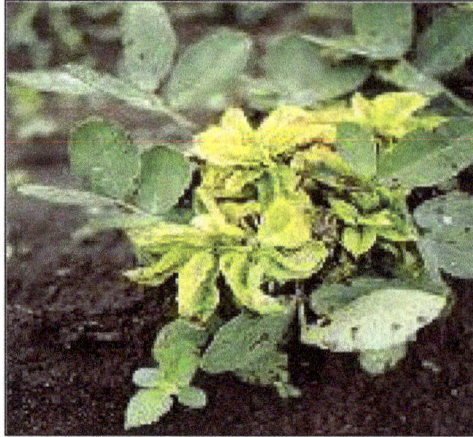

Figure 1.13: Groundnut Rosette Disease Symptoms on Leaf (Photo courtesy icrisat. agropedia.in).

2. Plant tolerant/resistant varieties, *e.g.* 'Asirya Mwitunde'.

3. Remove virus-infected plants after harvest, and volunteer plants that are primary source of infection.

References

Ainsworth, G. C., Sparrow, F. K. and Sussan, A. S. (1973). The fungi. A taxonomic review with keys. Vol.4, A and B: New York, US at Academic Press.

Boshou, L. (2005). A broad review and perspective on breeding for resistance to bacterial wilt, p. 225-238. In C. Allen, P. Prior, and A. C. Hayward (ed.), Bacterial Wilt Disease and the *Ralstonia solanacearum* Species Complex. APS Press, St. Paul, MN.

Boswell, T. E. (1984). Root - lesion nematodes. Pages 41- 42 in Compendium of peanut diseases (Porter, D. M., Smith, D. H. and Rodriguez-Kabana, R., Eds.). St. Paul, MN, USA: American Phytopathological Society.

Coutinho, T. A. (2005). Introduction and prospectus on the survival of *R. solanacearum*. Pages 29-38 in: Bacterial wilt disease and the *Ralstonia solanacearum* species complex. Allen, C., Prior, P., and Hayward, A. C., eds. APS press, St. Paul, M. N.

Denny, T. P. (2006). Plant pathogenic *Ralstonia* species. Pages 573-644 in: Plant-associated bacteria. S. S. Gnanamanickam, ed. Springer Publishing, Dordrecht, The Netherlands.

Dropkin, V. H. (1980). *Introduction to Plant Nematology*. New York, USA: John Willey. FAO STAT (2011). http://faostat.fao.org.

Gibbs, A. and Harrison, B. (1976). *Plant Virology: The Principles*. London, UK: Edward Arnold Publishers. 292pp.

Gitaitis, R. D. and Hansons, R. O. (1984). Bacterial wilt. Pages 36-37 in Compendium of peanut diseases (Porter, D. M., Smith, D. H. and Rodriguez-Kabana, R., Eds.). St. Paul, MN, USA: American Phytopathological Society.

Holt, B. L. and Reddy, D. V. R. (1984). Peanut clump. Pages 50-51 in Compendium of peanut diseases (Porter, D. M., Smith, D. H. and Rodriguez-Kabana, R., Eds.). St. Paul, MN, USA: American Phytopathological Society.

Kimmins, F. M., Busolo-Bulafo, C., Van Der Merwe, P., Naidu, R. A. and Subrahmanyam, P. (2002). Evaluation of Groundnut Rosette Resistant Varieties and Impact on Farmers' Livelihoods in the Teso System of Uganda pp. 47-48 In: Summary Proceedings of the Seventh ICRISAT Regional Groundnut Meeting for Western and Central Africa, 6-8 December 2000 Cotonou, Benin. Wailiyar, F. and Adomou, M. (Eds.).

Minton, M. A. (1984). Ring nematodes. Pages 43- 44 in Compendium of peanut diseases (Porter, D. M., Smith, D. H. and Rodriguez-Kabana, R. Eds.). St. Paul, M N, USA: American Phytopathological Society.

Nautiyal, P. C., Bandyopadhyay, A. B. and Zala, P.V. (2001). *In situ* sprouting and regulation of fresh-seed dormancy in Spanish type groundnut (*Arachis hypogaea* L.). *Field Crop.*

Pal, K. K., Dey, R., Singh, J. P. and Joshi, B. H. (2000). Biological control of groundnut Bruchid: A serious storage pest of groundnut, SARC, Newsletter, page No. 4 - 8 January-March, vol. 10, No. 1.

Rama Rao, D., Kiresur, V. R. and Sastry, K. R. (2000). Technological Forecasting of Future Oilseeds Scenario in India (ICAR-AP CESS Project). National Academy of Agricultural Research Management, Rajendranagar, Hyderabad.

Reddy, D. V. R. (1984[a]). Tomato spotted wilt virus. Pages 48-49 in Compendium of peanut diseases (Porter, D. M., Smith, D. H., and Rodriguez-Kabana, R., Eds.). St. Paul, MN, USA: American Phytopathological Society.

Reddy, D. V. R. (1984[b]). Groundnut rosette. Pages 48-49 in Compendium of peanut diseases (Porter, D. M., Smith, D. H. and Rodriguez-Kabana, R., Eds.). St. Paul, MN, USA: American Phytopathological Society.

Reddy, D. V. R. (1987). Techniques used for detection of plant viruses. Lecture note to VII International Training Course in Legume Pathology, 12-31, Jan 1987, ICRISAT Center, India. Patancheru, A.P. 502324, India: International Crops Research Institute for the Semi-Arid Tropics. 15pp. (Limited distribution).

Reddy, D. V. R. and McDonald, D. (1983). Pages i-viii in Proceedings of the National Seminar on Management of Diseases of Oilseed Crops, 21- 22, Jan 1983, Madurai, India (Narayanasamy, P. Eds.) Madurai, Tamilnadu, India: Tamilnadu Agricultural University, Agricultural College and Research Institute.

Reddy, D. V. R., Nightman, J. A., Beshear, R. J., Highland, B., Black, M., Sreenivasulu, P., Dwivedi, S. L., Demski, J. W., McDonald, D., Smith, Jr. J.W. and Smith, D. H. (1991). Bud necrosis: A disease of groundnut caused by tomato spotted wilt virus. Information Bulletin no. 31, Patancheru, A. P. 502324, India: International Crops Research Institute for the Semi-Arid Tropics. *Research,* **70**: 233-241.

Rodrigues Kabana, R. (1984[a]). Root - knot nematodes. Pages 38-41 in Compendium of peanut diseases (Porter, D. M., Smith, D. H. and Rodriguez-Kabana, R., Eds.). St. Paul, MN, USA: American Phytopathological Society.

Rodrigues Kabana, R. (1984[b]). Other nematodes. Pages 44 in Compendium of peanut diseases (Porter, D. M., Smith, D. H. and Rodriguez-Kabana, R., Eds.). St. Paul, MN, USA: American Phytopathological Society.

Saddler, G. S. (2005). Management of bacterial wilt disease. Pages 121-132 in: Bacterial wilt disease and the *Ralstonia solanacearum* species complex. Allen, C., Prior, P., and Hayward, A. C., eds. APS press, St. Paul, M. N.

Smith, D. H. (1984). Early and late leaf spots. Pages 5-7 in Compendium of peanut diseases (Porter, D. M., Smith, D. H. and Rodriguez-Kabana, R., Eds.). St. Paul, MN, USA: American Phytopathological Society.

Subrahmanyam, P. and Ravindranath, V. (1988). Fungal and nematode diseases. Pages 453-5 2 3 in Groundnut (Reddy, P. S., Eds.). New Delhi, India: Indian Council of Agricultural Research.

Subrahmanyam, P., McDonald, D. and Hammons, R. O. (1984). Rust. Pages 7-9 in Compendium of peanut diseases (Porter, D. M., Smith, D. H., and Rodriguez - Kabana, R., Eds.). St. Paul, MN, USA: American Phytopathological Society.

Subrahmanyam, P., Gibbons, R. W., Nigam, S. N. and Rao, V. R. (1980). Screening methods and further sources of resistance to peanut rust. *Peanut Science,* **7**: 253-256.

2 Diseases of Linseed and their Management

R.B. Singh, H.K. Singh, Mahesh Singh and Shiwangi

Introduction

Linseed is a multipurpose crop which is grown all over the world in tropical, subtropical and temperate areas mainly in India, Argentina, Canada, Egypt, Ethiopia, France, USA *etc*. In India it is cultivated generally for oil purpose, whereas in western countries it is grown especially for fibres. The oil content of the seed varies from 33 to 45 per cent. The residue cakes remaining after the oil extraction contain about 3 per cent oil and 36 per cent protein (Gill, 1987). Besides oil, it is known good manure and animal feed; it contains about 5 per cent nitrogen, 1.4 per cent phosphorus and 1.8 per cent potash. Every part of the linseed plant is utilized commercially either directly or after processing. There are different varieties of linseed meant for both purposes, which differ considerably in growth characters. The fibre has a tall slender stem, having a high ratio of good quality fibre and seeds with low oil content (Turner, 1987). Varieties having shorter stem and more branches usually bear larger seeds with higher oil content. Linseed straw produces good fiber which is used in manufacture of cigarette paper, linens, insulating material, book paper and fiber boards. Fiber obtained from stem and used in making rough as well as good clothes and inter decorating material (Gill, 1987). Flax is grown in the warmer regions of the temperate zones as a winter crop and in the cooler regions as a summer crop (Anonymus 2006). There are six species of *Linum*, grown in India *viz., L. angustifolium, L. usitatissimum, L. mysorense, L. grandiflorum, L. strictum,* and *L. perenne*. Some other species have been also introduced as flowering ornamental plants.

There are several diseases which affect linseed crop, amongst Alternaria blight and rust in Northern India and powdery mildew and wilt in central and peninsular regions are most damaging (Singh, *et al.*, 2014). Although, other diseases are of minor significance at present in India but they need regular monitoring so that do not become major problems in future.

Wilt

Distribution and Economic Importance

The wilt disease of linseed occurs in every country where this crop is grown. This disease was reported very severe and destructive in specially USA and Japan. In India it was first reported From Bihar in 1923. It was reported that the yield losses due to wilt disease in Rajasthan region of India is up to 70 per cent. Singh and Singh, (2011) have also reported considerable yield losses in different cultivars of linseed in U.P. under sick field condition. In severe cases complete crop can be destroy.

Symptoms

Fusarium oxysporum generally produces symptoms such as wilting, chlorosis, necrosis, premature leaf drop, browning of the vascular system, stunting, and damping-off. The symptoms of wilt may appear at any stage of plant growth. According to the stage of the plant, symptoms may be divided into two types, first at seedling stage and second at adult plant growth stage. At seedling stage the symptoms occurs before the third week after sowing when susceptible seedlings are grown at high temperatures. In case of very young seedlings the edge of cotyledons rolls inward and ultimately the whole cotyledon fall down. The infected seedlings droop over on the ground, become threads like and untimely plants die. In case of adult plants the symptoms appear as small brownish spots on the leaves and at later stage it turn into yellow colour. The vascular system of infected plants is covered with fungus mycelium. On pulling out and splitting the plants, the black streaks can be seen in the vascular tissues of root and stem. At maturity it is difficult to distinguish between healthy and diseased plants except that the later may turn brown and dry before maturity (Kolte and Fit, 1997 and Singh, *et al.*, 2014).

Pathogen

Fusarium oxysporum f. sp. *lini* (Bolley) Synder and Hensan. The pathogen belongs to division Eumycata, subdivision Deuteromycotina, class Hyphomycetes, orders Monilials and family Tuberculariaceae. Mycelium is septate, branched intracellular developing into thin stroma of various shades. Conidiophores are short, branched and hyaline. The conidia are 3-septate, slightly curved and measure 27-38 µ in length and 3-3.5 µ in width (Borlaug, 1945). Chlamydorpores are spherical to pyriform smooth or wrinkled usually 1-celled 5-13 µm in diameter, terminal and intercalary and are formed abundantly (Turner, 1987; Kolte and Fit, 1997; Singh, *et al.*, 2014).

Disease Cycle and Epidemiology

This is a soil as well seed borne disease. The pathogens survive in the soil for several years even in the absence of main host on several other wild hosts or when the contaminated seeds are sown, the seedling are being infected. Continuous cropping of the same crop in the same field helps to the reservoir of primary inoculum. Fungus enters into in the xylem vessels through the soft roots of the young plants. High temperature *i.e.* 25-28°C, high moisture and light sandy soil are important factors in the development of the disease (Broadfoot and Stakman, 1926). The secondary spread of the pathogen takes place by means of irrigation water and some species of nematode.

Disease Management

Flooding of the infected fields and 4-5 years crop rotation of rape with linseed are effective. Soil solarization along with soil amendments with neem cake or mustard cakes help in reduction of inoculum. Sowing 5 to 15 November was best for linseed cultivation in Uttar Pradesh (Singh and Singh, 2011). Cultivars Nagarkot, Jawar-23, Jeevan, Padmini were found as resistant and can be cultivated under high sick field condition (Singh *et al.*, 2012). Seed treatment with Bavistin and Thiram (1: 2) or Carboxin and Thiram (1: 2) @ 3 g/kg seed or slurry seed treatment with systemic fungicides Benlate T20 or Vitavax each @ 0.5-1.0 g/kg seed may be quite useful in controlling wilt pathogen (Sharma *et al.*, 2002., Singh *et al.*, 2005., Kishore and Singh, 2008).

Leaf extract of *Xanthium strumarium* (Karanj) and *Tribulus terrestris* showed maximum toxicity and inhibited 81.18 per cent and 77.66 per cent infection, respectively (Singh *et al.*, 2014). The disease can be managed up to some extent either soil and seed treatment with the leaf extract of *Tribulus terrestris* (10 per cent w/v) or seed and soil treatment with *Trichoderma harzianum* (2.5 kg/ha) under field condition (Duijff, *et al.*1999, Singh, *et al.*, 2014., Singh *et al.*, 2015).

Sowing of resistant varieties is the only reliable method to manage the disease. Some of the rust resistant varieties of linseed such as RR-9 (in U.P.) posses wilt resistance also. The cultivars like K-2, LC-185, LC-54, Himalini and Jawahar-552 are tolerant to wilt disease. Genotypes, namely RLS-1, RLS-5, RLC-6, RLC-18, RLC-19 and R-552 and varieties BR-9, BR-29, Indore-1, Malvi-1, NP-12, Bison-69, Bison-70 and Canadian Western No.1 have been found resistant to wilt. Among older cultivars, B-5128, NP-1, NP-124, NP (RR) 50, 10, 80 are resistant to flax wilt. Sixteen genotypes namely RSJ-1, Surabhi, Sl-5-21, Type-397, NDL-2002, NDL-4-5, Rashmi, LCK-9436, LCK-7034, LCK-7035, K-4515, Kl-160, EC-282809, EC-8363, EC-5449, and R-552 were found resistant to *Fusarium* wilt under sick field condition (Singh *et al.*, 2012).

Alternaria Blight

Distribution and Economic Importance

Dey (1933) from India first reported the occurrence of *Alternaria* blight affecting the flower buds of linseed of Kanpur area (Uttar Pradesh) in 1933. The loss due to the disease was estimated to vary from 26–60 percent (Chauhan and Srivastava, 1975 and Singh et al., 2003). Severe occurrence of the disease was further recorded at the farm of Indian Agricultural Research Institute (IARI) in March-April, 1949, in January 1964 in Punjab and in February-March 1974 in Jabalpur. The disease is still considered to be of economic importance in India. Similar kind of disease is reported to occur in Britain, Canada, France, Netherlands, Sweden and USSR. The blight of linseed is commonly occurs in eastern Uttar Pradesh and Madhya Pradesh. The disease is also common in other parts of the world where linseed is grown.

Symptoms

Initial symptoms in the field appear on the cotyledonary leaves as water soaked pinhead dots, later turning light-brown to dark-brown, and progressively enlarging from 1.5 to 5 mm diameter, dark- brown spots of oval to irregular shape. Under favourable host-environment situations, water soaked edges advance fast and finally blight cover over 50 per cent area singly or rarely in coalescence with neighbouring spots. Older spots have grey centres producing conidia profusely.

Symptoms on leaves appear quite independent of the cotyledonary leaves about one and half months or still later after the sowing. These too initially appear as water-soaked bluish-dots, later turning light to dark-brown as on cotyledonary leaves, solitary to numerous number. Dots grow as bigger spots progressing from spherical to oval and finally irregular in shape. But spots formed on leaf edges are crescent shaped, circular side extending fast towards the mid rib. Unlike these, spots formed close to the mid rib away from the edges, progress slowly but cause substantial damage by covering over 50-100 per cent leaf width on their side of the mid rib, blocking the flow of water, nutrients and photosynthates. These probably, coupled with pathogen toxins produced, turn the affected side chlorotic pale yellow and finally blighted and withered. Spots developed on the mid rib, generally constrict the leaves at the affected point. Infected leaves develop senescens and drop quickly in some varieties whereas in others these blight/curl up and keep attached to the plants for longer duration. Spots vary from 2-4 mm in diameter in spherical ones and 2-4 mm in width and up to 12mm in length in irregular ones. These tend to coalesce. Disease progresses from lower to the upper leaves.

On stem, light-brown linear spots of varying size of darker margins appear and progress to bigger dimensions with darker colour individually or through coalescens.

Initially symptoms on floral buds appear at the base of the calyx, the way these appear on leaves, enlarge, take light to darker-brown colour and cover part or whole of the bud. These progress to the pedicel, sometimes blighting them completely. A junction of the pedicel and capsule shrinks leading to collapse of whole flower bud (decapsulation). Petals, stamens and carpel take the infection fast, rot, shrink and serve as substrate to a variety of saprophytic fungi and bacteria. In post fertilization infections too symptoms appear but with low severity. Later infections, though allow formations of capsules, seeds remain undersized. These capsules exhibit minute, brown, coloured spots, on on occasions turn the whole capsule brown to dark-brown give burnt up appearance. The disease capsules either do not bear seed or give shrivelled light seed. Infection on the fully formed capsules too produced brown to dark-brown spots and discolouration but confined to capsule coat only. Such infections have minimal effect on seed size and weight (Singh and Singh, 2005).

Pathogens

Infected tissues yielded *Alternaria linicola* Groves and Skolko and *A. lini* Dey, former being predominant on the flower buds, capsules and cotyledonary leaves and the later on the leaves, branches and the stem. On artificial inoculation, these produced symptoms on all above ground plant parts but *A. linicola* caused severe disease only on reproductive parts. The mycelium is branched, septate; very light coloured and bearing conidiophores. The conidiophores are short, septate, sometimes branched bearing conidia terminally. The conidia are longitudinally and transversely septate, constricted at the septa, tapering towards the beak side, short beaked and almost rounded base.

Perpetuation and Epidemiology

The disease is seed borne which initiate primary infection in the host plants. The secondary spread is through wind borne conidia. Disease inoculums remains buried in the field soil along with previous seasons infected crops debris or reaches it through infected contaminated seed material. It multiplies there initially, infect cotyledonary or normal leaf nearest to the ground. Disease moves up as a result of secondary infection through conidia formed in the primary legions. Conidia later become air borne and infect all the aerial plant parts. Symptoms initiate on leaves between 23 Nov. and 5 Jan. on normal sown (15-20 Oct) crops. Thus, predisposing factors for the onset of the disease appear in the October sown crop, moderate temperature, maximum (29-24°C) and minimum (12-7°C), high relative humidity (74-92 per cent) min. (42-68 per cent) and irrigation or rainfall. Irrespective of date of initiation, highest disease intensity is reached between 15 February and 2 March indicating thereby that the most favourable environmental conditions for the blight development prevail in the month of February. Therefore, the predisposing factors are min and max temperature of 6-10°C and 22-28°C, respectively and humidity 16-51 per cent and 80-90 per cent. Highest disease intensity was recorded in 15 Oct. sown crops followed by gradually decline intensity in later sown once, lowest being 5 December sown. Dense canopy of the pure or companion crop, cloudy weather, frequent rains or irrigation intensify the disease. A combination of these factors in wider areas leads to epidemics (Singh and Singh, 2005).

Integrated Disease Management

Cropping System

Rotate the crop with *rabi* cereals or pulses to reduce the build up of the pathogen population in field soil and strengthen its nutritional base. Intercropping linseed with these crops reduced aerial spread of the disease and slows inoculums build up in the field. Linseed is credited with reducing the inoculum build up of the soil borne diseases of the companion crops through its roots exudates.

Field Sanitation and Clean Cultivation

Select upland, well drained fields for the commercial cultivation of the crops. Avoid saline-alkaline soils scrupulously as excess moisture and salt devitalized the root system and favours *Alternaria* blight development too. Collect and burn the infected mustard crop debris lying in the field post-harvest. Give deep summer ploughing to turn the remaining inoculums down top safer depths.

Date of Sowing

Sow the crop between 1 and 10 Nov. as early sown results in loss of yield and quality owing to high disease intensity and late sowing causes poor germination, slow growth, bud fly infestation and terminal heat. Crop sown on 5th November gives less blight incidence and high yield (Singh and Singh, 2011).

Healthy Seeds

Use certified seeds, preferably of recent releases. In case of recycling use clean seed obtained from a disease free crop and treat it with Thiram 75 WP @ 2.5g/kg seed. Both practices help in getting good crop stand and vigorous seedlings.

Irrigation

Frequent irrigations prolong vegetative phase of the plant, thus delaying its maturity and increasing disease severity at later stage. Give only 2 to 3 irrigations, depending upon soil type and crop requirement.

Nutritional Management

Apply farmyard manure or compost. Alternatively, practice green manuring using sunhemp or *dhaincha* (*Sesbania*). These in addition to nutrients, add organic matter to soil, which serve as substrate for the native bio-control microorganism. Apply 35 kg Murate of Potash and 25 kg Zinc sulphate in addition to normal dose of nitrogen and phosphorus. Additional to their normal function these imparts disease resistance to the plants.

Tolerant Varieties

Linseed lacks varieties resistant or moderate resistant against Alternaria blight Sheela, Gariam, Jeevan. Shweta, Kiran, Narendra Alsi-1 and Subra have moderate susceptibility against the disease. In a further screening of disease varieties, based on avoidable yield loss, Sheela and Garima was found highly tolerant Jeevan, Shweta and Kiran tolerant and Subhra moderately tolerant. Select varieties out of these as per need and location (Singh and Singh, 2011).

Chemotherapy

Spray the crop with aqueous suspension of Mancozeb 75 WP @ 0.25 per cent at the initiation of the disease. Repeat once or twice at 15 days interval, if necessary, depending upon the varietal susceptibility, crop growth stage and environmental conditions (Singh and Singh, 2004). Seed treatment with *Trichoderma viride* (4g/kg seed) followed by two sprays of mancozeb (0.25 per cent) decrease the blight severity (56.16 per cent) and increased the seed yield with higher test weight (Singh *et al.*, 2013).

Rust

Distribution and Economic Importance

Rust is also important disease of linseed and the cultivation of crop is greatly affected by this disease in most of the linseed (flax) growing areas of the world. It appeared as moderate to severe form in all linseed growing states especially Uttar Pradesh, Bihar, Jharkhand, Madhya Pradesh, Maharashtra, Rajasthan, Gujarat and other northern parts of India. The crop is sown in October and November and harvested in March or April. The rust appears by the end of October in north western Himalayas and in the month of February in other parts of north India (Singh, *et al.*, 1981).

The yield loss depends on early stage of crop infection. Linseed rust results in severe epidemics year after year causing 20 to 100 per cent yield losses and 13 per cent reduction in oil content in the seed harvested from heavily rusted plants (Singh *et al.*, 1981).

Symptoms

The affected plants are very conspicuous in the field because of the bright orange- coloured uredia occur on both surfaces of the leaves as well as on other aerial parts of the plant. At a later stage the necrosis becomes more general spots are mostly dark brown turning to black with crust like pustules and the leaves die prematurely. The pustules are almost round shaped and small on leaves but on the stem they are large, elongated and irregular. However, they are most common and prominent on the stem. Often, surrounded by the reddish brown telia are the orange yellow uredia on the stem (Lawrrence, 1988).

Pathogen

Rust of linseed is caused by *Melampsora lini* (Ehrenb.) Lev. The pathogen belongs to Kingdom- Fungi, Phylum- Basidiomycota, Class- Pucciniomycetes, Order- Pucciniales, Family- Melampsoraceae, Genus- *Melampsora* and Species- *lini*. The uredospores are small, spherical to oval, 15-30 × 13-18µ in diameter, with a colourless warty hyaline wall and equatorial pores. The spores germinate with germ tubes that readily infect linseed. The teleutospores are formed in a single row, closely packed with sub-epidermal teleutosori. It is single celled, cylindrical and polygonal in cross- section and measure 42-60 × 10-20 µ. The sporidia can infect linseed, forming pycnia and aecia. The small pycnia are subepidermal on both sides of the leaf. The pycniospores are minute, single celled, hyaline and measure 3-4 × 2-3µ. The aecia are cup-like and bright orange yellow in colour. They are devoid of a peridium and the chains of aeciospores are roundish, with hyaline, warty wall, measuring 21-28 × 19-27µ. The aeciospores germinate to produce germ tubes which can infect linseed to produce uredia. Thus all the four stages of the fungus are formed on the same host; hence it is an autocious rust.

Disease Cycle and Epidemiology

The pathogen is air borne in nature. It is autoceous pathogen because all the stages of fungus life cycle completed on the same host species. The thick walled teliospores survive on plant debris in the hills. These teliospores germinate and produce haploid basidiospores and these basidiospores reinfect to susceptible plants. After infection pycnia develops and produce pycniospores in a liquid exudate called nectar (Mc Fadden, *et al.*, 2001). Dikaryotic aeciospores produced in aecia infect the susceptible plants and after 8-10 days, uredosori are produced which have uredospores. Secondary spread of pathogens takes place by insects; water dripping *etc.* with temperature 15-25 °C and leaf wetness period is about 4-5 hours (Thind, 1998).

Disease Management

Early sowing of crops *i.e.* before third week of October helps to escaping of disease incidence. Avoid the excess use of nitrogenous fertilizers. Sowing of resistant cultivars is the cheapest and best method of disease management. The cultivars NP (RR) 9, 10, 56, 95, 218, 297, B, 279 K3, 368, 381, 389, 415 and 501 had been cited as resistant to the diseases. In Punjab and Himanchal Pradesh, LC-54, LC 115, K-2, Himalini and Jawahar-7 are grown as resisted varieties. Cultivars LC 216, LC 255, LC 256 are resistant to all races of the pathogen prevalent in the hills (MC Fadden *et al.*, 2001 and Saharan, *et al.*, 2005). Two spray of fungicides such as Tridemorph @ 0.05 per cent, Mancozeb @ 2 kg/ha or sulphur dust @ 17.5 kg/ha at 10 days interval have been found more effective against linseed rust.

Powdery Mildew

Distribution and Economic Importance

Powdery mildew of linseed has been identified as one of the serious problem particularly in late sown crop with rich soil under irrigated conditions in several areas. However, the disease is not important in North America, Europe and Asia. When powdery mildew appears in severe form at early stages of the plant growth, yield reduction is very high. Infected plants produce poor quality of seed and fibre (Singh *et al.*, 1989). The adverse effect of the disease on yield is mainly through reduction in seed number in capsules and also in seed size. The yield losses from powdery mildew disease is about, 12-38 per cent (Pandey and Mishra, 1992).

Pathogen

Powdery mildew disease of linseed is caused by *Oidium lini* Skoric and *Leveillula taurica* (Lev.) Arnaud. The fungus is an obligate parasite and belongs to division-Eumycota, subdivision- Ascomycota, class- Pyrenomycetes, order- Erysiphales and family- Erysiphaceae. Conidiophores and conidia are present in chains with septate mycelium. The size of conidia ranges from 12.20-39.99 × 12.19μ. Conidia borne singly on short hyphal branches, predominantly of two distinct shapes, cylindrical and radicular, varying in size. Redish to brown scattered perithecia develop with simple appendages. Ascospores are large, cylindrical to pyriform, sometimes slightly covered. Asci are stalked, clavate, hyaline and usually contain various oblong to ovate ascospores (Beale, 1991).

Disease Cycle and Epidemiology

The clestothecia of the fungus survive in soil when the favourable environmental condition come it becomes active and asci release ascospores which germinate and attack the crop. The spread of pathogen in Northern and central part of India is usually in the last week of February and up to the middle of March when temperature ranges between 20 to 25°C with 60 per cent relative humidity. (Saharan *et al.*, 2005).

Disease Management

Sow the crop as early as possible following the crop rotation practices. The adults and larva of *Coccinellid theacincta* (*Psylobora cincta*) are reported feeding on the powdery mildew of linseed and cleaning the infected part of the plants from the fungus. Cultivar Jawahar 23, R-552, JLS-1, Chambal, Garima, Sweta, Shubhra were moderately resistant to powdery mildew. The inheritance of resistance against powdery mildew disease in linseed was studied in F_1, BC_1, BC_2 and F_2 generations in three resistance varieties Polf 5, NDL 2004-05 and LCK 9406 and crossed with one susceptible genotype, PKDL-52. In the F1 generation of all the three crosses, LCK 9406 x PKDL 52, Polf-5 x PKDL 52 and NDL 2004-05 x PKDL 52, the plants were resistant to powdery mildew (Chauhan, *et al.*, 2016). Application of Sulfex or Wettable sulphur @ 0.3 per cent or Karathane @ 0.2 per cent or Calixin @ 0.1 per cent, Triademefon (0.1 per cent), Dinocap (0.1 per cent) and Myclobutanil (0.1 per cent) at 15 days interval are more effective management of powdery mildew disease of oilseed crops (Kushwah and Chand, 1971, Nene and Thapliyal, 1979, Singh and Singh, 2003, Singh *et al.*, 2016).

Stem Break and Browning

Distribution and Economic Importance

The disease was first reported in Ireland by Lafferty in 1921 and later it has been reported from several country of the world. Stem break and browning are two different phases of a disease caused by the seed borne and soil borne fungal pathogen. This disease has minor importance in India; however, some damage may occur.

Pathogen

Polyspora lini, the fungus which causes the browning and stem-break disease of flax.

Symptoms

The first visible symptom appears as stem break, due to development of a canker at the stem base, the plants become weak when the plants are still young, or at a later stage. Plants may remain alive after stem breakage. The browning phase is initiated by infections on the upper part of the stem that appear as small oval or elongated brown spots. These small spots may coalesce together, and leaves and stem turn brown, giving the disease name of browning (Thind, 1998). At the severe infection the fungus may penetrate the bolls as well as the seeds, or may produce spores on the seed surface.

Disease Cycle and Epidemiology

The primary infections in spring may start from spores produced on diseased plant debris. Infections may start during seedling emergence when seed coats of diseased seed are lifted above the ground, and the fungus produces the first cycle of spores of the season. Secondary spread takes place by means of wind and rain splashes.

Disease Management

Practice crop rotation at least 2-3 years. Early sowing of the crop may reduce injury since the crop could ripen before the disease becomes severe. Use of clean seed produced by healthy plants is one of the best control measures. Seed treatments with fungicides likes Thiram @ 3 g/kg seeds or Benlate T20 @ 0.5-1.0 g/kg seed may be controls surface-borne inoculum (Nene and Thapliyal, 1979).

Seedling Blight and Root Rot

Distribution and Economic Importance

This disease is distributed all over the linseed growing areas of the world particularly in sandy soils under warm and moist condition. The pathogens survive as a composite of strains that differ in host range and pathogenicity. Strains attacking sugar beets and legumes such as alfalfa and field peas, also attack flax. Yellow-seeded varieties are more prone to cracking which renders them more susceptible to seedling blight and root rot than brown-seeded varieties (Singh, and Singh, 2011).

Pathogen

Seedling blight and root rot caused by soil-borne fungi such as species of *Fusarium, Pythium* and *Rhizoctonia*. However, *Rhizoctonia solani* is the principal causal agent of the disease. (Singh and Singh, 2011).

Symptoms

Infected seedlings symptoms may appear as singly or in patches. The seedlings turn yellow, wilted and ultimately die. Roots of recently affected plants show red to brown lesions, and may turn dark and shrink at later. Root rot symptoms appear in plants after the flowering stage (Singh and Singh, 2011). Plants may wilt on warm days, and turn brown prematurely; plants with root rot usually set little or no seed.

Disease Management

Seedling blight and root rot disease can be controlled by integrated disease management practices. Proper field sanitation and other tillage practices should be done before sowing. Practice crop rotation at least for three years between flax crops and also avoid legumes and sugar beets in the rotation. Soil solarization should be done in summer season. Use certified seed of a recommended variety. Cracking of seed at the time of threshing should be avoided. Treatment of the seed with fungicides are effective measure.

Aster Yellows

Distribution and Economic Importance

Aster yellows is one of the important disease of flax caused by a mycoplasma-like organism. The pathogen has wide host range such as canola and sunflower, and several weeds. An epidemic was reported in 1957 in Western Canada caused widespread severe yield losses in flax and other crops. The aster leafhopper vector, *Macrosteles quadrilineatus*, is responsible to transmit yellows phytoplasma from plant to plant (Rashid *et al.*, 2008).

Symptoms

The main symptoms of aster yellows are yellowing of the upper part of the plants, conspicuous malformation of the flowers, and stunted growth. Affected leaves are somewhat narrower than healthy leaves. Plants may be stunted or with numerous secondary shoots. Small yellowish green leaves are seen on flowers, it become sterile and produce no seed (Rashid and Duguid, 2005). The severity of the disease depends on the stage at which plants become infected and severity of insect vectors.

Disease Transmission

This is not a soil or air borne disease it only survive inside host plants and insect vector, which transmits the mycoplasma-like organism in flax. Aster yellows phytoplasma inhabits the phloem vessels of infected plants and is carried from plant to plant by insect vectors known as the six-spotted leafhopper (*Macrosteles quadrilineatus*). The disease can be serious when dry weather forces leafhoppers to migrate from wild weeds to irrigated fields of susceptible plants.

Disease Management

Always keep in mind the relationship between cultivated hosts, insect vectors and wild or alternate hosts; and proper field sanitation practices should be done. The migrating leafhoppers should be avoided in mid and late season. Early crop showing also reduces the incidence and severity of aster yellows. Destroy affected plants in small areas as soon as they appear to be diseased. Chemical control is presently the most effective means of controlling the vector (Saharan, *et al.*, 2005). After first appearance of vector, growers should sprays routinely on a weekly basis chemical insecticide.

Anthracnose

Distribution and Economic Importance

Generally it is distributed wherever flax is growing but is most common in cool, humid, flax-growing areas. The disease was first described by Bolley (1903) in the United State as flax canker and named the causal organism *Colletotrichum lini* Bolley. Some researchers also reported the disease in Ireland and noticed that it was destructive to flax seedlings. The disease is more severe in several countries such as India, Argentina, Australia, Canada, China, and other parts of the world and causes some time severe damage to linseed crop. *C. linicolum* can produce economic losses (Saharan, *et al.*, 2005). Loss of 50 per cent in fibre yield in an epidemic was also reported and caused considerable loss in fibre strength after attacks of the pathogen.

Symptoms

This disease causes cankers on the primary leaves of flax. The circular, sunken brown spots appear on the leaves that spread under cool and moist conditions. Seedling blight takes place either before or after emergence. Under high moisture condition leaf spots and stem cankers are very common on growing season. Brown coloured spots are also formed on the seed.

Pathogens

Linseed anthracnose is caused by *Colletotrichum linicolum* (Pethyb and Laff). The acervuli of the fungus rapture the epidermis of the host. Setae are erect, septate, dark brown and measure 150× 4µ. At near-saturated humidities acervuli are formed on the host surface without setae. At lower relative humidity, acervuli with several setae are formed. Conidiophores are shorts, hyaline and mostly simple. Conidia are cylindrical, tapering towards both ends, slightly curved, hyaline, one celled and measures 11-21 × 4µ in size (Rashid, *et al.*, 2008).

Disease Management

Seed treatment is the most effective means of the disease control. Adequate crop rotation and removal of crop debris should also minimise the risk of the pathogen being picked up from the field. Early sowing may be effective in restricting the spread of the disease. Disease reduction may also be achieved by uniform sowing and avoidance of clumps of plants (Saharan, *et al.*, 2005).

References

Anonymus (2006). Annual progress report of linseed, AICRP on Linseed, Directorate of Oil Seed Research, Rajendra Nagar, Hydrabad.

Beale, R.E. (1991). Studies of resistance in linseed cultivars to *Odium lini* and *Botrytis cineria*. In production and protection of linseed. Edited by R.J. Fraud Williams. *Aspect Applied Biology Series. Association of Applied Biologists, Horticulture Research International*, Willisbourne, Waruick, U.K. Vol. **28**. pp. 85-90.

Borlaug, N.E. (1945). "Variation and variability of *Fusarium lini*", *Minn. Agr. Exp. Sta. Tech. Bul.*, p.168.

Broadfoot, W.C. and Stakman, E. C. (1926). Physiological specialization in *Fusarium lini*. Bolley. *Phytopath*. **16**: 84-85.

Chauhan, M.P., Singh, H.K., Rahul, V.P. and Yadav, J.K. (2016). Inheritance of resistance to powdery mildew disease of linseed (*Linum usitatissimum* L.). *Indian Phytopath*. **69** (4s) 218-220.

Chauhan, L.S. and Srivastava, K.N. (1975). Estimation of loss of yield caused by blight disease in linseed. *Indian J. Farm Science*. **31**: 107-109.

Duijff., B.J.; Recorbet, G.; Bakker, P.A.H.M.; Lopee, J.E. and Lemansceau, P. (1999). Microbial antagonism at the root level is involved in the suppression of Fusarium wilt by the combination on non-pathogenic *Fusarium oxysporum* F047 and *Psudomonas putida* ECS358. *Phytopath.* **89**: 1073.

Gill, K.S. (1987). Linseed, *Indian Council of Agriculture Research*, New Delhi, India.

Kishore, R. and Singh, J. (2008). Evaluation of Fungicides against *Fusarium oxysporum* f.sp. *lini* of Linseed. *Ann. Pl. Protec. Sci.* **16** (1): 165-167.

Kolte, S.J. and Fitt, B.D.L. (1997). Diseases of linseed and fibre flax. Pp.-247, Shipra Publications, Delhi.

Kushwah, U.S. and Chand, J.N. (1971). Efficacy of fungicide for the control of powdery mildew of linseed (*Linum usitatissimum*) caused by *Oidium lini*. *Indian Phytopath.*, **24** (1) 200.

Lawrrence, G.J. (1988). *Melampsora lini*- Rust of flax and linseed. *Advances plant Pathol*, **6**: 313-331.

Mc Fadden, H.G.; Lawrence, G.J. and Dennis, E.S. (2001). Differential induction of chitinase activity in flax (*Linium usitatissimum*) in response to inoculation with virulent or strains of *Melampsora lini* the cause of flax rust. *Aust. Plant Pathol.* **30** (1): 27.

Nene, Y.L. and Thapliyal, P.N. (1979). Fungicides in plant disease control. Sec. Ed. *Oxford and IBH Publishing Co. Bombay.* p. 507.

Pandey, R.N. and Mishra, D.P. (1992). Assessment of yield loss due to powdery mildew of linseed, *Indian Bot. Rev.*, **11**: 62-64.

Rashid, K. and Duguid, S. (2005). Inheritance of resistance to powdery mildew in flax. *Can. J. Pl. Pathol.*, **27**(3): 404-409.

Rashid, K.Y., Desjardins, M.L., and Duguid, S. (2008). Diseases of flax in Manitoba and eastern Saskatchewan in 2007. *Can. Pl. Dis. Survey.* p. 111.

Saharan, G.S., Mehta, N. and Sangawan, M.S. (2005). Disease of oilseed crops. *Indus Publishing Co., New Delhi.*

Sharma, R.C., Singh, B.P., Thakur, M.P. and Verma K.P. (2002). Chemical management of linseed wilt caused by *Fusariym oxysporum* f. sp. *lini. Ann. Pl. Protec. Sci.*, **10** (2): 290-291.

Singh, B.P.; Shukla, B.N. and Sharma, Y.K. (1981). Effect of rust infection on the oil and protein content of linseed (*Linum usitatissimum*). *JNKVV Res. J.*, **12**: 101.

Singh, H.K., Kumar, Ketan., Kumar, P., Singh, S., Singh, R.B., Maurya, K.N. and Chauhan M.P. (2016). Management of powdery mildew of rapeseed-mustard. *Indian Phytopath.* **69** (4s) 394-396.

Singh, J., Soyarmma, V. and Kerkhi, S.A. (2005). Effect of seed treatment for the control of linseed wilt. *Ann. Pl. Protec. Sci.*, **13** (2): 507-508.

Singh, N.K., Chauhan, Y.S., Kumar, K. and Gupta, R.P. (1989). Inheritance of powdery mildew resistance in linseed. (*Linum usititissimum* L). *Indian J. genetics and Pl. Breeding*, **49** (3): 421-422.

Singh, R.B and Singh, H.K. (2011). Date of sowing and varieties for the management of Alternaria blight of linseed (*Linum usitatissimum* L.). *Proc. Natl. Acad. Sci. India. Sect. - B. Biol. Sci.,* **81** (4): 375-380.

Singh, R.B and Singh, R.N. (2011). Date of sowing and varieties for the management of root-rot wilt complex of linseed (*Linum usitatissimum*). *Indian J. Agric. Sciences,* **81** (3): 287-289.

Singh, R.B. and Singh, R.N. (2003). Management of mildew of mustard. *Indian Phytopath.* **56**: 147-150.

Singh, R.B., Singh, A. K. and Srivastava, R.K. (2003). Assesment of yield loss due to Alternaria blight in linseed. *J. Oilseed Res.* **20** (1): 168-169.

Singh, R.B., Singh, H. K. and Parmar, A. (2014). Evaluation of mycotoxic potential of some higher plants against *Fusarium oxysporum* f.sp. *lini* causing Wilt in Linseed (*Linum usitatisimum*). *J. AgriSearch.,* **1**(1): 26-29.

Singh, R.B., Singh, H.K. and Parmar, A. (2013). Integrated management of Alternaria blight in linseed. *Proc. Natl. Acad. Sci. India. Sect. - B. Biol. Sci.,* **83** (3): 465-469.

Singh, R.B., Singh, H.K., Parmar, A. and Shiwangi (2015). Ecofriendly management of fusarium wilt in linseed (*Linum usitatissimum* L.). *Res. Environ. Life Sci.* **8** (1): 33-36.

Singh, R.B. and Singh, R. N. (2004). Occurrence and management of Alternaria blight of linseed in Eastern India. *Pl. Dis. Res.,* **19** (2): 120-124.

Singh, R.B. and Singh, R. N. (2005). Integrated management of Alternaria blight of linseed. *Indian Farming,* **54** (10): 26-28.

Singh, R.B., Singh. H.K. and Parmar, A. (2012). Identification of resistant sources to *Fusarium* wilt of linseed (*Linum usitatisium* L.). *Plant Archives.***12** (1): 329-330.

Thind, T.S. (1998). Diseases of field crops and their managements. *National Agricultural Technology Information Center,* Ludhiana, India.

Turner, J. (1987). Linseed law: a handbook for growers and Advisors BASF united Kingdom Limited, U.K. p. 356.

3 Major Diseases of Castor and their Management

V.S. Verma, Vishal Gupta, Rishu Sharma and Kavaljeet Kaur

Castor (*Ricinus communis* L.), belonging to the family Euphorbiaceae, is the most important non-edible oilseed crop of arid and semi-arid regions of India. It possesses exceptional oil features for the chemical industry. Castor oil is important in wide range of industries such as nylon, fibers, jet engine lubricants, hydraulic fluids, dyes, detergents, soaps, ointments, greases, varnishes, cosmetics, perfumes *etc.* (Pathak, 2003). Due to constant pressure for renewable fuels, castor has become a potential source of biofuel. In India castor is grown in the states of Gujarat, Andhra Pradesh, Tamil Nadu and Orissa. Castor crop is attacked by several pests and diseases causing huge losses in crop yield. Some of the commonly occurring diseases encountered in the castor crop are discussed here under:

Leaf Blight

Leaf blight of castor has been reported from different parts of India from time to time and is assuming serious proportions in the recent years, particularly in Bumbai area of Maharashtra State. In some other countries also, leaf spot is considered to be one of the serious diseases of castor. In humid years, extensive fructification of the fungus takes place on the capsules of castor plant giving black, sooty appearance, which on shaking gives rise to black clouds of conidia. In some fields about 70 per cent of the plants are reported to be affected with the disease causing serious losses in yield and oil content. When infection takes place in early stages of flower development, buds are killed and the inflorescence turns black. If the disease appears at a later stage, flowers shed without capsule formation, and in mild attack, only individual flowers dry up. The height and general vigour of plants are reduced due to the disease, which affects yield to a considerable extent.

Symptoms

The disease appears on leaves, stem, inflorescence and capsule. At seedling stage, light brown spots first appear on cotyledonary leaves, which become angular with age. The spots coalesce to form large patches of the foliage blight leading to premature defoliation. Severe infection results in the death of young seedlings. Symptoms on adult plant leaves are brown, zonate and variable in size and usually surrounded by a yellow halo. Extensive damage takes places due to infection on the inflorescence and capsules in the form of sooty growth during extended humid climate. Two types of symptoms may be observed on capsules, one involving sudden wilt, purple or dark brown discolouration, collapsing of pedicel or hang down, failure of normal dehiscence and the others characterized by a unilateral sunken area, which gradually enlarges to cover the whole pod. Sunken spots develop on capsules on one side, which gradually enlarge to cover the whole capsule with fungal growth. Such capsules are smaller in size and have underdeveloped or wrinkled seeds with little oil content. The capsules crack and seeds become infected. In heavily affected castor crop, all the young racemes and even flower primordia are killed (Stevenson, 1945).

Pathogen

Causal Organism: *Alternaria ricini*

The conidia of *Alternaria ricini* are produced abundantly on the diseased portion under moist conditions and are borne in chains on the conidiophores. The conidiophores are straight, erect or irregularly bent, solitary, sometime in fascicles and are olivaceous in colour. The conidia are obclavate, light olive in colour, but become darker in mass. They are segmented into 5-16 cells having transverse and longitudinal septa with and without beak. The beak is narrow, colourless, long and unbranched. The conidia formed in culture are usually smaller in size than those formed on the host. The disease is carried over through the seed, both externally and internally (Pawar and Patel, 1957). When diseased seeds are sown, the disease causes pre- and post-emergence damping-off and seedling as well as foliage blight. Presence of a susceptible variety, high atmospheric humidity (85-90 per cent) and low temperature (16-20°C) are favourable for disease development. The pathogen survives on hosts like *Jatropha pandurifloria* and *Bridelia hamiltoniana*. The causal pathogen is externally as well as internally seed-borne and causes primary infection. The secondary infection occurs through air-borne conidia of the pathogen.

Disease Management

Removal of reservoir host is very important. Seed treatment with fungicides is effective in averting the disease in initial phase of the blight (Stevenson, 1945). Seed treatment alone may not be effective. Foliar application of mancozeb (0.2 per cent) at an interval of 15 days starting from appearance of the disease is beneficial and helps contain the disease. Judicious use of nitrogenous fertilizers also helps in reducing the disease development.

Grey Mould

Grey mold of castor was first reported in the USA during the summer of 1918, mainly in Florida and other southern states (Godfry, 1923). Later, the disease was found in all castor growing countries (Kolte, 1995) and now it is distributed worldwide. In India, a major castor producing nation, grey mould is found in Andhra Pradesh and Tamil Nadu, where weather conditions are more favourable for disease development and an epiphytotic of grey mould occurred in 1987 (Dange *et al.*, 2005).

Symptoms

In the development stage, inflorescence and capsules are primary targets of the fungus, (Araujo *et al.*, 2007). The first symptoms are visible as bluish spots on the inflorescence, on both female and male flowers (before anthesis) and on the developing fruits. On fruits, the symptoms can evolve circular or elliptic, sunken, dark coloured spots that may lead to rupturing of the capsule (Araujo *et al.*, 2007). These symptoms are usually more frequent when a period of low relative humidity, unfavourable to fungal sporulation, occurs soon after the fungus penetrates the host tissues. Depending on weather conditions (long periods with high relative humidity soon after the fungus penetrates the host), the occurrence of yellow ooze at the point of infection is frequent as a result of the rapid enzymatic tissue degradation (Dange *et al.*, 2005). The symptoms on the male flowers, before anthesis, are small, pale brown, necrotic spots, which can evolve to larger brown spots with a darker edge. The infected flowers and young capsules become softened due the fungal colonization and mycelial growth is pale grey at first, later becoming dark olivaceous. A profuse sporulation is usually observed at such stage. When infection starts on immature capsules, they become rotten; if the infection starts later, with fully developed capsules, the seeds usually become hollow with coat discoloration and weight loss (Dange *et al.*, 2005). Infection can lead to complete destruction of the raceme, particularly if it reaches the main stem and the weather conditions are favourable for the disease. Several other plant parts *e.g.* leaves, petioles and stem can also be infected, mainly due to the deposition or fall of infected material from the inflorescence or racemes. On leaves, the lesions are usually irregular, but can assume an elliptical or circular pattern, the size is very variable, sometimes coalescing and resulting in foliar blight. On petioles and stems, necrotic, sunken lesions are usually formed, which can cause strangulation and consequently death of plant parts above the infection point (Dange *et al.*, 2005).

Pathogen

Causal Organism: *Botryotinia ricini*

The pathogen penetrates directly through the host cuticle (Godfrey, 1923). After penetrating cuticle, the fungus quickly spreads over the host tissues leading to a complete disorganization and breakdown. Enzymatic action has also been found responsible in the penetration process, without tissues dissolving, prior to infection. Thomas and Orellana (1963) found that it was not possible to verify the direct germ-tube penetration of *B. ricini* through the cuticle or stomata, on castor capsules, before tissue maceration by pectic enzymatic action, suggesting that the fungus first degraded the cuticle and later penetrated the host tissues (Orellana and Thomas, 1962). Probably, *B. ricini* uses both mechanical and chemical processes to penetrate the undamaged host tissue; however, no further studies have been done to clarify these questions. It is likely that enzymes, such as lipases and cutinases play an important role in the infection process similar to several other *Botrytis*-host interactions (Kars and Van Kan, 2007). Hoffmann *et al.* (2004) extracted α and β esterase and superoxide dismutase from *B. ricini*. It is important to highlight great distinction between the penetration process of a fungus under controlled and highly favourable conditions in contrast with the natural process in the field. In the latter case, all aerial parts of the host are potential targets for the deposition and penetration of *B. ricini*, because not only the conidia, regarded as the major propagative unit, usually responsible for the epidemic outbreak, but also the ascospores, sclerotia and mycelial fragments can give rise to infection, as observed in *Botrytis* spp. (Jarvis, 1978). Besides direct penetration, probably natural openings and wounds also serve as a portal of entry for the fungus. Growth of the fungus on the host surface and consequently its penetration in the host tissues will depend on factors such as inoculum type, free water and nutrient availability, cuticle features, presence of exudates on floral organs and other glands, besides the abundance of natural openings and the size and age of wounds (Holz *et al.*, 2007).

Epidemiology

A temperature around 25°C and high relative humidity are highly favourable for the disease development (Godfrey, 1923). The minimum and maximum temperature for mycelial growth was found to be 12 and 35°C, respectively (Godfrey, 1923). Some other studies have confirmed that temperatures around 25°C are favourable to fungal growth and disease development (Sussel, 2009). At temperatures below 20°C, the disease is little expressed and highly dependent on long periods of high relative humidity. There is a high correlation between the temperature and duration of leaf wetness with the disease incidence and severity (Sussel, 2009). At 25°C and relative humidity near saturation, under controlled conditions, the incubation period of *B. ricini* can vary from 4 to 88 hours and the latent period from 72 to 144 hours (Soares *et al.*, 2010)

Disease Management

Protection of inflorescences and immature capsules is crucial to avoid heavy yield losses. Fully developed capsules are less susceptible to pathogen attack and the severity levels are usually lower when compared with infection in young capsules or on the inflorescence. There is no single measure to keep the disease under acceptable levels. As the pathogen has a very short incubation period, easily wind-dispersed, its destructive potential is very high. Different management practices need to be integrated judiciously for keeping the disease losses below economic threshold.

(i) Cultural

Cultural practices are usually applied with the aim to prevent the introduction of inoculum into the field, reducing its survival, spread, build-up or rendering the host less prone to disease attack (Termorshuizen, 2001). The use of varietal resistance is regarded as a better method for disease management. However, there are no varieties with satisfactory resistance levels to grey mold (Milani *et al.*, 2005). Several workers have recommended the use of healthy seeds, removal of plant debris, adequate choice of planting area and growing season, and use of less susceptible cultivars (Galli *et al.*, 1968; Massola and Bedendo, 2005). It is also recommended to use plant spacing in such a way as to allow maximum aeration (Kolte, 1995; Lima *et al.*, 2005). The use of healthy, disease-free seeds, including seed treatment with fungicides is the best options for raising a healthy crop. Elimination of alternate and reservoir hosts (euphorbiaceous hosts), as well as removal and destruction of inoculum persisting in plant residues, are effective practices for the management of grey mold and usually result in lower disease levels. Avoidance of the long wet periods is recommended. (Kolte, 1995; Dange *et al.*, 2005).

(ii) Chemical

Seed treatment with chemicals has been most frequently recommended (Gonçalves, 1936; Batista *et al.*, 1996; Massola and Bedendo, 2005; Sussel, 2009) mainly to avoid introduction of the pathogen into new areas. After disease establishment, spraying of fungicide is adopted to reduce the disease progress. Spraying of systemic fungicides soon after appearance of the first symptoms has been found to reduce disease progress and avert or delay the epiphytotic (Araujo *et al.*, 2007). Two prophylactic sprays with carbendazim (0.05 per cent) - the first at 50 per cent of flowering and the second, when the first disease symptoms appear have been found helpful in checking the disease. Several new active fungicides with distinct modes of action, and usually with high specificity, have provided satisfactory levels of control of many plant diseases. Preliminary studies under controlled conditions have shown that carbendazim and azoxystrobin are effective against the grey mold pathogen (Bezerra, 2007).

(iii) Biological

Biopesticides like *Trichoderma* and *Clonostachys rosea* have been used for the control of grey mold of castor, and promising results have been obtained (Raoof *et al.*, 2003; Tirupathi *et al.*, 2006; Bhattiprolu and Bhattiprolu, 2006; Demant *et al.*, 2006; Chagas, 2010). However, much needs to be done before permanent recommendations regarding the use of biological control agents can be authenticated for grey mold management.

Rust

Rust of castor occurs in southern part of India and has been observed in Mumbai, Deccan districts, Coimbatore and Nagpur. It usually appears in Mumbai between November and February on castor sown in June. The damage caused by this disease has been very severe in moist localities and at places where the disease appeared quite early. In Hyderabad the disease appears only in December when the capsule formation has already started so that little damage is done to the crop.

Symptoms

Minute, orange-yellow coloured, raised pustules appear with powdery masses on the lower leaf surface and the corresponding areas on the upper leaf surface are yellow. The uredeopustules are grouped in concentric rings and coalesce together to form bigger patches. The pustules burst at a later stage exposing the powdery mass of orange yellow uredospores.

Pathogen

Causal Organism: *Melampsora ricini*

The pathogen produces only uredosori on castor plants and other stages of the life cycle are unknown. Uredospores are of two kinds, one is thick-walled while the other is thin-walled. They are elliptical to round, orange-yellow coloured and finely warty. The fungus survives in the self sown castor crops in the off season. It can also survive on other species of *Ricinus*. The fungus also attacks *Euphorbia obtusifolia* and *E. geniculata*.

Etiology

The hyphae of the causal organism (*Melampsora ricini*) collect beneath the epidermis of the leaf and form a minute cushion. The uredospores arise beneath the epidermis on the tips of branched hyphae. Stout, club-shaped paraphyses are also formed in the sorus. At maturity the epidermis ruptures releasing clouds of spores. The uredospores are round or elliptic, warty and orange in colour and measure 25-29 x 19-25µ. They germinate by germ tube, which comes out from the pores and infects fresh leaves of castor. It was observed that the rust culture could not be maintained on host plants under Delhi conditions beyond the beginning of April because of rise in temperature. *M. ricini* is pathogenic to *Euphorbia obtusifolia, E. ipecacuanha, E. giniculata and E. marginata*, besides 24 strains and varieties of *R. communis*. Isolates of *M. euphorbiae* from 5 species of *Euphorbia* including *E. peplus* failed to infect castor plant.

Disease Management

Self-sown castor plants and other weed hosts should be rouged out. Spraying of mancozeb (2 kg/ha) or propiconazole (1 litre/ha) has been found effective against rust disease.

Brown Leaf Spot

Brown spot disease of castor plants has been reported from Bihar, Uttar Pradesh and Andhra Pradesh and is probably present in many other parts of the country. It causes considerable injury to the leaves and is the source of loss of food for the Eri-silkworm which is maintained on castor plant.

Symptoms

The disease appears as minute, brown specks surrounded by a pale green halo. The spots enlarge to greyish-white centre having deep brown margins. The spots may be 2-4 mm in diameter and when several spots coalesce, large brown patches appear which are restricted by veins. Infected tissues often drop off leaving shot-hole symptoms. In severe infections, the older leaves may be blighted and withered.

Pathogen

Causal Organism: *Cercospora ricinella*

The disease is caused by *Cercospora ricinella*. The hyphae of the causal organism collect beneath the epidermis and form very small stomata. Clusters of conidiophores, usually in groups of 10-20, emerge through any part of the leaf tissue and form the fructification of the fungus. The conidiophores are brown below and lighter towards the tip and are septate, unbranched and measure from 24-70 x 3 -6.5μ. The upper part of the conidiophores is characteristically knobby or flexed. The conidia are elongated, colourless, tapering above and truncated below, straight or slightly curved and multi-septate (up to 7 transverse septa). The pathogen remains as dormant mycelium in plant debris and disease usually spread and cause secondary infection through wind borne conidia.

Disease Management

Use of resistant varieties is the most effective method for combating the disease. However, spraying castor crop with Bordeaux mixture (1 per cent) may help contain the disease, but spraying is not desirable where the cultures of Eri-silkworm are reared on castor plants. Spraying twice with mancozeb (0.25 per cent) or carbendazim (500g/ha) at 10-15 days' interval reduces the disease incidence. Treatment of castor seed with thiram or captan (2g/kg seed) before sowing has been found to check the disease.

Powdery Mildew

Powdery mildew of castor has been reported to be prevalent during November to March at Coimbatore in India.

Symptoms

The symptom of the disease comprise of typical mildew growth usually confined to the under-surface of the leaf. When the infection is severe, the upper-surface is also covered by the whitish fungal growth. Light green patches, corresponding to the diseased areas on the under surface appear on the upper surface they are clearly visible when the leaves are held against light.

Pathogen

Causal Organism: *Levillula taurica*

The disease is caused by *Leveillula taurica*. The pathogen is endophytic and consists of hyphae which are intercellular and occupy the spongy parenchyma of the mesophyll. The haustoria penetrate into some of the parenchymatous cells. The conidiophores of the fungus are branched and usually emerge through stomata in aggregation. The conidia are hyaline, varying in shape, bear minute papilla-like projection at the broad end, and are borne singly at the tip of each branch. These germinate readily in water producing a germ tube from one end. The fungus is also reported to produce powdery mildew disease on *Cyamopsis tetragonoloba*, *Capsicum annuum*, *Medicago sativa* and *Vinca pusilla*.

Disease Management

When weather is comparatively dry, spraying twice with wettable sulphur (0.2 per cent) at 15 days' interval, starting from 3 months after sowing has been found effective against powdery mildew disease. Spraying the castor crop with hexaconazole (0.1 per cent) or dinocap (0.05 per cent) at fortnightly intervals has also been found effective against powdery mildew. The variety Jawala of castor is resistant to this disease.

Seedling Blight

Seedling blight of castor plants is known to exist in India since 1909 when it was reported from Pusa, Bihar. It generally appears during rainy season, *i.e.*, about the end of June and continues up to September. The disease is severe in low lying and badly drained fields and it is in such areas that it destroys nearly 30-40 per cent seedlings, particularly those which are 6-8 inches high. The disease has also been reported from Hyderabad in 1947 and from Uttar Pradesh in 1948.

Symptoms

The disease first makes its appearance on both the surfaces of the cotyledonary leaves in the form of roundish patch of dull green colour which soon spreads to the point of attachment causing the leaf to rot and hang down. The infection further spreads to the stem with the result that the seedling is killed either due to the destruction of growing point or by the collapse of stem. The true leaves of seedlings and the very young leaves of older plants may also be affected, but ordinarily not much injury is caused. The leaf spots turn yellow and then brown and concentric zones of lighter and darker brown colour are formed. The outer border is not well defined and is greenish above and brownish-grey below. The disease spots coalesce at a later stage covering almost the entire leaf. The presence of veins, even in the mature leaves, does not obstruct the spread of the disease. The affected leaves shed prematurely. Under moist conditions, a very fine whitish haze is found on the under-surface of the leaf spots. In case of mature plants also the disease may spread from young leaves to the stem through the petiole. When the older leaves are attacked the pathogen however remains localized on the leaf blade. Direct attack of stem is not very common, and if it is attacked, the disease does not cause much damage. This has been considered to be due to the presence of cuticularised epidermis in older stem, which resists penetration. Infection of flowers and green fruits in nature has not been reported though under laboratory conditions these parts get infected.

Causal Organism: *Phytophthora colocasiae*

The causal organism, *Phytophthora colocasiae*, consists of inter- and intra-cellular mycelium which develops inside the host tissue. After a few days of growth numerous branches emerge from the lower epidermis of the leaf, generally through stomata as sporophores, either singly or in twos or threes. A single colourless ovoid or roundish sporangium is borne at the tip of sporophore. A ripe sporangium liberates zoospores when put in water. The number of zoospores varies from 5 to 45 in each sporangium. The zoospores germinate readily by one or rarely two germ tubes. These are formed freely during hot and dry months when sporangia are scanty and retain the power of germination for many months. Oospores are also produced on artificial cultural media. The oospores remain viable for a long period and it is considered possible that these might serve as a source of carrying the parasite over from one crop to the next. Secondary infection spreads rapidly through sporangia provided the weather conditions are favourable. The sporangia are easily disseminated by the wind and germinate readily on the leaves producing zoospores which penetrate by means of germ tubes either through stomata or directly and produce diseased spots within 24 hours and the next crop of sporangia appear in about two days. The fungus also causes infection on young potato, tomato and brinjal plants and seedlings of several garden annuals. It also produces infection on the leaf of sesame. These plants might serve as collateral or alternative hosts of the fungus and help in the perpetuation of the disease.

Disease Management

Damp and low lying localities should be avoided for sowing of castor. Sanitation of the field, including destruction of badly affected plants, should also be practiced. The best control is through cultivation of resistant varieties. Seed dressing with *Trichoderma viride* formulation (4g) or metalaxyl (4g/kg seed) can reduce disease incidence. Soil drenching with copper oxychloride (3g/litre) is also useful. To protect older leaves spraying with Bordeaux mixture or other fungicides may be recommended.

Bacterial Leaf Blight

Bacterial leaf blight of castor is prevalent in USA, Korea, Japan, Brazil, Sudan, South Africa and Uganda, besides India. Castor crop to the extent of 25 per cent was lost due to bacterial blight disease in Texas and Oklahoma in USA. The disease was first reported in India from Gujarat (Patel *et al.*, 1951) and has thereafter been reported from Tamil Nadu, Andhra Pradesh, Rajasthan, Maharashtra and Delhi. However, reliable information regarding losses due to this disease in India is not available.

Symptoms

The disease first appears on the cotyledons as water-soaked lesions, which gradually turn dark brown to black. Small, round, water soaked spots appear on young leaves, which may enlarge to 2-5 mm in diameter and turn somewhat angular. These spots later turn brown to black, dry up and become brittle. These lesions coalesce in favourable weather conditions and the infected leaves turn yellow, dry and result in defoliation. Black lesions may also appear on petioles, stem and branches. Bacterial ooze in the form of shining beads may be observed on the affected leaves in the morning hours.

Pathogen

Causal Organism: *Xanthomonas campestris* pv. *ricini*

The cells of the bacterium are rod-shaped, formed in chains and capsules but no spore and the cells are motile by a single polar flagellum (monotrichus). The bacterium is gram-negative and aerobic. The pathogen enters the host through stomata, hydathodes or wounds. Long distance dissemination occurs through seed and local spread takes place by rains. Maximum disease development occurs when there are frequent showers with humidity ranging from 84-85 per cent or a temperature above 28°C.

Disease Management

Seed infection can be eradicated by hot water treatment of seeds at 58 to 60°C for 10 minutes. Seed treatment with streptocycline and Sandoz 6334 has been reported effective in the management of the disease (Singh *et al.*, 1996). Proper prophylactic sprays of combination of Paushamycin (0.025 per cent) + Copper oxychloride (0.3 per cent) have been found very effective in reducing the disease with increased yield. Removal and destruction of diseased plant debris is recommended to reduce the soil borne inoculums. Deep ploughing after harvest buries the infected stalks and this reduce survival ability of the bacterium in soil. Destruction of possible alternate and collateral hosts is also essential.

References

Araújo, A. E., Suassuna, N. D. and Coutinho, W. M. (2007). Doenças e seu Manejo. In: O Agronegócio da Mamona no Brasil (Eds: D.M.P. Azevedo and N.E. de M. Beltrão), Embrapa Informação Tecnológica, Brasília, Brazil, pp. 283-303.

Batista, F. A. S., Lima, E. F., Soares, J. J. and Azevedo, D. M. P. (1996). Doenças e pragas da mamoneira (*Ricinus communis* L.) e seu controle. Embrapa Algodão, Campina Grande, Brazil, p. 53.

Bhattiprolu, S. L. and Bhattiprolu, G. R. (2006). Management of castor grey rot disease using botanical and biological agents. *Indian Journal of Plant Protection*, **34**: 101-104.

Bezerra, C. S. (2007). Estrutura genética e sensibilidade a fungicidas de *Amphobotrys ricini* agente causal do mofo cinzento da mamoneira. MSc Dissertation (Genetic and Molecular Biology), Universidade Federal do Rio Grande do Norte, Natal, Brazil, pp. 1-47.

Chagas, H. A., Basseto, M. A., Rosa, D. D., Zanotto, M. D., and Furtado E. L. (2010). Escala diagramática para avaliação de mofo cinzento (*Amphobotrys ricini*) da mamoneira (*Ricinus communis* L.). *Summa Phytopathologica*, **36**: 164-167.

Dange, S. R. S., Desal, A. G. and Patel, S. I. (2005). Diseases of Castor. In: Diseases of Oilseed Crops (Eds: G.S. Saharan, N. Mehta and M.S. Sangwan), Indus Publishing Co., New Delhi, India, pp. 211-234.

Demant, C. A. R., Furtado, E. L., Zanotto, M. and Chagas, A. A. (2006). Controle do mofo cinzento com o uso de *Trichoderma*, *Proccedings of 2nd Congresso Brasileiro de Mamona*.04.06.2010.Availablefrom: http://www.cnpa.embrapa.br/produtos/mamona/publicacoes/trabalhos_cbm2/043.

Galli, F, Tokeshi, H, Carvalho, P. C. T, Balmer, E., Cardoso, C. O. N. and Salgado, C. L. (1968). *Doenças da Mamoneira*. In: *Manual de Fitopatologia: Doenças das Plantas e seu Controle*, Editora Agronomica Ceres, São Paulo, Brazil, pp. 292-297.

Godfrey, G. H. (1923). Gray mold of castor bean. *Journal of Agricultural Research*, **23**: 679-715.

Gonçalves, R. D. (1936). Mofo cinzento da mamoneira. *O Biológico* **2**: .232-235.

Hoffmann, L. V., Coutinho, T. C., Duarte, E. A. A., Bandeira, C. M. and Suassuna, N. D. (2004). Cultivo de *Amphobotrys ricini* e detecção das enzimas málica, superóxido dismutase e esterase, *Proccedings of 1st Congresso Brasileiro de Mamona*, 14.06.2010. Available from: http: //www.cnpa.embrapa.br/produtos/mamona/publicacoes/trabalhos_cbm1/126.

Holz, G., Coertze, S. and Williamson, B. (2007). The ecology of *Botrytis* on plant surfaces. In: *Botrytis*: Biology, Pathology and Control. (Eds: Y. Elad, B. Williamson, P. Tudzynski and N. Delen), Springer, Dordrecht, Netherlands, pp. 9-28.

Jarvis, W. R. (1978). Epidemiology. In: The Biology of *Botrytis*. (Eds: J. R. Coley-Smith, K. Verhoeff, W. R. Jarvis), Academic Press, London, UK, pp. 219-250.

Kars, I. and Van Kan, J. (2007). Extracellular enzymes and metabolites involved in pathogenesis of *Botrytis*, In: *Botrytis*: Bi-ology, Pathology and Control. (Eds: Elad Y., Williamson, B., Tudzynski, P, and Delen, N.) Springer, Dordrecht, pp. 99-118.

Kolte, J. S. (1995). *Castor: Diseases and Crop Improvement*. Shipra Publications, Delhi, India, pp. 1-119.

Lima, V. P. T., Graça Leite, E. A, Botrel, E. P., Fraga, A. C. and Castro Neto, P. (2005). Avaliação de ataque de mofo cinzento da mamoneira, variedade Al Guarany 2002, em diferentes espaçamentos. *Proccedings of 2nd Congresso Brasileiro de Plantas Oleaginosas, Óleos, Gorduras e Biodiesel* 12.06.2010. Available from: http: //oleo.ufla.br/anais_02/artigos/t183.

Massola J. R. and Bedendo, I. P. (2005). *Doenças da Mamoneira (Ricinus communis L.)*. In: Manual de Fitopatologia: Doenças das Plantas Cultivadas (Eds: H. Kimati, L. Amorim, A. Bergamin Filho, L. E. A. Camargo and J. A. M. Rezende), Agronomica Ceres, São Paulo, Brazil, pp. 497-500.

Milani, M., Nóbrega, M. B. M., Suassuna, N. D. and Coutinho, W. M. (2005). Resistência da mamoneira (*Ricinus communis* L.) ao mofo cinzento causado por *Amphobotrys ricini*. Embrapa Algodão, Campina Grande, Brazil, pp. 1-22.

Orellana, R. G. and Thomas, C. A. (1962). Nature of predisposition of castor beans to *Botrytis* nd relation of leachable sugar and certain other biochemical constituents of the capsule to varietal susceptibility. *Phytopathology*, **52**: 533-538.

Patel, M. K., Kulkarni, Y. S. and Dhande, G. W. (1951). Bacterial lef blight of castor. *Current Science* **20**: 20.

Pathak, H.C. (2003). Emerging trends in castor seed development. In: Proceedings of National Seminar on castor seed, castor oil and its value added products. 22 May, 2003. Sovent Extract Association of India, Ahmedabad, India, pp. 54-62.

Pawar, V. H. and Patel M. K. (1957). Alternaria leaf spot of *Ricinus communis* L. *Indian Phytopathology*, **10**: 110.

Raoof, M. A. Yasmeen, M. and Kausar, R. (2003). Potential of biocontrol agents for the management of castor grey mold, *Botrytis ricini* Godfrey. *Indian Journal of Plant Protection*, **31**: 124-126.

Singh, S., Majumadar, D. K., Rehan, H. M. S. (1996). Evaluation of anti-inflammatory potential of fixed oil of *Ocimum sanctum* (Holybasil) and its possible mechanism of action. *Journal of Ethnopharmacology*, **54**: 19-26.

Soares, D.J., Fernandes J.N and Araujo, A.E. (2010). Componentes monociclicos do mofo cinzento (Amphobotrys ricini) em diferentes genotipos de mamoneria, *Proceedings of 4 th Congresso Brasileiro de Mamona*, 12.07.2011. Available from http: //www.cbmamona.com.br/pdfs/FTT-01.pdf.

Smith, E. F. and Godfrey, G. H. (1921) Bacterial wilt of castor beans (*Ricinis comminis* L.). *Journal of Agricultural Research*, **21**: 255-261.

Stevenson, E.C. (1945). *Alternaria ricini* (Yoshii) Hansfor, the cause of a serious disease of the castor bean plant (*Ricinus communis*) L.). *Phytopathology*, **35**: 249.

Sussel, A. A. B. (2009). Epidemiologia e Manejo do Mofo-cinzento-da-mamoneira. Embrapa Cerrados, [Documentos 241, Brasília, Brazil, pp. 1-27.

Termorshuizen, A. J. (2001). Cutural Control. In: Plant Pathologist's Pokcket Book. (Eds: J. M. Waller, J. M. Lenné and S. J. Waller), CABI Publishing, Wallingford, UK, pp. 318-327.

Thomas, C. A. and Orellana, R. G. (1963) Biochemical tests indicative of reaction of castor bean to *Botrytis*. *Science*, **139**: 334-335.

Tirupathi, J. Kumar, C. P. C. and Reddy, D. R. R. (2006). *Trichoderma* as potential biocontrol agents for the management of grey mold of castor. *Journal of Research, ANGARU*, **34**: 31-36.

4 Diseases of Safflower and their Management

P. Kishore Varma and Mohan V. Totawar

Safflower (*Carthamus tinctorius* L.) is an important *rabi* rainfed oilseed crop. It is grown in *rabi* season preceding either fallow kharif or short duration fallow pulses such as greengram, blackgram in Andhra Pradesh, Maharashtra and Karnataka or soybean in Madhya Pradesh. The crop is known for its healthy oil with high PUFA (76 per cent Linoleic acid) content.

India occupies premier position in safflower in the world as it was cultivated over an area of 364 thousand hectares (50 per cent of world area) and had a production of 229 thousand tons (27 per cent of world production) during 2005-06. However, after attainment of the peak area in 1988 (69 per cent of world area) and peak production in 1994 (69 per cent of world production), the area and production have been in continuous decline. This decline has been attributed by Nimbkar (2008) due to the following reasons:

1. Higher remuneration than safflower obtained for competing crops like sorghum and gram over the years.
2. Low oil content of 30 per cent or less.
3. Import of cheap palm oil.
4. Susceptibility to various biotic and abiotic stresses such as aphids (*Uroleucon compositae* Theobald), wilt caused by *Fusarium oxysporum* f. sp. *carthami*, moisture stress and low nutrients.

Safflower production in India can be affected by many diseases caused by various fungi, bacteria and viruses as well as by disorders from environmental conditions. Diseases may be a problem in safflower, especially in years of above normal rainfall with extended periods of high humidity. Among the biotic factors, about 50 diseases are known to be incited by fungi, bacteria, viruses and mycoplasma. However, in India only a few of them have economic importance based on the weather prevailing in a particular region. American Phytopathological society has given the list of safflower diseases in its website www.apsnet.org which is as follows:

Bacterial Diseases	
Bacterial leaf spot and stem blight	*Pseudomonas syringae* van Hall

Fungal Diseases	
Alternaria leaf spot	*Alternaria carthami* Chowdhury
Botrytis hard rot	*Botrytis cinerea* Pers.: Fr.
Cercospora leaf spot	*Cercospora carthami*
Charcoal rot	*Macrophomina phaseolina* (Tassi) Goidanich
Damping-off	*Pythium* spp., *Phytophthora* spp.
Fusarium wilt	*Fusarium oxysporum* Schlechtend.: Fr. f. sp. *carthami* Klisiewicz and Houston
Phytophthora root and stem rot	*Phytophthora* spp.
	P. cactorum (Lebert and Cohn) J. Schröt.
	P. cryptogea Pethybr. and Lafferty
	P. drechsleri Tucker
	P. nicotianae Breda de Hann var. *parasitica* (Dastur) G.M.Waterhouse = *P. parasitica* Dastur
Pythium root rot	*Pythium* spp.
Powdery mildew	*Erysiphe cichoracearum* DC. (anamorph: *Oidium asteris-punicei* Peck)
Ramularia leaf spot	*Ramularia carthami, R. cercosporelloides*
Rhizoctonia blight, stem canker	*Rhizoctonia solani* Kühn (teleomorph: *Thanatephorus cucumeris* (A.B. Frank) Donk)
Rust (foliage and hypocotyl)	*Puccinia calcitrapae* DC. var. *centaureae* (DC.) Cummins
	= *P. carthami* Corda
	P. verruca Thuem.
Sclerotinia stem rot and head blight	*Sclerotinia sclerotiorum* (Lib.) de Bary
Septoria leaf spot	*Septoria carthami*
Verticillium wilt	*Verticillium dahliae* Kleb.

Virus Diseases	
Alfalfa mosaic	Alfalfa mosaic virus
Cucumber mosaic	Cucumber mosaic virus
Lettuce mosaic and necrosis	Lettuce mosaic virus
Turnip mosaic and necrosis	Turnip mosaic virus

Mollicutes	
Phyllody	Phytoplasma

Safflower diseases can be categorized as soil borne and foliar diseases. Safflower suffers severely from soil pathogens, which may attack seed, germinating seed, and young seedlings or at the time of seed formation, causing directly or indirectly yield and quality losses. Fusarial wilt of safflower is the more devastating soil borne diseases of safflower under irrigated conditions and wherever monocropping is practiced.

The foliar diseases have been serious in areas where high relative humidity prevails from flowering to maturity. In India, more serious and wide spread foliar disease is the leaf spot caused by *Alternaria carthami*. The other leaf spots incited by Cercospora, Ramularia, Botrytis, Erysiphe and Septoria are of localized concern. A brief description of important diseases of safflower is presented in this chapter.

Fusarium Wilt

Wilt incited by *Fusarium oxysporum* f.sp. *carthami* is a major soil borne disease in safflower particularly under irrigated conditions. It has become endemic in major safflower growing areas in the country. Fusarium infects the plant right from seedling stage to flowering and can cause upto 100 per cent yield loss under severe incidence (Prasad *et al.*, 2006).

Symptomatology

The initial symptoms are seen in cotyledonary leaves as small brown spots either scattered or arranged in a ring on the inner surface. The infected cotyledons may be shrivelled or rolled or curved as the infection progress. Above ground symptoms become apparent when plants are in 6-10 leaf stage. Initial symptoms often include a dull green appearance of the lower leaves that precedes a loss of turgor pressure and wilting. Wilting is followed by chlorosis of the leaves followed by necrosis. The wilting generally starts with the older leaves and progresses to the younger foliage. The prominent field symptoms appear when the plant putforth capitula. Plants die

Figure 4.1: Wilt Affected Safflower Plants.   **Figure 4.2: Partial Wilt.**

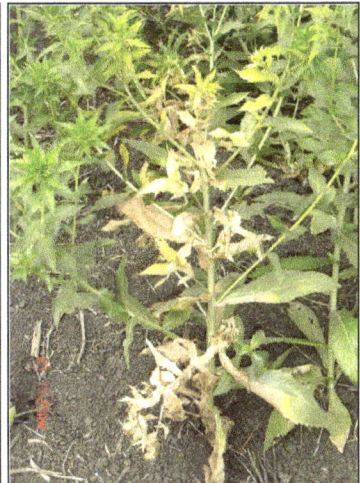

in patches under conditions of sufficiently high inoculum density coupled with cultivation of very susceptible varieties like JSF-1, Bhima and Nira.

The primary diagnostic symptom of *Fusarium* wilt is a discoloration of the vascular system (xylem), which can be observed readily in longitudinal or cross section of roots or stems. In older plants the lateral branches on one side may be killed while the remainder of the plant remains free from the disease (partial wilt). Partial wilting of the plant is a definite indication of *Fusarium* wilt and distinguishes from *Phytophthora* root and stem rot that kills the whole plant. Infected plants produce small sized flower heads which are partially blossomed. Most of the ovaries fail to develop seeds or they may form blackish, small, distorted chaffy seeds.

Causal Organism

The disease is caused by sporodochial fungus, *Fusarium oxysporum* Schlechtend.: Fr. f.sp. *carthami* Klisiewicz and Houston. The fungus belongs to the class Hyphomycetes, order Tuberculariales, and family Tuberculariaceae of the sub-division, Deuteromycotina. The pathogen is known to produce chlamydospores, macroconidia and microconidia. Several studies revealed morphological, cultural and pathogenic variability among isolates of *Fusarium oxysporum* f. sp. *carthami*. The colony diameter of the isolates ranged from 51.6 to 84.4 mm after 7 days of incubation at 25 ±1°C on Potato Dextrose Agar. The dry mycelial weight of the isolates ranged from 56.0 to 252.0 mg. The size of micro- and macro-conidia ranged from 8.0 to 14.8 x 2.7-2.9 µm and 17.5 to 26.8 x 4.0-4.5 µm, respectively, whereas the diameter of chlamydospores ranged from 5.8 to 9.5 µm. The number of microconidia ranged from 4.8 to 28.2 x 10^4/mm², whereas the number of macroconidia ranged from 1.40 to 6.8 x 10^4/mm² on PDA at 15 days after incubation at 25±1°C (Prameela *et al.*, 2005).

Twenty Indian isolates were categorized into five groups based on their reaction to the varieties, Bhima, HUS-305, Manjira and Nira (Raghuwanshi *et al.*, 2008). The first group consisted of those isolates which had incompatibility with the genotype HUS-305 only. The second group had the isolates which showed compatibility with all cultivars. The third group had the isolates which were incompatible with Nira. The fourth group isolates were incompatible with Bhima, HUS-305 and Manjira. Group five included only one isolate which was incompatible with HUS-305 and Nira.

Murumkar and Deshpande (2009) reported that most of the isolates from Maharashtra showed moderately resistant (1-10 per cent wilting) and tolerant (10-20 per cent wilting) reaction on safflower variety NARI-6 as well as on hybrids *viz.*, NARI-NH-1 and DSH-129. However, all the *Fusarium oxysporum* f. sp. *carthami* isolates were pathogenic (>50 per cent wilting) to safflower variety Nira and Bhima.

Survival and Spread

The fungus survives in seed, soil and infected plant debris. The primary spread is by soil-borne chlamydospores and also by seed contaminant. The secondary spread in the field is through irrigation water and implements.

Management

Resistant cultivars are of prime importance in the management of Fusarium wilts; however, there are a number of options that can help in reducing the severity of the disease.

Host Resistance

Use of resistant varieties for the management of *Fusarium* wilt is the best and most economical method of control. Pallavi *et al.* (2007) has reported wild safflower species like *Carthamus glaucus, C. creticus, C. lanatus, C. turkestanicus* and *C. oxycantha* as novel sources of resistance to fusarial wilt. Many lines from crosses of *Oxycantha-tinctorius* and *tinctorius-turkestanicus* and germplasm lines GMU 1502, 1870 and 1973 showed immunity to wilt (Prasad *et al.*, 2006). Avoid cultivating susceptible varieties like JSF-1 and Nira in problematic areas (Murumkar *et al.*, 2008c)

Crop Rotation

The same crop should never be grown in succession on the same land as successive cropping can cause the pathogen to build up to high levels leading to mass mortality. The pathogen survives as chlamydospores in the soil, crop rotation with non-host crops like chickpea, cowpea and pigeonpea reduce wilt incidence and ensure better yields

Cropping Systems

Cropping system is known to influence the fusarial wilt incidence in safflower. Studies at DOR, Hyderabad has revealed that cropping system involving sorghum in *kharif* followed by safflower in *rabi* for successive years increased the incidence of wilt. Chickpea followed by safflower resulted in low disease incidence. The system sorghum + chickpea-sorghum + safflower were found most profitable with a high B: C ratio (Prasad *et al.*, 2006).

Soil Solarization

Soil solarization or slow soil pasteurization is the hydro/thermal soil heating accomplished by covering moist soil with polyethylene sheets as soil mulch during summer months for 4-6 weeks. Soil solarization for six weeks in ploughed and irrigated fields reduces the pathogen propagules in the soil; however the adoption of technology by resource poor farmer is very difficult (Kalpana Sastry and Chattopadhyay, 1999).

Seed Treatment

Seed treatment with the fungicides (Captan, Mancozeb and Carbendazim) and leaf extracts of *Becopa monniera* and *Adathoda vasica* were found extremely effective in reducing safflower wilt and improved seed germination, seedling vigour and plant stand under greenhouse and field conditions and also effectively suppressed the wilt up to flowering (Govindappa *et al.*, 2011; Saroj Singh *et al.*, 2006).

Biological Control

Role of antagonists like *Trichoderma viride*, *Trichoderma harzianum* and *Trichoderma virens* in the management of safflower wilt was well established (Chattopadhyay and Sastry, 1997). Seed treatment with *T. viride/T. harzianum* @10g/kg seed was recommended for the management of the disease (Prasad, 2007). The bioagents *Aspergillus fumigatus* (9.26 x 10[7] spores/ml) and *Trichoderma harzianum* (8.72 x 10[7] spores/ml) significantly reduce the disease incidence of safflower in inoculated soil in pots under greenhouse conditions (Gaikwad and Behere, 2001).

Integrated of Management Practices (IDM)

IDM studies revealed that carbendazim + *Pseudomonas fluorescens* seed treatment along with soil application of recommended potash is the best for safflower wilt management (Prasad *et al.*, 2006). The wilt disease can also be managed effectively by adopting a moderately susceptible variety like A-1 and application of *Trichoderma* species either to the soil or on the seed (Prasad and Anjani, 2008).

Root Rot

Root rot caused by *Macrophomina phaseolina* (Sclerotial stage: *Rhizoctonia bataticola*) is a sporadic disease of safflower in some areas. Moisture holding capacity of 50 per cent was conducive for maximum disease development and survival of the pathogen (Lukade and Rane, 1993).

Symptoms

The first symptom is general wilting of the plant during the middle of hot days followed by a recovery in the evening as temperature declines. The stems of infected plants eventually take on a gray discoloration at the base and finally, the vascular bundles may become covered with microsclerotia of the fungus. Since charcoal rot restricts the flow of water and nutrients to the upper parts, ultimately reduce the seed sizes. Disease incidence and severity are often more when maturing plants are stressed by drought and high temperature which leads to premature plant death. In some cases, the pathogen kills up to 25 per cent of the plants in commercial fields of safflower (Govindappa *et al.*, 2005).

Figure 4.3: Root Rot.

Causal Organism

The disease is caused by a pycnidial fungus *Macrophomina phaseolina* (Tassi) Goidanich. The fungus belongs to the class Coelomycetes, order Sphaeropsidales, and family Sphaeropsidaceae of the sub-division, Deuteromycotina. *Macrophomina phaseolina* is highly variable, differing in size of sclerotia and the presence or absence of pycnidia. The mycelium is superficial or immersed, hyaline to brown, branched, septate, often dendroid in form. The pycnidial stage is uncommon, however, some strains produce pycnidia on culture media. Pycnidia are dark brown, solitary, immersed, 100-200µm in diameter, becoming erumpent opening by apical ostioles. The pycnidial wall is multicellular with heavily pigmented thick walled cells on the outermost side. Conidiophores are hyaline, short, obpyriform to cylindrical, 5-13 x 4-6 µm in size bearing hyaline, ellipsoid to obovoid conidia, 14-30 x 5-10 µm in size.

Survival of Pathogen

In cultures, sclerotia are most common. On the host these structures are seen within roots, and stems. Sclerotia are black, smooth and hard, 100µm to 1mm in diameter. Sclerotia of the fungus can survive free in soil or embedded in host tissues. Survival is longer in dead host tissues than in free condition in the soil. Dry soils prolong survival of these sclerotia. In wet soil sclerotia cannot survive for more than seven to eight weeks while the mycelium cannot survive for more than seven days.

The disease symptoms appear at temperatures of 28 to 35°C. Seedling blight is seen in tropical countries only where soil temperatures are 30°C or above at planting time. Soil moisture stress further enhances disease severity.

Management

Management strategies to control charcoal rot in safflower include crop rotation, lower plant density and scheduling planting date and irrigation to reduce the effect of mid-season drought stress (Smith and Carvil, 1997). Planting resistant cultivars is the most permanent and practical way for the control of the disease, because above mentioned strategies fail to provide adequate control.

☆ Grow resistant or moderately resistant varieties like AKS-152, AKS-68, AKS-216 and AKSF-13. Avoid growing highly susceptible varieties like Bhima and A-1 in endemic areas (Ingle *et al.*, 2004).

☆ Application of Farm yard manure (40 cartloads per hectare) or wheat straw, paddy straw and press-mud cake reduces the incidence of root rot by triggering the antagonistic microorganisms in the soil (Lukade, 1992; Lukade and Rane, 1994).

☆ Treatment of the seed with *Trichoderma harzianum* and *Pseudomonas fluorescens* (@10g/kg) enhanced the seed germination and growth parameters against root-rot disease and they also induced systemic resistance and/or physiological changes leading to plant defense mechanisms. Biocontrol agents have triggered defense related enzymes involved in phenyl proponoid pathways and phenols. Higher activity of peroxidase, phenylalanine ammonia-lyase, chitinase, polyphenol oxidase and beta -1,3-glucanase was observed in *P. fluorescens* and *T. harzianum* treated safflower plants after challenge inoculation with *M. phaseolina* (Govindappa *et al.*, 2010).

☆ Treat the seed with carbendazim @2g/kg seed (Prashanthi *et al.*, 2000)

☆ Spot drench with 0.1 per cent Benomyl (Mathur and Tyagi, 1988).

Alternaria Leaf Spot

Alternaria blight is a serious disease problem in safflower (*Carthamus tinctorius* L.) in Canada, Ethiopia, India, Israel, Kenya, Tanzania, U.S.S.R., Zambia and possibly also Rhodesia, Rumania and U.S.A (Ellis and Holliday, 1970). This disease is the main constraint in safflower production in India under humid conditions. The disease occurred in epidemic form during 1997 in all safflower growing areas of Maharashtra, Andhra Pradesh and Karnataka states of India.

In India, the disease is reported to cause 25-60 per cent yield loss every year (Singh and Prasad, 2005). Survey and surveillance in different parts of the country indicate 5-40 per cent and 58-70 per cent incidence in Ranga Reddy district of Andhra Pradesh and Marathwada region of Maharashtra respectively (Kishore Varma *et al.*, 2007; Relekar *et al.*, 2010). Survey on intensity of Alternaria leaf blight of safflower in northern India revealed 27-90 per cent yield loss when the disease appears at early stage of crop growth (Krishna Prasad, 1988).

Symptomatology

The initial symptoms appear during seedling stage as dark brown lesions on hypocotyls and cotyledons. Symptoms appear before flowering on all parts, but especially on the leaves. Isolated light brown to dark brown lesions appear on lower leaves which gradually progress upwards. The centre of the spot is light brown with a dark brown margin. The spots increase in size under continuous cloudy and wet weather conditions developing concentric pattern in the mature spots. Shot-holes may develop and coalescence of the spots leads to large, irregularly outlined lesions and breakup of the lamina. Elongated black lesions can be seen on the petiole and stem. The infection spreads to the flower buds under high humid conditions leading to drying and shedding of capitulum. Dark sunken lesions are also seen on the testa of seed making them shriveled and empty. Seed yield and oil content is reduced (Ellis and Holliday, 1970). The pathogen is reported to produce a phytotoxin known as zinniol (Prasad *et al.*, 1998).

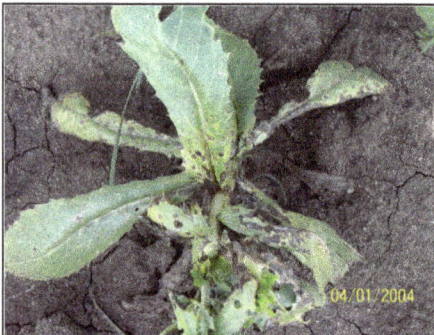

Figure 4.4: Alternaria Leaf Spot.   **Figure 4.5: Concentric Pattern in Mature Lesion.**

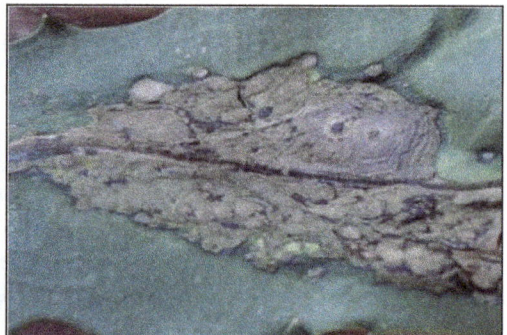

Portals of Entry

Spines present on the leaf margin in safflower (*Carthamus tinctorius*) cultivars were observed to be a site of infection by the fungal pathogen *A. carthami*. The infection through spines is known to occur whenever the pathogen was externally seedborne, serving as a primary source of inoculum. An opening diameter of 120μ at the apex of individual spines was a pre-requisite for infection through spines. Spine apex openings varied with the location of the spine on the leaf margin and, in general, increased at a rate of 7μ/day until they reached 120μ. Therefore, the relationship between the position of the spines on the leaf margin and infection was governed by the diameter of the opening at the spines apex (Borkar, 1997).

Causal Organism

The causal agent, *Alternaria carthami* was first isolated and described by Chowdhury in 1944 from safflower leaves. The fungus belongs to the class Hyphomycetes, order Moniliales, and family Dematiaceae of the sub-division, Deuteromycotina. The fungus produces simple, erect, straight or flexuous, brown or olivaceous brown conidiophores bearing brown and multi-septate conidia. Conidia are solitary or in very short chains, straight or curved, obclavate, sometimes constricted at the septa, 60-100 x 15-26 μm in size with 7-11 transverse septa and upto 7 longitudinal or oblique septa. The conidia are straight or curved, light brown and translucent in shade, and possess a long beak (Chowdhury, 1944). The conidia measure 36 to 171 μm (with beak) and 36 x 99 μm (without beak) in length, and 12 to 28 μm in width with 3 to 11 transverse septa and upto 7 longitudinal septa. The optimum growth *in vitro* was found between 25 to 30°C at pH 6-7.

Survival and Spread

The pathogen survives in seed (Externally seed borne) and in infected plant debris. Primary infection develops from infected seed. Secondary infection takes place through airborne conidia. The pathogen is also known to infect and survive on snow lotus (*Saussurea involucrata*) in China (Zhao *et al.*, 2008)

Favourable Conditions

Available reports indicate that the rains received coupled with high RH above 80 per cent and temperature in the range of 21-33°C favoured the primary infection of the crop (Murumkar *et al.*, 2008a). The disease becomes particularly severe in irrigated crops and in warmer areas where periods of dew or frequent rainfall occur.

Management

Cultural

The pathogen is externally seed borne; hence, use of disease free seed is important in reducing the primary inoculum. The pathogen is also known to survive on infected plant debris in soil. So, the infected plant debris needs to be collected and destroyed. Advancing planting time to August exposes safflower crop to more congenial weather conditions that favour *Alternaria* disease development, there by causing maximum yield loss. Hence, adopt recommended planting time in problematic areas and avoid sowing safflower in low lying areas. Proper fertilization with 60 kg N/ha+20 kg P/ha recorded lowest *Alternaria* leaf spot incidence compared to recommended dose of fertilizers (Tripathi, 2000)

Host Resistance

Sources of resistance to *Alternaria* are not available in cultivated safflower gene pool. The wild safflower species, *Carthamus palaestinus, C. lanatus, C. creticus* and *C. turkestanicus* were immune to *A. carthami* and can be used in resistance breeding programme to introgress the resistance genes into susceptible high yielding cultivated safflower genotypes (Prasad and Anjani, 2008). Many interspecific derivatives having resistance to the disease were identified at DOR, Hyderabad. A variable degree of resistance is observed in safflower lines EC 32012, NS 133, 1015, 1016, 1021 and 181866A, GMU lines 822, 1921, 163, 487, 624, 645, 1824 and 1825. The hybrids DSH-129, NARI-NH-1 and variety NARI-6 are moderately resistant to the disease (Prasad *et al.*, 2006). Three genotypes, *viz.*, GMU-3706, GMU-4538 and GMU-5133 were found to be tolerant at Solapur (Murumkar *et al.*, 2009).

Chemical Management

As the pathogen is externally seed borne, the seed has to be treated with common seed protectants like Thiram and Captan @ 2.5 g/kg seed. Seed treatment with mefenoxam+thiram and difenoconazole+mefenoxam were found to increase total seed yield and oil yield per hectare over untreated seed (Jacobsen *et al.*, 2008). Two sprays with difenaconazole@0.05 per cent starting from first appearance of the disease on lower leaves were found effective in the management of the disease (Sumitha and Nimbkar, 2009). Spraying thrice with carbendazim@0.1 per cent immediately after disease appearance (generally at rosette stage, *i.e.*, 25 DAS) followed by need based second and third sprays at 15 days after first spray and during flowering/seed setting stage is recommended for effective and economical management of Alternaria leaf spot (Murumkar *et al.*, 2008b). A single application of azoxystrobin applied at first flower increased yields and oil content equal to two or three applications of other fungicides applied at different stages of growth, *viz.*, first sign of disease, first flower initiation and 14 days after that (Bergman and Jacobsen, 2005).

Plant Extracts

Bulb extract of *Allium sativum* L. and leaf extract of *Datura metel* L., *Datura innoxia* L. and *Nerium indicum* L. have inhibited the mycelium growth of *A. carthami* under laboratory conditions (Ranaware *et al.*, 2010; Shinde *et al.*, 2008). Flavonoid, cabruvin (200, 500 and 1000 mg/ml) extracted from roots of *Clerodendrum infortunatum* exhibited good inhibition of spore germination of *A. carthami* (Roy *et al.*, 1996).

Ramularia Leaf Spot

Ramularia leaf spot is one of the major diseases of safflower under irrigated conditions in Maharashtra, which can cause a yield loss of 18-23 per cent (Prasad *et al.*, 2006). Epiphytotic occurrence of the disease was observed in Jhansi of Uttar Pradesh during 1983.

Symptoms

Two species were reported on safflower, *viz.*, *Ramularia carthami* and *R. cercosporelloides*, of which *R. carthami* is the most important in India. High incidence of the disease is favoured by temperatures greater than 28°C with high relative humidity.

Ramularia carthami: Numerous circular to irregularly shape brown lesions, 3-10 mm in diameter, are produced on both sides of leaves and flower bracts. Severe infection results in stunting of plants and reduction in seed yield (Hostert *et al.*, 2006).

Ramularia cercosporelloides: Leaf spots were initially pale which become sunken, and necrotic with age. Under continuous presence of dew, new infections occur and the number of lesions increase, coalesce and eventually the entire leaf and plant turned brown and dried up. The fungus appears as a whitish mould covering the lesions (Huerta-Espino, 2006).

Causal Organism

The fungus belongs to the class Hyphomycetes, order Moniliales, and family Moniliaceae of the sub-division, Deuteromycotina.

Ramularia carthami: Mycelium is hyaline, septate, branched, 1.5 to 3µm wide, forming small to moderately large stromata in the leaf spots. Conidiophores are produced in small to moderately rich fascicles emerging through the stomata or rarely through the cuticle. Conidiophores are simple, rarely branched, continuous or septate, hyaline, smooth bearing ellipsoid to ovoid, subcylindric to fusoid conidia. The conidia are 0 to 3 septate, hyaline, smooth to faintly rough, ends rounded or attenuated with somewhat thickend and darkened hilum.

Ramularia cercosporelloides: Mycelium is internal; hyphae hyaline or subhyaline, septate, sparsely branched, forming small to well developed substomatal stromatic hyphal aggregations. Conidiophores arising from stromata form small to moderately large loose fascicles. Conidiophores are erect, geniculate to sinuous, simple, 20-90 x 3-6µm in size, continuous to septate, hyaline or with a faintly yellowish or greenish tinge and smooth with thickened, darkened and refractive conidial scars. Conidia formed singly or in short chains. Conidia are hyaline or with fainty yellowish or greenish tinge, obovoid, cylindrical, 20-45 x 3.5-10µm with 0-3 septa which are partly constricted at the septa with rounded or occasionally truncate apex.

Survival and Spread

The pathogen can survive as conidia in plant debris (83 days) and also as sclerotia in infected leaves and seedlings for more than 8 months (Patil and Hegde, 1990). The pathogen is disseminated through wind borne conidia. High temperature, normal R.H with dry rainless days was found to favour disease development (Kumar and Joshi, 1995).

Management

☆ In Mexico, four linoleic varieties (CIANO-LIN, RC-1002, RC-1005 and RC-1033) and a new oleic variety were found highly tolerant to *Ramularia carthami* (Montoya Coronado, 2008). In India, GMU-822, 827 and 828; C1609, B-3-8-7, 195925-1, B-445, C-1611, -278383, B461, EC-170274, NS-99a, 199935C and BLY-1080 were identified as resistant lines (Prasad *et al.*, 2006).

☆ Spray thrice with 0.2 per cent mancozeb and 0.05 per cent carbendazim at 15 days interval starting from 55 days after sowing (Patil and Hegde, 1989). Chitosan (3.4g/L), an elicitor of plant defenses and also an inhibitor of fungal growth, can also be used for the management of Ramularia leaf spot (Quintana-Obregon *et al.*, 2011)

Rust

Safflower rust is an obligate parasite with an autoecious life cycle. The disease is reported from Africa (Egypt, Ethiopia, Kenya, Sudan); Asia (India, Iran, Iraq, Israel, Japan, Pakistan, Syria, Turkey, U.S.S.R.); Europe (Austria, Bohemia, Cyprus, France, Germany, Italy, Malta, Portugal, Rumania, Spain) and N. America (Canada, U.S.A.).

Symptoms

The pathogen chiefly infects leaves, but is also prevalent on the foot, root and tender stems. Seedlings and older plants are equally susceptible. Affected plants initially show discolouration of leaves and this is later followed by drooping or wilting. In certain cases seedlings have been known to die without showing any above ground symptoms. Infection on the hypocotyl causes hypertrophy and in older plants girdling of invaded areas is very characteristic. Severe infection leads to reduction in stands and serious damage to safflower production (Punithalingam, 1968).

Causal Organism

Safflower rust is caused by *Puccinia carthami* (Hutz.) Corda. The fungus belongs to the class Teliomycetes, order Uredinales, and family Pucciniaceae of the sub-division, Basidiomycotina. The fungus is an obligate parasite with autoecious life cycle in safflower. Pycnia are amphigenous, flask shaped or spherical, 80-100 μm in diameter, seen in small groups. Aecia are uredinoid, amphigenous, chestnut brown in colour and are associated with pycnia in clusters. Uredia are amphigenous, upto 0.5mm in diameter and are scattered. Uredospores are single celled, light brown coloured and echinulate and measure 21-28 x 19-25μ. Teliospores are globose to broadly ellipsoid, two celled, chestnut brown in colour, thick walled with hyaline pedicels and measure 28-39 x 21-26 μm. The uredospores germinate readily at 18-20°C and remain viable for an year or longer.

Three other rusts are recorded on Carthamus. *Puccinia carduncelli* Syd. is recorded from Silicy can be distinguished from *P. carthami* in having urediniospores with two superequitorial pores and teliospores of variable size and shape. *Puccinia veruca* Theum. is a microcyclic species with compact telia in large groups and teliospores with thickened apex. *Melampsora ricini* Noroha has been reported on safflower from Ethiopia and can be distinguished by the paraphysate uredia and hyaline to pale yellow urediospores.

Epiphytotics occurred in USA during 1949 and 1950. A severe outbreak also occurred in the South of France in 1951. Infection takes place entirely by teliospores and the optimum temperature for germination is 12-18°C. The soil temperature has a marked influence on the pathogenicity of soil and seed borne inocula. Low temperature (15 - 25°C) appears to have a favourable effect on the foot and root phase of the disease while high temperature (30-35°C) seems to check disease development.

Disease Cycle

The fungus remains on the seeds and infected crop debris in the soil as teliospores for more than a year. Seed and soilborne teliospores account for the widespread occurrence of the disease. The fungus also produces uredial and telial stages in the collateral host *Carthamus oxyacantha* and this also serves as primary source of infection in addition to dormant teliospores in soil. The secondary spread occurs through wind-borne uredospores.

Survival and Spread

Seed and soilborne teliospores account for the widespread occurrence of the disease. Teliospores remain viable for 12 months under field conditions and on infected safflower straw stored at 5 °C for a period of 45 months. The pathogen also infects wild safflower species like *Carthamus glaucus, C. lanatus, C. oxycantha, C. syriacus* and *C. tenuis*. Secondary spread of the disease is through wind borne uredospores.

Management

☆ Resistant genotypes play a major role in the management of safflower rust. The genotypes SF1954, SF1956-1, SF1946-1, SF1951, SF1960-1, SF1971, G-2219-10, APR-1 to 5 and EC-11972 were found to be rust resistant in India.

☆ Seed treatment with Chlorothalonil @ 3g/kg seed is found effective in controlling rust disease (Prasad *et al.*, 2006).

Cercospora Leaf Spot

The leaf spot is caused by *Cercospora carthami* and the disease is seen in India, Philippines, Israel and East Africa.

Symptoms

Leaf lesions are circular to irregular with a diffuse or a darker margin surrounded by a yellow brown halo. These lesions are individually 0.2 to 0.5 cm in diameter, discrete or confluent, especially seen at the leaf margins. These small brown lesions coalesce contributing to general necrosis. Severely infected leaves are misshapen and the diseased tissue disintegrates producing a shot hole effect. Young flower heads may be destroyed. High relative humidity during November-December and unseasonal rainfall favours disease development

Figure 4.6: Cercospora Leaf Spot.

Causal Organism

Leaf spot is caused by *Cercospora carthami* (H. and P. Sydow) Sundararaman and Ramakrishnan. The fungus belongs to the class Hyphomycetes, order Moniliales, and family Dematiaceae of the sub-division, Deuteromycotina. The fungal mycelium is hyaline initially and gradually turns to smoky brown. The pathogen produces dark brown stromata upto 80µm in size beneath the epidermis. Conidiophores are amphigenous, but more abundant on upper leaf surface, also occurring on bracts, capitula and stems under severe infections. The conidiophores are fasciculate and are produced in groups of 12 to 40, more or less straight or flexuous, geniculate, unbranched, uniformly, pale brown, smooth, 2-4 septate mostly in the lower part, 100-150 x 3.5-4.5 µm in size. Conidiogenous cells polyblastic, integrated, terminal, sympodial with 1 to 4 prominent thickened scars. Conidia solitary, subulate, straight or curved hyaline, truncate at the base with a prominent thickened scar, 80-220 x 2.5-4 µm in size, gradually tapering to 1 µm wide at the apex.

Survival and Spread

The pathogen may survive on seed and in crop debris (Waller and Sutton, 1979) and is disseminated through wind borne conidia.

Management

☆ Grow disease tolerant germplasm lines like GMU-645, 1553, 1629, 1948, 2053, 2240, 2516, 3448, 5826, 5833, 5929, 6938, 6943 and 6979 (Prasad *et al.*, 2006)

☆ Spray copper oxy chloride (0.3 per cent) or mancozeb (0.25 per cent) or carbendazim (0.1 per cent) at first appearance of the disease.

Viral Diseases

Mosaic

Mosaic has not been a serious disease in India, and has been confined to isolated pathches. According to recent survey in Karnataka, incidence of safflower mosaic ranged from 4-16 per cent (Kulkarni and Byadgi, 2004). The pathogen associated with safflower mosaic was found to be a strain of cucumber mosaic virus. In its severe form mosaic results in necrosis of terminals, stems or parts of leaves. The virus is also known to infect *Helianthus annuus, Chenopodium quinoa, Dolichos lablab* [*Lablab purpureus*], *Pisum sativum, Cucumis sativus, Cucurbita maxima, Luffa acutangula, Solanum melongena, Abelmoschus esculentus, Nicotiana tabacum, Capsicum annuum* and *Amaranthus* sp. (Kulkarni and Byadgi, 2005).

The virus is transmitted by various aphids, *viz., Aphis craccivora, A. gossypii, Myzus persicae, Rhopalosiphum maidis* and *Uroleucon compositae.* However, safflower aphid, *Uroleucon compositae* is the most efficient vector for transmission of safflower mosaic virus (Kulkarni and Byadgi, 2004). Apterous adults were better transmitters of the virus than nymphs and alate adults (Ravinder *et al.,* 1980). Spraying of monochrotophos @1.6ml/lt was suggested for the management of Safflower aphid and thereby the disease transmission (Ali *et al.,* 1994).

Tobacco Streak Virus

Rao *et al.* (2003) have first reported the existence of *Tobacco streak virus* on safflower cv. NARI-6 in Maharashtra. The symptoms include veinal and leaf necrosis, necrotic streaks on the stem, necrosis of the terminal bud and plant death.

References

Ali, S., Ramchandra Reddy, K. and Govinda Rao, N. 1994. Field efficiency of insecticides against safflower aphid *Uroleucon compositae* Theobald and safflower mosaic disease. *Journal of Research APAU* **22** (1): 45-48.

Anonymous, 1998. Annual Progress Report on Safflower (1997-98). Directorate of Oilseeds Research, Hyderabad, p.5.

Bergman, J. W. and Jacobsen, B. J. 2005. Control of *Alternaria* blight (*Alternaria carthami* Choud.) of safflower (*Carthamus tinctorius* L.) in the United States Northern Great Plains region. In: Proceedings of the VIth International Safflower Conference, Istanbul-Turkey, 6-10 June, 2005, pp. 215-221.

Borkar, S.G. 1997. Spines, a new avenue for infection by *Alternaria carthami* on safflower. *Journal of Mycology and Plant Pathology* **27** (3): 311-316.

Chattopadhyay, C. and Sastry, K.R. 1997. Effect of bioagents on safflower wilt caused by *Fusarium oxysporum* f.sp. *carthami.* IVth International Safflower Conference, Bari. 2nd -7th June, 1997, pp. 299-302

Chowdhury, S. 1944. An *Alternaria* disease of safflower. *J. Ind. Bot. Soc.* **23:** 59-65.

Ellis, M.B. and Holliday, P. 1970. *Alternaria carthami.* [Descriptions of Fungi and Bacteria]. IMI Descriptions of Fungi and Bacteria 25, Sheet 241.

Gaikwad, S. J. and Behere, G. T. 2001. Biocontrol of wilt of safflower caused by *Fusarium oxysporum* f. sp. *carthami.* In: Safflower: a multipurpose species with unexploited potential and world adaptability. Proceedings of the 5th International Safflower Conference, Williston, North Dakota and Sidney, Montana, USA, 23-27 July, 2001, pp.63-66.

Govindappa, M., Lokesh, S. and Rai, V.R., 2005. A new stem splitting symptom in safflower caused by *Macrophomina phaseolina*. *Journal of Phytopathology* 53: 560-561.

Govindappa, M., Lokesh, S., Rai, V. R., Naik, V. R. and Raju, S. G. 2010. Induction of systemic resistance and management of safflower *Macrophomina phaseolina* root-rot disease by biocontrol agents. *Archives of Phytopathology and Plant Protection* **43** (1): 26-40.

Govindappa, M., Rai, V. R. and Lokesh, S. 2011. *In vitro* and *in vivo* responses of different treating agents against wilt disease of safflower. *Journal of Cereals and Oilseeds* **2** (1): 16-25.

Hostert, N. D., Blomquist, C. L., Thomas, S. L., Fogle, D. G. and Davis, R. M. 2006. First report of *Ramularia carthami*, causal agent of ramularia leaf spot of safflower, in California. *Plant Disease* **90** (9): 1260.

Huerta-Espino, J., Constantinescu, O., Velasquez, C., Herrera-Foessel, S. A. and Figueroa-Lopez, P. 2006. First report of *Ramularia cercosporelloides* on *Carthamus tinctorius* in Northwestern Mexico. *Plant Disease*, **90** (12): 1552.

Ingle, V.N., Deshmukh, V.V., Jiotode, D.J., Chore, N.S. and Dhawad, C.S. 2004. Screening of safflower varieties against root rot (*Rhizoctonia bataticola*). *Research on Crops*, **5** (1): 113-114.

Jacobsen, B.J., Bergman, J.W. and Flynn, C.R. 2008. Comparison of safflower fungicide seed treatments. In: Safflower: unexploited potential and world adaptability. 7th International Safflower Conference, Wagga Wagga, New South Wales, Australia, 3-6 November, 2008. pp.1-4.

Kishore Varma, P., Patibanda, A.K. and Bhadru, D. 2007. Survey on diseases of safflower in Ranga Reddy district of Andhra Pradesh. In: Extended summaries: ISOR National seminar on "Changing global vegetable oils scenario: Issues and challenges before India. Jan 29-31, 2007. Indian society of oilseed Research, Hyderabad, pp. 198-199.

Krishna Prasad, N. V. 1988. Studies on the *Alternaria* leaf blight of safflower (*Carthamus tinctorius* L.) Ph.D., Thesis, Benaras Hindu University, Varanasi. p. 168.

Kulkarni, V. R. and Byadgi, A. S. 2004b. Studies on transmission of safflower mosaic disease. *Karnataka Journal of Agricultural Sciences* **17** (4): 841-842.

Kulkarni, V. R. and Byadgi, A. S. 2004a. Incidence of safflower mosaic disease in North Karnataka. *Karnataka Journal of Agricultural Sciences* **17** (4): 836-837.

Kulkarni, V. R. and Byadgi, A. S. 2005. Studies on host range of safflower mosaic disease. *Karnataka Journal of Agricultural Sciences* **18** (1): 169-171.

Kumar, A. and Joshi, H.K. 1995. Development of leaf spot caused by *Ramularia carthami* and reaction of safflower cultivars. *Journal of Agricultural Science*, **125** (2): 223-225.

Lukade, C. M. and Rane, M. S. 1994. Effect of organic and inorganic soil amendments on pre and post emergence of root rot and yield of safflower. *Madras Agricultural Journal*, **81** (1): 3-4.

Lukade, G. M. and Rane, M. S. 1993. Saprophytic activity of *Rhizoctonia bataticola* causing root rot of safflower. *Madras Agricultural Journal*, 80: 421-424.

Lukade, G.M. 1992. Effect of organic soil amendments on root rot incidence of safflower. *Madras Agricultural Journal*, **79** (3): 179-181.

Mathur, A. K. and Tyagi, R. N. S. 1988. Translocation of Vitavax and Benlate in safflower seedlings. *Indian Phytopathology*, **41** (3): 484-486.

Montoya Coronado, M. C. L. 2008. Mexican safflower varieties with high tolerance to Ramularia carthami. In: Safflower: unexploited potential and world adaptability. 7th International Safflower Conference, Wagga Wagga, New South Wales, Australia, 3-6 November, 2008. pp. 1-3.

Murumkar, D. R. and Deshpande, A. N. 2009. Variability among isolates of *Fusarium oxysporum* f. sp. *carthami*. *Journal of Maharashtra Agricultural Universities*, **34**(2): 178-180.

Murumkar, D. R., Gud, M. A. and Deshpande, A.N. 2008c. Screening of released varieties and hybrids of safflower against wilt (*Fusarium oxysporum* f. sp. *carthami*). *Journal of Plant Disease Sciences*, **3** (1): 114-115.

Murumkar, D. R., Gud, M.A., Indi, D.V., Shinde, S.K. and Kadam, J.R. 2008a. Development of leaf spot of safflower (*Alternaria carthami*) in relation to weather parameters. *Journal of Plant Disease Sciences*, **3** (2): 201-205.

Murumkar, D.R., Gud, M.A., Akashe, V.B., Shinde, S.K. and Kadam, J.R. 2009. Identification of sources of resistance to major diseases and pests of safflower. *Journal of plant disease sciences*, **4** (1): 107-109.

Murumkar, D.R., Indi, D.V., Gud, M.A. and Shinde, S.K. 2008b. Field evaluation of some newer fungicides against leaf spot of safflower caused by *Alternaria carthami*. In: Safflower: unexploited potential and world adaptability. 7th International Safflower Conference, Wagga Wagga, New South Wales, Australia, 3-6 November, 2008, pp. 1-6.

Nimbkar, N. 2008. Issues in safflower production in India. In: Safflower: unexploited potential and world adaptability. 7th International Safflower Conference, Wagga Wagga, New South Wales, Australia, 3-6 November, 2008.

Pallavi, M., Prasad, R.D. and Anjani, K. 2007. Novel sources of resistance to *Fusarium* wilt in *Carthamus* Genus. In: Extended summaries: ISOR National seminar on "Changing global vegetable oils scenario: Issues and challenges before India. Jan 29-31, 2007. *Indian society of oilseed Research*, Hyderabad, pp. 59-60.

Patil, M. S. and Hegde, R. K. 1990. Perpetuation of *Ramularia carthami* on safflower crop. *Journal of Maharashtra Agricultural Universities* 15 (2): 252-253.

Patil, M. S. and Hegde, R. K. 1989. Efficacy of fungicides in controlling leaf spot of safflower caused by *Ramularia carthami*. *Journal of Oilseeds Research*, **6** (1): 118-122.

Diseases of Oilseed Crops and their Management

Prameela, M., Rajeswari, B., Prasad, R. D. and Reddy, D. R. R. 2005. Variability in isolates of *Fusarium oxysporum* f.sp. *carthami* causing wilt of safflower. *Indian Journal of Plant Protection*, 33 (2): 249-252.

Prasad, N. V. K., Chaudhary, K. C. B. and Rao, G. R. K. 1988. Production of phytotoxin by *Alternaria carthami* incitant of leaf blight of safflower. *International Journal of Tropical Plant Diseases* 6 (2): 251-255.

Prasad, R. D. and Anjani, K. 2008. Exploiting a combination of host plant resistance and *Trichoderma* species for the management of safflower wilt caused by *Fusarium oxysporum* f. sp. *carthami* Klisiewicz and Houston. *Journal of Biological Control*, 22 (2): 449-454.

Prasad, R.D. 2007. Integrated management of safflower wilt. In: Extended summaries: ISOR National seminar on "Changing global vegetable oils scenario: Issues and challenges before India. Jan 29-31, 2007. Indian society of oilseed Research, Hyderabad, pp. 229-230.

Prasad, R.D. and Anjani, K. 2008. Sources of resistance to *Alternaria* leaf spot among *Carthamus* wild species. In: Safflower: unexploited potential and world adaptability. 7th International Safflower Conference, Wagga Wagga, New South Wales, Australia, 3-6 November, 2008.

Prasad, R.D., Singh, V. and Singh, H. 2006. Crop protection, In: AICRP on Safflower, DOR, 2006. Research achievements in safflower. All India Coordinated Research Project on safflower, Directorate of oilseeds Research, Hyderabad, India. P. 111.

Prashanthi, S. K., Kulkarani, S., Sangam, V. S. and Kulkarni, M. S. 2000. Chemical control of *Rhizoctonia bataticola* (Taub.) Butler, the causal agent of root rot of safflower. *Plant Disease Research*, 15 (2): 186-190.

Punithalingam, E. 1968. *Puccinia carthami*. [Descriptions of Fungi and Bacteria]. IMI Descriptions of Fungi and Bacteria 18, Sheet 174.

Quintana-Obregon, E. A., Plascencia-Jatomea, M., Sanchez-Marinez, R. I., Burgos-Hernandez, A., Gonzalez-Aguilar, G. A., Lizardi-Mendoza, J. and Cortez-Rocha, M. O. 2011. Effects of middle-viscosity chitosan on *Ramularia cercosporelloides*. *Crop Protection* 30 (1): 88-90.

Raghuwanshi, K. S., Dake, G. N., Mate, S. N. and Naik, R. M. 2008. Pathogenic variations of *Fusarium oxysporum* f. sp. *carthami*. *Journal of Plant Disease Sciences*, 3 (2): 241-242.

Ranaware, A., Vrijendra Singh. and Nimbkar, N. 2010. *In vitro* antifungal study of the efficacy of some plant extracts for inhibition of *Alternaria carthami* fungus. *Indian Journal of Natural Products and Resources*, 1 (3): 384-386.

Rangaswami, G. 1996. Diseases of crop plants in India, Third edition. Prince-Hall of India Private, Limited, New Delhi. Pp.498.

Rao, S. C., Rao, R. D. V. J. P., Kumar, V. M., Raman, D. S., Raoof, M. A. and Prasad, R. D. First report of tobacco streak virus infecting safflower (*Carthamus tinctorius*) in Maharashtra, India. *Plant Disease* 87 (11): 1396.

Ravinder, T., Rao, N. G. and Sastry, K. S. 1990. Relationship of safflower mosaic with the aphid, Uroleucon compositae Theobald. *Journal of Insect Science,* **3** (2): 177-179.

Relekar, N., Khalikar, P.V. and Nikam, P.S. 2010. Survey and Surveillance of Alternaria blight of Safflower Caused by *Alternaria Carthami* in Marathwada Region. *Journal of Plant Disease Sciences* 5: 2.

Roy, R., Pandey, V. B., Singh, U. P. and Prithiviraj, B. 1996. Antifungal activity of the flavonoids from *Clerodendron infortunatum* roots. *Fitoterapia,* **67** (5): 473-474.

Sastry K. R. and Chattopadhyay, C. 1999. Effect of soil solarization on *Fusarium oxysporum* f.sp. *carthami* populations in enemic soils. *Indian Phytopathology,* **52** (1): 51-55.

Shinde, A. B., Hallale, B.V. and Vaidya, A.P. 2008. Antifungal activity of leaf extracts against *Alternaria* blight of safflower. *National Journal of Life Sciences,* **5** (2): 203-206.

Singh, S., Kumar, S. and Suryawanshi, D. S. 2006. Evaluation of different treatments for the management of safflower insect pests and diseases under rainfed conditions. *Indian Journal of Entomology,* **68** (3): 286-289.

Singh, V. and Prasad, R.D. 2005. *Integrated management of pests and diseases in safflower.* Directorate of Oilseeds Research, Hyderabad, India, pp. 49.

Smith, G.S. and Carvil, O.N., 1997. Field screening of commercial and experimental soybean cultivars for their reaction to *Macrophomina phaseolina*. *Plant Disease,* **81**: 363-368.

Sumitha, R. and Nimbkar, N. 2009. Efficacy of fungicides against Alternaria leaf spot disease in safflower. *Annals of plant protection sciences,* **17** (2): 507-508.

Tripathi, A. K. 2000. Impact of nitrogen and phosphorus on the incidence of Alternaria leaf spot and yield of safflower. *Flora and Fauna,* **6** (2): 111-112.

Waller, J. M. and Sutton, B. C. 1979. *Cercospora carthami.* [Descriptions of Fungi and Bacteria]. IMI Descriptions of Fungi and Bacteria 63, Sheet 626.

Zhao, S., Xie, G., Zhao, H., Li, H. and Li, C. 2008. First report of a leaf spot on snow lotus caused by *Alternaria carthami* in China. *Plant Disease* **92** (2): 318.

5 Major Diseases of Sesame and their Management

Dinesh Rai and H.K. Singh

Sesame (*Sesamum indicum* L., 2n = 26), which belongs to the genus *Sesamum* of the family *Pedaliaceae*, is one of the oldest oilseed crops and is cultivated in tropical and subtropical regions of Asia, Africa and South America (Ashri, 1998; Anilakumar, *et al.*, 2010). Its cultivation history can be traced back to between 5,000 and 5,500 years ago in the Harappa Valley of the Indian subcontinent (Bedigian and Harlan, 1986). The total area of sesame harvested in the world is currently 7.8 million hectares, and annual production is 3.84 million tons (2010, UN Food and Agriculture Organization data) Asia contributes 68 per cent in sesamum growing area of the world. The major sesamum growing countries in Asia are India, China, Burma and to a lesser extent Afghanistan, Pakistan, Bangladesh, Indonesia and Srilanka. India ranks first in area (29 per cent), production (26 per cent) and export (40 per cent) of sesame in the world (AICRP, 2012). In India, the major growing states are Uttar Pradesh, Rajasthan, Madhya Pradesh, Tamil Nadu, Maharashtra, Andhra Pradesh, Karnataka and Gujarat. It is grown to extract edible oil used in cooking, salad and manufacturing of marginarine. Its oil forms the basis of most of the fragrant and scented oils and perfumes. Its cake is used as feed for farm animals. The overall productivity of this crop in India is 3.35q/ha (Hegde, 2002).

The main reason of low productivity of this crop is the attack of various fungal, bacterial and viral diseases. Seventy two fungi, seven bacteria, one *phytoplasma* and one virus has been reported in India (Vyas *et al.*, 1984). Out of these about thirty two diseases (14 major and 18 minor) occur in India (Table 5.1). At present chemical fungicides are the first choice for the farmers to combat diseases because of their easy adaptability and immediate therapy. Due to health risk and pollution hazards by use of chemical fungicides in plant disease control, it is considered appropriate to minimize their use. Since sesame seed and oil are in high demand for export

Diseases of Oilseed Crops and their Management

Table 5.1: Distribution of Fungal and Bacterial Diseases of Sesame in India

Disease	Cause	Distribution
Phytophthora blight	*Phytophthora parasitica* var. *sesami*	Madhya Prasesh, Uttar Pradesh, Gujarat, Punjab, Haryana, Maharashtra, Andhra Pradesh, Tamil Nadu
Charcoal rot	*Macrophomina phaseolina*	Madhya Prasesh, Rajasthan, Uttar Pradesh, Gujarat, Punjab, Haryana, Maharashtra, Andhra Pradesh, Tamil Nadu, West Bengal, Bihar, Karnataka, Kerala and Orrisa
Fusarium wilt	*Fusarium oxysporum* f. sp. *sesami*	-do-
Root rot/stem rot	*Rhizoctonia solani*	
Alternaria leaf spot	Alternaria sesame	Madhya Prasesh, Rajasthan, Punjab, Haryana, Andhra Pradesh, Karnataka, Kerala and Orrisa
Cercospora leaf spot	*Cercospora sesame*	-do-
Angular leaf spot	*Cercospora sesamicola*	-do-
Powdery mildew	*Oidium erysiphoides*	-do-
	Sphaerotheca fuliginea	
	Leveillula taurica	
	Erysiphe cichoracearum	
Corynespora blight	*Corynespora cassiicola*	-do-
Brown angular spot	*Cylindrosporium sesame*	-do-
Areal stem rot	*Helminthosporium sesame*	-do-
Leaf spot	*Alternaria* spp.	-do-
	Alternaria tenuis	
	Ascochyta sesame	
	Botrybasidium rolfsii	
	Cercospora sp.	
	Cladosporium sp.	
	Colletotrichum sp.	
	Curvularia sp.	
Wet rot of seedlings	*Choanephora cucurbitarum*	-do-
Damping off	*Pythium aphanidermatum*	-do-
	P. ultimum	
Bacterial		
Bacterial leaf spot	*Pseudomonas syringae* pv. *sesami*	-do-
Bacterial blight	*Xanthomonas campestris* pv. *sesami*	M.P., Rajasthan, U.P.
Bacterial wilt	*Pseudomonas solanacearum*	
	=Ralstonia solanacearum	

Source: Verma, *et al.*, 2005.

due to their high unsaturated fat and methionine content, focus has been shifted out safer alternatives to chemical fungicides in recent years. Biological control had attained importance in modern agriculture to curtail the hazards of intensive use of chemicals for disease control. The major diseases and technology generated for their management is being described in this chapter.

Phytophthora Blight

Phytophthora blight of sesame was first reported from India by Butler, 1918. Now it has been reported from all major sesame growing areas of the country such as Madhya Prasesh, Rajasthan, Uttar Pradesh, Gujarat, Punjab, Haryana, Maharashtra, Andhra Pradesh, Tamil Nadu, West Bengal, and Bihar (Verma, 2002a; Vasudeva, 1961). This disease causes 66.8 percent losses in Gujarat (Kale and Prasad, 1957), 78.8 percent in Madhya Pradesh (Singh *et al.*, 1976) and 52.8 per cent in Assam (Kalita *et al.*, 2000). It may even 100 per cent loss under most favourable conditions when infection occurs severely at seedling stage. The infected seeds lose their viability.

Symptoms

The disease affects the all aerial parts of plants. The first symptom is the appearance of water-soaked brown spots on leaves and stems. The spots gradually encrease in size. Under favourable weather conditions the brownish discoloured spots spread rapidly both upwards and down wards and also around the stem. The brownish area later turns deep brown and becomes black with the spread of the infection. The pods are also affected. In humid weather the white woolly growth of the fungus can be seen on the surface of affected pods (Vyas *et al.*, 1983). The pods are poorly formed and the seeds in affected pods are shrivelled and become brown.

Causal Organism

Phytophthora parasitica var. *sesami* Dastur (1918) reported that *Phytophthora parasitica* from castor attacks the seedlings of sesame causing curling of leaves and considered the species *Phytophthora* of castor and sesame to be same. Kale and Prasad (1957) found that some isolates of *P. parasitica* to be highly specialized on sesame crop and proposed the name *P. parasitica* var. *sesami*.

The fungus produces non-septate, hyaline mycelium. The sporangiophores are hyaline and branched sympodially and bear sporangia. The sporangia are hyaline and spherical with a prominent apical papilla and measure 25 to 50 × 20 to 35 μm in size. The oospores are smooth, spherical and thick walled.

Disease Cycle

The pathogen can survive in mycelial form up to 50°C temperature, and culture having chlamydospores may survive up to 52°C (Sehgal, 1963). Viability of the culture-can be kept in a refrigerator for one year at 5°C. These studies suggest that the fungus can survive in soil during the summer and winter where temperature never rises beyond 50°C or drops below 5°C (Prasad, *et al.*, 1970). The fungus survives in soil during the unfavorable period in the form of dormant mycelium and/or in the form of chlamydospores. In addition to soil, seed also appears to play an important role in the recurrence and spread of the disease. In seed the mycelium has been located in the embryo (Sehgal and Prasad, 1966). Maiti *et al.* (1988) reported that the fungus reduces the seed viability but is not seed borne.

The reoccurrence of the disease takes place in new soil with the seed but in the old soil with seed soil and plant debris. The disease first appears in certain places in the field and from infection foci. When the rains occur and humidity persists for longer period due to cloudy weather the zoospore are formed and secondary spread of disease takes place from one plant to other covering whole field giving the appearance of blighting.

Epidemiology

Early sowing with the onset of monsoon is less favourable for disease development. Seedling stage is more susceptible than the adult plant stage. Heavy rains for at least two weeks and high humidity (above 90 per cent) for three weeks or more favor the development of the disease. The moderate nitrogenous fertilizers application lead to more incidence of phytophthora blight (Verma and Bajpai, 2001). This disease is more severe in heavy soils (Verma, 2002b). It is observed that the initial development of the disease is much earlier when the soil temperature is 28°C, while the initial appearance of the disease is delayed with an increase in the soil temperature up to 37°C (Gemawat *et al.*, 1964 and Prasad *et al.*, 1968). It is further reported that incidence of the Phytophthora blight of sesame shows a close parallelism to the growth of the fungus.

Management

☆ Use resistant/tolerant varieties like TKG-22, TKG-55 and JTS-8.

☆ Intercropping system, sesamum + pearl millet (3: 1) should be followed.

☆ Rathaiah (1985) reported more disease in May-June sown crop than the early August sown crop in Assam. It has been observed that fertilizers application of N30, P20, and K10 shows 36.22 per cent disease and 55 per cent in N60, P30, and K20 as compared to control (without fertilizers) with 0.5 per cent disease. The application of FYM alone or neem cake with inorganic fertilizers (N60, P4O, K20) reduces the disease as compared to without FYM.

☆ It has been observed that dense population (3.3 lakhs plant/ha) shows 30 per cent incidence as compared with (11.0 per cent) 2.2 lakhs plant/ha.

☆ Choi *et al.* (1984) reported that the cultivation of sesame in 0.2mm wide ridge in plots mulched with black vinyl reduces the spread of disease by at least 30 per cent and increased yield by 22 per cent compared with mulching.

☆ Soil amendment with biological control agent like *Trichoderma harzianum* and *Trichoderma viride* or seed treatment with *T. harzianum/T. viride/Bacillus subtilis* (0.4 per cent) reduces the incidence of disease.

☆ Erwin (1983) reported that species of *Pseudomonas, Bacillius* and *Streptomyces* which are most active at 25-27°C at field capacity moisture level can be suppressive to *Phytophthora* species in soil. *Trichoderma* species which are common fungi are also antagonistic to the *Phytophthora* species.

☆ Application of phosphorus solublizing bacteria (PSB) along with neem cake or 50 per cent NPK+FYM 100 per cent NPK +PSB reduces the disease (Verma and Bajpai), 2001.Verma 2002[a,b] reported that *Trichoderma viridae*, *T.harzianum* and *Pseudomonas fluorescence* as seed treatment not only reduces the disease significantly but substantially increases the yield.

☆ Treat seed before sowing with Apron 35 SD (0.3 per cent) or Thiram (0.3 per cent).

☆ For root/stem infection, drench soil 2 to3 times with Ridomil MZ (0.25 per cent) along with diseased plants at 7 days interval.

☆ Spray crop 3 times with Ridomil MZ (0.25 per cent) or copper oxychloride (0.25 per cent) at an interval of 10 days from the initiation of disease.

☆ Kalita *et al*. (2000) reported that spraying of 0.3 per cent copper oxychloride at 20, 40 and 60 days after sowing significantly reduced the disease incidence and yield loss.

Macrophomina Rot

The disease has been reported from all the sesame growing areas throughout the world. In India it is widely distributed in sesame growing areas of Madhya Prasesh, Rajasthan, Uttar Pradesh, Gujarat, Punjab, Haryana, Maharashtra, Andhra Pradesh, Tamil Nadu, West Bengal, Bihar, Karnataka and Kerala. The losses from the diseases are 5-100 per cent in farmers and experimental field. Sundararaman (1933) observed 36.6 per cent due to *M. Phaseoli* leading to 57 per cent loss in yield. Murugesan *et al*. (1978) found loss of 110 and 111 kg/ha in monsoon and summer crop respectively. You and Park (1980) reported that *M. phaseolina* causes severe reduction in seed germination and seedling stand.

Symptoms

The disease symptom starts as yellowing of lower leaves, followed by drooping and defoliation. The stem portion near the ground level shows dark brown lesions and bark at the collar region shows shredding. The sudden death of plants is seen in patches. In the grown-up plants, the stem portion near the soil level shows large number of black pycnidia. The stem portion can be easily pulled out leaving the rotten root portion in the soil. The infection when spreads to pods, they open prematurely and immature seeds shriveled and become black in colour. Minute pycnidia are also seen on the infected capsules and seeds. The rotten root as well as stem tissues contains a large number of minute black sclerotia. The sclerotia may also be present on the infected pods and seeds.

Causal Organism

The pathogen is *Macrophomina phaseolina* (Tassi) Goid. The pathogen produces dark brown, septate mycelium showing constrictions at the hyphal junctions. The sclerotia are minute and dark black in colour. The pycnidia are dark brown with a prominent ostiole. The conidia are hyaline, elliptical and single celled.

Disease Cycle

The fungus remains dormant as sclerotia in soil as well as in infected plant debris in soil. Sudaraman (1931) found that the sclerotia of this fungus remain viable at room temperature for about 54 months in a pot experiments. The infected plant debris also carries pycnidia. The fungus primarily spreads through infected seeds which carry sclerotia and pycnidia. Singh and Singh (1990) examined seeds of sesame from 18 districts of Rajasthan and found 108 samples infected with 1-67 per cent incidence. Microsclerotia are present in the seeds. Gupta and Cheema (1990) positively correlated the number of microsclerotia on sesame seeds with plant infection and negatively with seed germination, dry matter production and root/shoot length of seedlings. The fungus also spreads through soil-borne sclerotia. The secondary spread is through the conidia transmitted by wind and rain water.

Epidemiology

Lower pH favours the fungal growth and development. Jain and Kulkarni (1965) reported that 35°C favours maximum sclerotial production and causes more infection at 100 per cent soil moisture as compared to 60 per cent. Rodriguez and Zambrano (1985) reported that root and stem rot disease is affected by plant age, soil humidity and temperature. Phosphorus application in soil checks the disease severity (Khare and Jharia, 2002).

Management

☆ Sesame lines RT 54, RT 103,RT 125, RT 127, RT 46, B 67, TKG 55, JTS 8, AT 9, HT 6 RT 16, ST 58, Gwalior 5, ES 25, ES 56, IS 190, SI 249, Rajeshwari, and SI 254 are reported resistant against *M. phaseolina* (Jharia *et al.*, 2003).

☆ Avoid continuous cropping of sesamum in the same field.

☆ Remove and destroy infected plant debris.

☆ Daftri and Verma (1975) found that mixed cropping of sesame in alternate rows with black gram, green gram, cowpea, cluster bean or moth reduces the incidence of diseases.

☆ It has also been observed that disease is low in 30x15cm spacing (2 lakhs plant/ha) as compared to 30x10(3 lakhs plant/ha).

☆ Sesame mixed or intercropped with green gram in 1: 1 gave less incidence of Macrophomina stem and root rot and also provided a higher sesame seed yield equivalent as compared to sole sesame Rajpurohit (2002).

☆ Sirry *et al.* (1980) observed that higher phosphorus level reduces the disease.

☆ Gemawat and Verma (1971) reported mustard cake (16.6 per cent)effective in comparison to Farm Yard manure(FYM), froundnut cake and sesame cake in comparison to untreated plot (83.3 per cent). Similarly, Mishra and Mukherjee (1982) found that soil mixed with 2000kg/ha sesame oil cake

3 weeks before sowing followed by light irrigation reduces the incidence of disease.

☆ Chattopadhyay (1999) reported that soil solarization to be effective in the minimizing of disease and found his experiment significant reduction of *Macrophomina phaseolina* population *viz.*, 84.5 and 100 per cent at 5 cm and 37.3 and 5.3 per cent at 15 cm depth.

☆ Apply farm yard manure or green leaf manure at 10t/ha or neem cake 150 kg/ha.

☆ Treat the seeds with *Trichoderma viride* at 4g/kg.

☆ Captan (0.2 per cent), thiram (0.2 per cent), metalaxyl (0.4 per cent) gives complete control of seed borne infection of *M. phaseolina* when used as seed dressers.

☆ On appearance of the disease, drench soil with Carbendazim @1g/lit.

☆ Soil application with *P. fluorescens* or *T.viride* @ 2.5 kg/ha mixed with 50 kg/ha FYM reduces the incidence of disease.

Fusarium Wilt

The disease is quite serious where ever the crop is grown. In India, it has been reported from all the sesame growing areas such as Madhya Pradesh, Maharashtra, Andhra Pradesh, Rajasthan, Haryana, Punjab *etc.* The disease is serious when appears in early stages of the crop. The disease can be devastating on susceptible varieties of sesame, but many local varieties have been found to have some degree of resistance to local races of the fungus. Epiphytotic occurrence of the disease was reported in 1961 and 1964 in the U.S.A (Rivers *et al.*, 1965) and in 1959 in Venezuela (Malaguti, 1959).

Symptoms

Plants get infected at any stage of the crop development. Yellowing of the leaves is the first noticeable symptom of the wilt in the field. Leaves become yellowish, droop, and desiccated. Sometimes such leaves show inward rolling of the edges and eventually dry up. The terminal portion dries up and becomes shrunken and bent over. In a severe infection the entire plant becomes defoliated and dry. In a less severe infection or when mature plants are infected, only one side of the plant may develop symptoms, resulting in partial wilting. A blackish discoloration in the form of streaks appears on infected plants. Discoloration of the vascular system is conspicuous in the roots. Roots in the later stages show rotting, wholly or partially corresponding with that side of the plant showing disease symptoms. Numerous, pink, pin-head sized sporodochia (containing macroconidia of the fungus) may be seen scattered over the entire dried stem. The capsules of wilted plants also show numerous sporodochia.

Causal Organism

Fusarium oxysporum f. sp. *sesami* (Schelt.) Jacz. Butler (1926) from India identified the fungus as a strain of *Fusarium vasinfectum* Synder and Hansen and suggested that *Fusarium* causing wilt in sesame be treated a form *of F.oxysporum*. Castellani (1950) identified the fungus as *Fusarium oxysporum* f. sp. *sesame*. The fungus produces profuse mycelial growth on potato-dextrose-agar. The mycelium is arid, hyaline, septate, and richly branched, turning light pink when old. The microconidia are formed abundantly. They are hyaline, ovoid to ellipsoid, unicellular, and measure 8.5 × 3.25 µm in size. The macro conidia are also produced abundantly in sporodochia and size in the range of 35-49x4.5 µm. The chlamydospores are globose to sub globose, smooth or wrinkled and about 7-16 µm in diameter. The pathogen grows at the temperature range of 10-35°C with an optimum of 26°C and pH 5.6.

Disease Cycle and Epidemiology

The pathogen is reported seed borne as well as soil-borne, and it may persist for many years in the soil. Gaikwad and Pachpande (1992) reported that early sown crop *i.e.*, mid of April, showed higher disease as compared to the late sown crop. The pathogen penetrates the host through root hairs (Buldeo and Rane, 1978). It is also observed that minimum available soil moisture decrease the disease incidence (Salazar-Huerta and Ortiz, 1992). The Fusarium wilt of the sesame is reported to be associated with nematode attack as well as other pathogens (Buldeo and Rane, 1978). The culture filtrate of *F. oxysporum* f. *sesami* has been reported to have an inhibitory effect on sesame. Shoot and root growth is also inhibited by culture filtrate of the fungus, indicating the production of toxic substances by the pathogen. High soil temperature to a depth of 5 to 10 cm and 17 to 27 per cent water-holding capacity during dry periods is favorable for the development of the disease.

Management

- ☆ Sanitation and clean cultivation should be followed as additional measures to manage the disease.
- ☆ Deep summer ploughing during the months of May and June helpful in eradicating the pathogen.
- ☆ Follow five year rotation if soil is heavily infested.
- ☆ Soil amendment with farm yard manure @ 12.5 tonnes/ha is helpful in reducing the incidence of the disease.
- ☆ Destroy the diseased plant debris by burning or burying in the soil.
- ☆ Seed treatment with *T. viride* @ 4g/kg or *P. fluorescens* @ 10g/kg of seed.
- ☆ Application of *Trichoderma harzianum* and *T.viridae* in field significantly reduces the wilt incidence (Wuik *et al.*, 1998).
- ☆ Soil application of *P. fluorescens*/*T. viride* 2.5 kg/ha with 50 kg FYM reduces incidence of the disease.
- ☆ Treat the seed with Carbendazim or Thiram 2g/kg of seed.
- ☆ Drench the soil with Carbendazim 1g/lit.
- ☆ Bavistin (0.1 per cent) has been observed to be completely inhibitory to the pathogen in vitro conditions (Ahmed *et al.*, 1989).

Alternaria Leaf Spot

This disease of sesame was first described by Kavashnina (1928) from the North Caucasus, USSR. Kawamura (1931) from Japan identified the pathogen *Macrosporium sesame*. In India it was reported that pathogen *Alternaria sesami* is closely related to *M. sesami* but different on the basis of catenulation on the spores (Mohanty and Behera, 1958).The Alternaria leaf spot is now reported to occur in most of the tropical and subtropical areas of the world. The disease is reported to occur quite widely in Ethiopia, Nigeria, India, and USA (Kolte,1985; Bhale *et al.*, 1998). Recently, it has been reported from Kenya (Ojiambo *et al.*, 1999).

In India, it is reported from all those states wherever sesame is grown. However, it is serious in Karnataka, Andhra Pradesh, Maharashtra and Gujarat. The amount of damage to the sesamum plant is dependent on the stage of growth and environmental conditions. Epiphytotic occurrence of the disease has been reported from the Stoneville area in Mississippi in 1962, the Tallahassee area in Florida in 1958, the coastal area of the Orissa in 1957, and Maharashtra in India in 1975. The actual effect of the disease on yield has, however, not been investigated in detail. It is however, reported that about 0.1 to 5.7 g seeds/100 fruits are lost due to the disease under Karnataka conditions in India (Siddaramaiah *et al.*, 1981). It has been reported that Alternaria blight may cause a loss of about 20 to 40 percent in U.P. (Kumar *et al.*, 1992).

Symptoms

The pathogen attacks all parts of the plant at all stage. Initially small, circular, reddish brown spots (1-8mm) appear on leaves which enlarge later and cover large area with concentric rings. On the lower surface, the spots are lighter brown in color. Such spots often coalesce and may involve large portions of the blade, which become dry and shed. Dark brown, spreading, water-soaked lesions can be seen on the entire length of the stem. The lesions also occur on the midrib and even on veins of leaves. In very severe attacks, plants may be killed within a very short period after symptoms are first noted, while milder attacks cause de-foliation. The appearance of the disease at the seedling and young plants can cause pre- and post emergence damping-off.

Causal Organism

The pathogen is *Alternaria sesami* (Kawamura) Mohanty and Behera. The mycelium of the fungus is dull brown and septate and produce large number of pale grey-yellow conidiophores which are straight or curved. The conidia are produced singly or in chains of two. They are straight or slightly curved, obclavate, yellowish- brown to dark or olivaceous brown in color, and measure 30 to 120 x 9 to 30µm (excluding the beak). The conidia have 4 to 12 tranverse septa and 0 to 6 longitudinal septa at which they are slightly constricted, and terminate in a long hyaline beak measuring 24 to 220 x 2 to 4µm in (Ellis and Holiday, 1970). The beak may be simple or branched.

Disease Cycle and Epidemiology

Pathogen can survive through infected seed (Jain and Siddiqui, 1981). The spores of the fungus adhered to the seeds or capsule may serve to carry and disseminate the pathogen. The pathogen penetrates the seed coat where it remains viable until the germination of the seed. The infected diseased debris also plays an important role in perpetuation of the pathogen. The optimal temperature for growth of the fungus is in the range of 20 to 30°C, and the optimum pH for growth is 4.5 (Mohaptra *et al.*, 1977; Samuel and Govinddaswamy, 1972). High humidity favours the disease development (Ngabala and Zambettakis, 1970). Delay in sowing, increases the disease intensity. The appearance of the disease at the seedling stage is quite devasting (Tripathi *et al.*, 1998).

Management

☆ Grow resistant/tolerant variety like Krishna.

☆ Destruction of crop residues and weeds.

☆ Early planting *i.e.* immediately after onset of monsoon.

☆ Follow intercropping system of sesamum + sunflower (3: 1).

☆ Seed treatment with Bavistin (0.2 per cent) followed by three foliar applications of Bavistin (0.1 per cent) is most effective in reducing the disease (Swain *et al.*, 1989).

☆ Four sprays of Mancozeb (0.2 per cent) at 15 days interval are best in managing the diseases (Karunanithi *et al.*, 1996).

☆ Shekharappa and Patil (2001) reported three application of mancozeb (0.25 per cent) at 10 days intervals starting from the 40 days of sowing is best in managing the disease.

☆ Jeyalakshmi *et al.* (2013) reported that the soil application of *neem* cake (250 kg/ha) along with seed treatment and soil application (2.5 kg/ha) of *T. viride* followed by foliar spray of azadirachtin @ 3 ml/l on 30 and 45 DAS was found to be significantly effective by recording the minimum incidence of root rot and powdery mildew (2.01 per cent) and root rot (2.54 per cent) and Alternaria leaf blight (2.48 per cent) in IPM programme.

☆ Rajpurohit and Nema (2013) reported spraying of Carbendazim + Iprodion @ 0.1 per cent, Carbendazim + Mancozeb @ 0.2 per cent and propiconazole @ 0.1 per cent effective and economied for the management of disease.

Cercospora Leaf Spot or White Leaf Spot

The disease has been reported from all the sesame growing areas such as Karnataka, Maharashtra, Tamil Nadu, Andhra Pradesh and Gujarat and cause considerable loss in yield of sesame in India and Venezuela (Chowdhury,1945; Muller and Texera, 1941). In India it has been reported to cause about 20 per cent loss in yield (Mohanty, 1958)

Symptoms

Generally, the symptoms of the disease appear at the time of flowering but disease may also appear after 30-40 days after sowing. Disease appears as small circular spots are scattered on both leaf surfaces. At first they are minute and later they increase in size to become 5 mm in diameter with whitish center surrounded by a blackish-purple margin. The spots may enlarge rapidly, coalesce into irregular blotches which often become about 4 cm in diameter, and are concentrically zoned. Under humid conditions, the disease becomes severe involving premature defoliation. The symptoms on the petioles are visible as elongated lesions, whereas on capsules, it is more or less circular, brown to dark brown in colour. The disease causes defoliation particularly in early maturing varieties.

Causal Organism

The pathogen is *Cercospora sesami* Zimmerman. The hypha of the fungus is irregularly septate, light brown and thick walled. Conidiophores are produced in cluster and are 1-3 septate, hyaline at the tip and light brown coloured at base. Conidia are elongated, 7-10 septate, hyaline to light yellow, broad at the base and tapering towards the apex.

Disease Cycle and Epidemiology

The fungus is externally and internally seed-borne. The fungus also survives in plant debris. Primary infection may be from the seeds and infected debris. The secondary spread is through wind-borne conidia. Early sown crop shows less disease as compared to the late sown crop. The disease intensity is higher at closer spacing as compared to the wider spacing of 20-30 cm (Tripathi *et al.*, 1998). Enikuomehin *et al.* (2002) from Nigeria reported that crop sown in early July shows less disease as compared to crop sown from mid August to September.

Management

- ☆ Grow resistant/tolerant variety like TKG-21.
- ☆ Variety like BIC-7-2, Sidhi 54, Rewa 114 and Seoni Mala showed stable resistance to the disease (Tripathi *et al.*, 1996).
- ☆ Early planting *i.e.* immediately after onset of monsoon.
- ☆ Follow intercropping system of sesamum + pearl millet (3: 1).
- ☆ Foliar sprays of Zineb (0.25 per cent) or mancozeb (0.25 per cent) has been reported to be effective for the management of disease.

Powdery Mildew

The disease is of common occurrence especially in south india. Powdery mildew caused by many species/strains of the devastative fungal pathogen, *Oidium* sp. is common all over the sesame growing areas, especially in Andhra Pradesh and Tamil Nadu and causes as high as 50 per cent yield losses under favourable conditions. Four different fungi have been reported to cause powdery mildew of sesamum. These are described below.

Symptoms

Oidium erysiphoides Fr.

The symptoms of the disease appear only on the upper surface of leaves as chalky white spots which ultimately cover the whole surface. The petioles are also affected. Usually the disease appears when the crop is about 45 to 60 days old. The severely infected leaves drop off. The affected plants produce shriveled seeds, and the yields are reduced. The mycelium of *O. erysiphoides* is ectophytic, hyaline, septate, granular, 16.5 to 8.4 µm in width. The conidiophores are erect, cylindrical, becoming gradually broader at the top, rounded at the apex, and measure 74 to 132 x 12 to 15 µm. The conidia measure 23 to 36 x 14 to 20 µm in size.

Sphaerotheca fuliginea (Schlecht) Pollacci

Symptoms of the disease start as small whitish spots on the upper surface of the leaves. The spots coalesce to form a single spot, finally covering the entire leaf surface with dirty white fungal growth. Generally, the mildew is confined to the upper surface of the leaves. In some susceptible varieties it is observed on both surfaces. The mycelium of *S. fuliginea* is septate, superficial, branched, 4.2 to 5.1 µm in thickness. Conidiophores erect, simple, septate, bearing single-celled hyaline oval-to-elliptical conidia.

Leveillula taurica (Lav) Trnaud

Mycelium of the fungus is septate, branched, irregular in diameter, mostly internal, growing among the parenchyma cells. The conidiop-hores, single or in small clusters, emerge through stomata. They are hyaline, septate, branched and produce a single terminal conidium.

Erysiphe cichoracearum DC

Initially greyish-white powdery growth appears on the upper surface of leaves. When several spots coalesce, the entire leaf surface may be covered with powdery coating. In severe cases, the infection may be seen on the flowers and young capsules, leading to premature shedding. The severally affected leaves may be twisted and malformed. In the advanced stages of infection, the mycelial growth changes to dark or black because of development of cleistothecia. The Pathogen produces hyaline, septate mycelium which is ectophytic and sends haustoria into the host epidermis. Conidiophores arise from the primary mycelium and are short and non septate bearing conidia in long chains. The conidia are ellipsoid or barrel-shaped, single celled and hyaline. The cleistothecia are dark, globose with the hyaline or pale brown myceloid appendages. The asci are ovate and each ascus produces 2-3 ascospores, which are thin walled, elliptical and pale brown in colour.

Disease Cycle and Epidemiology

The Pathogen is an obligate parasite and disease perennates through cleistothecia in the infected plant debris in soil. The ascospores from the cleistothecia cause primary infection. The secondary spread is through wind-borne conidia. The early sown crop (August had higher disease intensity than the September sown crop (hazarika, 1998). Whereas Rao and Rao (2001) from A.P. reported that in the *rabi* season crop, early sown crop (Ist week of Jan) contracted less disease intensity as compared to the late sown crop.

Management

☆ A white seeded variety Rajeshwari has been reported to be tolerant to powdery mildew (Gangakishan *et al.*, 1989)

☆ Karunanithi and Dinakaran (1996) also reported that TNAU-17, US117, US9003 and DPI 1588 are tolerant to the disease.

☆ Remove the infected plant debris and destroy.

☆ Follow intercropping system of sesamum + pearl millet (3: 1).

☆ Powdery mildew can be managed by three to four foliar sprays of Wettable sulphur (0.2 per cent). Give first spray as soon as disease appears in the field. The sprays should be given between 10 to 15 days intervals depending upon the severity of disease.

☆ Two sprays of Wettable sulphur (0.2 per cent) at 15 days significantly reduced the disease (Karunanithi, 1996).

Corynespora Blight

Wei (1950) first time reported *Corynespora* on sesame leaves.

Symptoms

Dark, irregularly shaped spots appear on leaves and stems. They enlarge, become brown with light centers, and coalesce forming a blotchy configuration. Extensive defoliation occurs and the affected plants die. On the stem, brown specks are formed which develop into elongated, scattered, irregular lesions. These lesions on the stem increase in size both ways up to 10 to 15 cm. Affected stems are bent irregularly on the lesions. Cankers of various sizes also appear on the stem. The center of the cankers is warty and straw-colored. In mature plants, the infected stem cracks length and breadth-wise.

Causal Organism

The pathogen is *Corynespora cassiicola* (Berk and Curt.) Wei. Earlier the fungus was described under the name *Helminthosporium cassiicola* (Berk and Curt.) Berk. The my-celium is hyaline, becoming brown, septate, and branched. Conidiophores are single or in groups of 2 to 6, up to 20-septate, and measure 44 to 380 x 6 to 11 µm. Conidia are 10 to 15 septate, may be in chains. The conidial length is variable. The range of length is 39 to 280 µm with a mean of 153 µm. The spore germinates from the polar cells with germ tube about the length of the conidia. Lateral cell germination may also be noticed.

Disease Cycle and Epidemiology

The Pathogen perpetuates either through disease debris or infected seeds. The Pathogen is also known to survive more than ten months under field conditions

(Shukla *et al.*, 1987). According to Navas and Subero (1995) the pathogen become inactive within ten months when seeds are stored at 26 °C with 50 per cent relative humidity.

Management

☆ Sanitation and clean cultivation should be followed.

☆ Destruction of weed and crop residues.

☆ Early planting *i.e.* immediately after onset of monsoon.

☆ Follow intercropping system of sesamum+pearl millet (3: 1).

☆ Treat the seed with Thiram (0.15 per cent)+Bavistin (0.05 per cent) in 1: 1 ratio.

☆ Spray the crop with Mancozeb @ 0.2 per cent or 2 kg fungicide/1000 liters water/hectare if disease is noticed on the crop and repeat spraying at 15 days depending upon severity of disease.

☆ Spraying of crop with Dithane M-45 (0.15 per cent) at 10 -15 days interval is recommended (Yadav *et al.*, 1988).

Angular Leaf Spot

Mohanty (1958) first reported the occurrence of angular leaf spot of sesame from the Orissa state of India. Now it has been reported from Uttar Pradesh, Madhya Pradesh, Bihar and in other sesame growing areas but not in severe form.

Symptoms

The disease appears initially minute chlorotic spot on the upper surface of the leaves; later, when the affected tissues become necrotic, the color of the spots changes to dark brown, whereas on the corresponding lower surface of the leaves, the color of the spots remains olivaceous brown. These spots become angular and bound by leaf veinlets. The fruiting bodies of the fungus might become visible on both surfaces of the leaves but chiefly on the lower surface.

Causal Organism

The pathogen is *Cercospora sesamicola* Mohanty. The stromata dark brown, subglobular, 20 to 46 µm in diameter, fascicles closely packed, conidiophores olivaceous, brown, simple, 0 to 2 septate. Conidia hyaline, cylindric, straight to mildly curved, base truncate to sub-truncate, tip obtuse, indistinctly 2 to 7 septate, 20 to 120 x 2 to 2.8 µm.

Disease Cycle

The pathogen is reported to be seed-borne. Rathiaiah and Pavgi (1975) reported that pathogen perpetuates only through viable sclerotia in crop debris. The sclerotia are able to produce slender germ hyphae. It is reported that conidia and sclerotia of the pathogen are resistant to heat and desiccation. The infected seed also serve as primary source of inoculums (Ferrer, 1960).

Management

☆ The clean cropping practices and crop rotation are useful for minimizing the disease incidence.

☆ Hot water treatment of seed at a temperature of 53 °C for 30 minute gives good management of disease (Nusbaum,1941).

☆ Seed with systemic fungicide like Carbendazim (0.15 per cent) is reported to be effective in management of disease.

Bacterial Diseases

Bacterial Leaf Spot

Bacterial leaf spot of sesame was first reported by Malkoff (1903) in Bulgaria, with *Xanthomonas syringae* pv. *sesami* described as the causal agent. Now the bacterium is identified as *Pseudomonas syringae* Van Hall pv. *sesami* (Malkoff) Young, Dye, and Wilkie. (Young *et al.*, 1978; Bradbury, 1981).

The disease is worldwide in distribution and has been reported from Brazil, China, India, Japan, Somalia, Korea, Mexico, South Africa, Sudan, Tanzania, Turkey, Uganda, USA, Venezuela (Kolte, 1985 and Vyas *at al.*, 1984). The disease is reported to cause up to 27 per cent losses in India (Vajayat and Chakravarti, 1977; 1978). A capsule loss of about 60 per cent has been reported from severely infected plants (Bremer *et al.*, 1947).

Symptoms

Symptoms appear on all above-ground parts of the plant. The disease appears as water-soaked yellow specks on the upper surface of the leaves. They enlarge and become angular as resticted by veins and veinlets. The colour of spot may be dark brown with shiny oozes of bacterial masses. Badly affected leaves soon dry up and shed. Infection of the stem is initiated at the point of attachment of the petiole as a dark-brown lesion which rapidly spreads along the stem. Severe infection on the stem results in death. Spots on the capsules are usually slightly sunken, shiny, and purplish in color. The capsules may turn black when infected early and do not form seed.

Causal Organism

The pathogen is *Psedomonas syringae* Van Hall pv. *sesami* (Malkoff) Young, Dye, and Wilkie (= *P. sesami* Malkoff). The bacterium is a gram negative rod, single or in pairs, forms no capsule, is motile by 2 to 5 polar flagella and measures 0.6 to 0.8 x 1.2 to 3.8 μm. Colonies are white opalescent, circular, entire, smooth, flat, and striate. A green fluorescent pigment is formed on agar plates. The optimum temperature for growth is 30°C (Brabdbdury, 1981).

Disease Cycle and Epidemiology

The bacterium remains viable in the infected plant tissues. It is internally seed borne and secondary spread through rain splash and storms. It survives in sterilized soil for the duration of 45 to 96 days and in unsterilized soil for 7 to 30 days only. The bacterium may survive for about eleven months on seed (Vajayant and Chakravarti, 1978). Primary infection appears to be only through the infected seed. Secondary infection occurs through secondary growth of the bacteria as it becomes visible in the form of bacterial ooze on the infected tissue. It enters the host through wound and stomata, and invasion occurs through parenchyma tissue.

The disease becomes more serious under conditions of high rainfall or where humidity persists for long periods. The disease is less severe when sesame is grown in more arid areas under furrow irrigation, but when flood-irrigated, standing water encourages the development of a severe occurrence of the disease (Kolte, 1985).

Management

☆ Use resistant varieties like Rajeshwari and ES 227 as reported by (Gangakishan *et al.*, 1967; Shadakshari *et al.*, 1989).

☆ Destruction of crop residues.

☆ Steep the seed in Agrimycin-100 (250 ppm) or Streptocycline suspension (0.05 per cent) for 30 minutes for good management of the disease.

☆ Hot water treatment of seed at 52°C for 10 minutes found effective to controlling the seed borne infection (Singh, 1970).

☆ Foliar spray of Streptocycline (500 ppm) just after the appearance of disease.

Bacterial Blight

Bacterial blight of sesame is recognized as the most threatening problem affecting sesame in India. Barcenas (1960) reported *X. campestris* pv. *sesami* causing sesame blight in Columbia, USA. At the same time, bacterial blight was described as one of the most damaging diseases of sesame in Sudan (Sabet and Dawson, 1960). In India, Rao reported the bacterial blight of sesame in 1962. The disease is known to cause a yield loss of about 20 per cent from Jabalpur in MP (Shukla *et al.*, 1972)

Symptoms

The disease affects all the plant parts and can be seen at all the stages of crop growth. Initially water-soaked spots appear on the undersurface of the leaf and then on the upper surface. They increase in size, become angular and restricted by veins and dark brown in color. Several spots coalesce together forming irregular brown patches and cause drying of leaves. The reddish brown lesions may also occur on petioles and stem. Capsules are also infected by the bacterium and symptoms appear as oval, slightly raised, dark brown spots.

Causal Organism

Xanthomonas campestris (Pamel) Dowson pv. *sesami* (Sabet and Dowson) Dye. The bacterium is rod-shaped, produces capsules, is motile by a single polar flagellum, and measures 0.4 to 0.6 x 0.8 to 1.6 μm. The colour of the bacterium on medium is yellow, smooth, slightly viscoid and shining (Kolte, 1985).

Disease Cycle and Epidemiology

The bacterium survives in the infected plant debris and in seeds. The secondary spread is by rain water. Seed can carry the bacterium up to a period of eleven months (Rao and Durgapal, 1966). Bacterium may survive on dry and hanging leaves on the weed plants until the next season (Nayak and Sharma, 1980). The bacterium enters the host primarily through stomata and quickly becomes vascular. The secondary spread is by spattering rains. High temperature and humidity favour the disease. Seedling infection of sesame is most severe at soil temperature of 20°C. Infection does not take place when soil temperature is 40°C. The disease also becomes severe when the soil moisture is 30 to 40 per cent and relative humidity is 75 to 87 per cent (Kolte, 1985). The reduction in temperature, RH and Rain fall reduces the disease (Srivastava *et al.*, 1997).

Management

☆ Use resistant variety like T-58

☆ Early planting *i.e.* immediately after onset of monsoon.

☆ Remove and burn infected plant debris.

☆ Steep the seed in Agrimycin-100 (250 ppm) or Streptocycline suspension (0.05 per cent) for 30 minutes.

☆ Foliar spray of Streptocycline (500 ppm) as soon as symptoms are noticed and continue two more sprays at 15 days interval if necessary.

☆ Spray Streptomycin sulphate or Oxytetracycline hydrochloride or Strephocyclin at 100g/ha.

☆ Hot water treatment (52°C for 15 minutes) gives the best control of the disease (Singh, 1970).

☆ Rai and Srivastava (2003) reported that seed treatment with Streptocycline (100ppm) followed by three spray at 10 days interval of Streptocycline + Copper oxychloride (100ppm + 500ppm) is best in managing the disease.

Phytoplasma Disease: Phyllody

Sesame phyllody (SP), caused by phloemlimiting phytoplasmas, is primarily distributed in the tropical countries including Asia and Africa (McCoy *et al.*, 1989; Nakashima *et al.*, 1999) The occurrence of this disease was first reported from Burma (Mc Gibbon, 1924) and later in India in 1930 (Kashiram,1930) with the incidence of the disease from 15 to 100 per cent. The disease is now reported to occur in several countries *viz.*, India, Iran, Israel, Sudan, Tanzania, Thailand, Turkey, Uganda and Venezuela (Saran *et al.*, 2005). This disease is very common around Bangalore (Nagaraju *et al.*, 2000) and Andhra Pradesh (Prasad and Reddy, 1997)). The disease has been named as "green flowering disease" or "pothe" in Burma, Sepaloidy, and Stenosis in India, and "Phyllomania" or "green flowering" in Africa.

Affected plants remain partially or completely sterile, resulting in total loss in yield. As much as 10 to 100 per cent incidence of the disease has been recorded in the sesame crop in India. The yield loss due to phyllody in India is estimated to about 39 to 74 per cent (Sahambi, 1970). The losses in plant yield, germination, and oil content of sesame seeds may be as high as 93.66, 37.77 and 25.92 per cent, respectively (Vyas, 1981). It is estimated that a 1 per cent increase in phyllody incidence decreases the sesame yield by 8.4 kg under Coimbatore conditions in India. Up to 50 per cent yield loss was reported due to phyllody from Warangal and Karimnagar district of Andhra Pradesh (Prasad and Reddy, 1997).

Symptoms

The disease is characterized by virescence, phyllody, yellowing, floral sterility and stem proliferation of infected plants (Akhtar *et al.*, 2008). Affected sesame plants express symptoms, depending on the stage of crop growth and time of infection. A plant infected in its early growth remains stunted to about two thirds of a normal plant, and the entire plant may be affected. The entire inflorescence is replaced by a growth consisting of short, twisted leaves closely arranged on a stem with very short internodes. However, when infection takes place at laterstages, normal capsules are formed on the lower portion of the plants and phylloid flowers are present on the tops of the main branches, and on the new shoots that are produced from the lower portions.

The most characteristic symptom of the disease is transformation of flower parts into green leaf-like structures followed by abundant vein clearing in different flower parts. The calyx becomes polysepalous and shows multicostate venation compared to its ga-mosepalous nature in healthy flowers. The sepals become leaf like but remain smaller in size. The phylloid flowers become actinomorphic in symmetry and the corrola becomes polypetalous. The corrola may become deep green, depending upon the stage of infection. The veins of the flowers become thick and quite conspicuous. The stamens retain their normal shape, but they may become green in color. Sometimes the filaments may, however, become flattened, showing its tendency to become leaf-like.

Causal Organism

The Causal agent is a Phytoplasma (*mycoplasma*) like organism (PLO or MLO). The mycoplasmas like bodies are reported to be present in phloem sieve tubes of affected plants. Electron microscopy has revealed that the big mycoplasma bodies, ranging from 100 nm diameter to 625 nm diameter are present in the sieve tubes. Generally the PLOs are round, but some may be 1500 nm long and 200 nm wide. Bodies with beaded structures can also be noticed. The PLOs are bounded by a single unit membrane as is typical for the PLO, and show ribosome -like structure and DNA-like strands within (Klein, 1977).

Disease Cycle and Epidemiology

The pathogen has a wide host range and survives on alternate hosts like *Brassica campestris* var. *toria*, *B. rapa*, *Cicer arietinum*, *Crotalaria* sp., *Trifolium* sp., *Arachis hypogaea* which serve as source of inoculum. The disease is transmitted by jassid, *Orosius albicinctus*. Optimum acquisition feeding period of vector is 3-4 days and inoculation feeding period is 30 minutes. The incubation period of the pathogen in leaf hoppers may be 15-63 days and 13-61 days in sesame. Nymphs are incapable of transmitting the phytoplasma. Vector population is more during summer and less during winter months.

The pathogen undergoes an incubation period of about 3 weeks, both in the insect and the host, during summer months (May-July) under Delhi conditions in India, and this incubation period is considerably increased during winter months (October-January) due to low temperature. Among the weather factors, the night temperature (minimum tem-perature) prevailing from the 30th to the 60th day after sowing is found to have a greater increase of disease incidence. A fall in the night temperature by 1°C at this time brings about increase by 5 to 7 per cent in incidence of phyllody under Tamil Nadu conditions in India. The minimum acquisition feeding period has been observed to be 8 hr, while the minimum infection feeding period is 30 min during May and June. Both male and female insects are equally efficient in transmitting the pathogen. The nymphs of the insect are capable of acquiring the pathogen but they are unable to transmit it, as by the time the incubation period is completed, they reach the adult stage. Once the leaf-hoppers have picked up the pathogen and become infective, the adult leaf hoppers remain so throughout the remainder of their lives without replenishment of the pathogen from infected plants (Kolte, 1985).

Management

- ☆ Use intercropping system, sesamum + pigeon pea (6: 1).
- ☆ Remove all the reservoir and weed hosts and destruction of diseased plants.
- ☆ Three sprays of Dimethoate (0.03 per cent) at 30, 40 and 60 days after sowing are beneficial to control the vector.
- ☆ Avoid growing sesamum near cotton, groundnut and grain legumes.
- ☆ Spray Monocrotophos or Dimethoate @ 500ml/ha to control the jassids
- ☆ Soil treatment with Thimet 10G @ 10 kg/ha at the time of sowing.
- ☆ Reduced population of the vector during growth period of sesame crop is perhaps important in keeping the disease under check(Mathur and Verma, 1972)
- ☆ A possibility of biochemical control by spraying manganese chloride has been indicated (Purohit and Arya, 1980). It appeared that manganese chloride oxidizes the phenol and protects or inhibits the enzymes, bringing the auxin level to normal. Once hyper auxin is oxidized, the plant can grow to its normal condition.

Non-Parasitic Diseases

Nutrient Deficiencies

General Chlorosis, with or without withering of plants, appearing initially on older plants.

Nitrogen Deficiency

Plants light green, stalk slender, branching absent and leaves erect. Lower leaves become lemon-yellow turning to orange. Discolored leaves are shed.

Phosphorus Deficiency

Branching suppressed, stalks slender, lower leaves dull dark grayish-green, necrosis of lower or majority of leaves is followed by defoliation.

Magnesium Deficiency

Lower leaves develop interveinal chlorosis, light-yellow in color becoming orange later. Green color persists in midrib and veins giving a characteristic pattern. Chlorosis appearing as mottling with or without necrosis.

Potassium Deficiency

Plants become dwarf, and margins of lower leaves become wavy and cupped upward. Light lemon-yellow chlorotic mottling appears which later turns bright orange and finally copper-colored. Defoliation does not occur. Effects localized in the growing region, terminals die back.

Calcium Deficiency

Terminal dies out following distortion of the tips and bases of young leaves. Hooking downward of the young leaf tips is followed by twisting and puckering.

References

Ahmed, Q., Mohammad, Abu, Ahmad, Q. and Mohammad, A. (1989). Screening of fungicides against sesamum wilt. *Pesticides* **23**: 28-29.

Akhtar, K.P., Dickinson M., Sawar, G., Jamil, F.F.and Haq, M.A. (2008). First report on the association of a 16Sr II *Phytoplasmas* with sesame phyllody in Pakistan. *Plant Pathology,* **57**: 771.

Anilakumar, K.R., Pal, A., Khanum, F. and Bawa, A.S. (2010). Nutritional, medicinal and industrial uses of sesame (*Sesamum indicum* L.) seeds an overview. *Agric Conspec Sci.,* **75**: 159-168.

Ashri, A. (1998). Sesame breeding. In *Plant Breeding Reviews*. Edited by Janick J. Oxford: Oxford UK. 79-228.

Bedigian, D. and Harlan, J.R. (1986). Evidence for cultivation of sesame in the ancient world. *Econ Bot.,* **40:** 137-154.

Bhale, M.S, Bhale, U. and Khare, M.N. (1998). Diseases of important oilseed crops and their management. In: Pathological problems of economic crop plants and their management. (S.M. Paul Khurana, Ed.), Scientific Pub. India, Jodhpur. pp. 251-279.

Bradbury, J.F. (1981). *Pseudomonas syringae* pv. *sesame.* CMI Description of Pathogenic Fungi and Bacteria No. 696. CMI, Kew, Surrey, London.

Bremer, H., Ismen, H., Karel, G., Ozkan, H. and Ozkan, M. (1947). Contribution to knowledge of the parasitic fungi of Turkey. *Rev. Fac. Sci. Univ. Istanbul. Ser. B.* **13**: 122.

Buldeo, A.N. and Rane, M.S. (1978). Fusarium wilt of sesamum. *J. Maharashtra Agric. Univ,.* **3**: 167-170.

Butler, E. J. (1918). Fungi and diseases in plants. Thacker Sprink and Co. Calcutta. 547p.

Butler, E. J. (1926). The wilt diseases of cotton and sesamum in India. *Agric. J. India.* **21**: 268.

Castellani, E. (1950). L. avvizzimento del sesame. *Olearia.* **4**: 20.

Chattopadhyay, C. and Sastry, R.K. (1999). Effect of soil solarization on the sesame stem-root-rot pathogen *Macrophomina phaseolina* population *Sesame and Safflower Newsletter.* No. 14 pp. 81-90.

Choi, S.H., Cho, E.K. and Cho, W.T. (1984). Epidemiology of sesame Phytophthora blight in different cultivation types. *Res. Reports. Rural Dev. Korea Republic* 26: 64-68.

Chowdhury, S. (1945). Control of Cercospora blight of til. *Indian J. Agric. Sci.,* **15**: 140.

Daftari, L.N. and Verma, O.P. (1975). An integrated approach to control the root and stem rot of sesamum caused by *Macrophomina phaseoli. Symp. Plant Disease Problems, Soc. Mycol. Plant Pathol.,* Sept. 18-20th 1975. Udaipur, p. 37(Abstr.)

Dustur, J.F. (1913). *Phytophthora parasitica* a new disease of castor oil plant. *Mem. Dep. Agric. India, Bot. Surv.,* **8**: 177-231.

Ellis, M. P. and Holliday, P. (1970). *Alternaria sesami.* CMI Description of Pathogenic Fungi and Bacteria. No.250, CMI. Kew, England.

Enikuomehin, O. A., Olowe, V.IO., Alao, O.S. and Atayese, M.O. (2002). Assessment of Cercospora leaf spot disease of sesame in different planting dates in South Western Nigeria. *Moor. J. Agric. Res.,* **3**: 76-82.

Erwin, D.C. (1983). *Phytophthora* its biology, taxonomy, ecology and pathology. APS press. *American Phytopathological Soc.* pp. 237-257.

Ferrer, J. B. (1960).The occurrence of angular leaf spot of sesamum in Panama. *Plant Dis. Reptr.* **44**: 221.

Gaikwad, S.J. and Pachpande, S.M. (1992). Effect of temperature on wilt of sesame caused by *Fusarium oxysporum* f.sp. *sesami. J. Maharashtra Agric. Univ.*17: 76-78.

Gangakishan, A., Jagadishwar, K., Raju, C.S. and Sundarshanam, A. (1989). Rajeshwari a high yielding white seeded variety of sesame for Andhra Pradesh. *Indian Fmg.* **39**: 7-8.

Gemawat, P.D. and Verma, O.P. (1971). Diseases of sesamum (*Sesamum indicum* L.) in Rajasthan- A note on the control of charcoal rot. *Madras Agric. J.,* **58**: 321-323.

Gupta, I.J. and Cheema, H.S. (1990). Effect of microsclerotia of *Macrophomina phaseolina* and seed dressers on germination and vigour of sesame seed. *Seed Res.* **20**: 27-30.

Hazarika, D.K. (1998). Influence of sowing of dates varieties on development powdery mildew of sesame in Assam. *J. Phytological Res.*11: 73-75.

Hegde, D.M. (2002).Oilseed scenario with particular reference to sesame and niger. In: Integrated crop management of sesame and niger. (Duhoon, S.S., A.K. Tripathi, H.K.Sharma, Eds.), P.C. Uni, S and N. Jabalpur (M.P.) pp. 1-7

Jeyalakshmi, C., Rettinassababady, C. and Nema, S. (2013). Integrated management of sesame diseases. *J. Biopesticides*, **6**: 68-70.

Jharia, H.K., Khare, M.N. and Duhoon, S.S. (2003). Current status of resistance to pathogens causing diseases in sesame. In: ISOR 2003 Extended Summaries: *National Seminar on stress management for oil seeds for attaining self reliance in vegetable oils. Indian Soc. Oil seed Res.*, Hyderabad. pp. 101-103.

Kadian, O.P. (1972). Testing *Sesamum oriente* L. seeds for disease in Haryana. *HAU. J. Res.* **2**: 41-44.

Kale, G.B. and Prasad, N. (1957). Phytopthora blight of sesamum. *Indian Phytopathol.* **10**: 38-47.

Kalita, M.K., Pathak, K., and Barman, U. (2000). Yield loss in sesamum due to Phytophthora blight in Barak Valley of Assam. *Indian J. Hill Farming*, **13**: 42-43.

Kang, S. W. and Kim, H. K. (1989). *Glioclaidum virense* a potential biocontrol againgst damping off and Fusarium wilt of sesame (*Sesamum indicum* L.) in problem field in Korea. *Res. Reptr. Rural development administrative crop Prot.* Korea Republic, **31**: 19-26.

Kang, S.W., Cho, D. J. and Le, Y. S. (1985). Incidence of Fusarium wilt of sesame in relation to air temperature. *Korean J. Plant Prot.* **24**: 123-27.

Karunanithi, K. (1996). Efficacy of fungicides in the control of powdery mildew of sesamum caused by *Oidium acanthospermi. Indian J. Mycol. Pl. Pathol.* **26**: 229-230.

Kawamura, E.1931. New fungi on *Sesamum indicum* L. Fungi. *Nippon Fungological Soc.* **1**: 26.

Khare, M.N. and Jharia, H.K. (2002). Biostress and their management in sesame. In: Integrated crop management of sesame and niger. (S.S. Duhoon, A.K. Tripathi and H.K. Jharia, Eds.) pp. 82-94.

Kolte, S.J.1985. *Diseases of annual edible oilsees crops*. Vol. II Rapeseed Mustard and Sesame Diseases. CRC Press, Inc. Boca, Raton, Florida, pp.83-122.

Maiti, S., Hegde, S.R. and Chattopadhyay, S.B. (1988). Handbook of annual oilseed crops. Oxford and IBH Publishing Co. Pvt. Ltd. New Delhi. pp.109-137.

Malaguti, G. 1959. Epiphytotics of Fusarium wilt in sesame. *Agron. Trop. Maracy.* **8**: 145.

Malaguti.G. (1953). A stem rot of sesame (*Sesamum indicum* L.) caused by *Phytophthora*. *Agron. Trop. Maracy.* **2**: 201-102.

McCoy RE, Caudwell A, Chang CJ, Chen TA, Chiykowski LN, Cousin MT, Dale JL, De Leeuw GTN, Golino DA, Hackett KJ, Kirkpatrick BC, Marwitz R, Petzold H, Sinha RC, Sugiura M, Whitcomb RF, Yang IL, Zhu BM, Seemüller E. (1989). Plant diseases associated with mycoplasma-like organisms. In: Whitcomb RF, Tully JG (Eds.) The Mycoplasmas. New York NY. Academic Press. pp. 545-560.

Mishra, C. B. P and Mukerjee, N. (1982). Soil amendment with oil cakes in controlling the root- rot olitorius jute. *Jute Development J.,* **2**: 11-12.

Mohanty, N.N and Behera, B.C. (1958). Blight of sesame (*Sesamum indicum* L.) caused by *Alternaria sesami*. *Curr. Sci.,* **27**: 492.

Mohanty, N.N. (1958). Cercospora leaf spot of sesame. *Indian Phytopath.* **11**: 186.

Muller, A.S. and Texera, D.A. (1941). The white spot of Sesame. *Agric.Venez.* **5**: 47.

Murugesan, M. Shanmugam, M.M., Menon, P.P.V., Arokiaraj, A., Dhamu, K.P. and Kochubabu, M. (1978). Statistical assessment of yield loss of sesamum due to insect pests and disease. *Madras Agric. J.* **65**: 290-295.

Nagabala, A.Z. and Zambettakis, C. (1970). *Alternaria sesami* (Kaw.) Mohanty and Behera. *Rev.Mycol.* **35**: 1.

Nakashima, K., Chaleeprom, W., Wongkaew, P., Sirithorn, P. and Kato, S. (1999). Analysis of phyllody disease caused by phytoplasma in sesame and *Richardia* plants. *J. Japan International Research Center for Agricultural Science.* **7**: 19-27.

Navas, M. and Subero, I.J. (1995). Survival of *Corynespora cassiicola* in sesame seeds under two storage temperature. *Anales de Bot. Agricola.* **2**: 16-19.

Nayak, M.L. and Sharma, R.K. (1980). A new weed host of bacterial blight of sesame. *Indian Phytopath.* **33**: 482.

Nusbaum, C.J. (1941). The role of hot water seed treatment in the control of Cercospora mlight of benne, *Phytopathology,* **31**(8): 770.

Ojiambo, P.S., Ayiecho, P.O. and Nyabundi, J.O. (1999). Effect of plant age on sesame infection by Alternaria leaf spot. *African Crop Sci. J.* **7**: 91-96.

Rai, M. and Srivastava, S.S.L. (2003). Field evaluation of some agro chemical for management of bacterial blight of sesame. *Farm Sci. J.* **12**: 81-82.

Rajpurohit, T.S. (2002). Influence of Intercropping and Mixed Cropping with Pearl Millet, Green gram and Mothbean on the Incidence of Stem and Root rot (*Macrophomina phaseolina*) of Sesame. *Sesame and Safflower Newsletter,* 17-20, 2002.

Rajpurohit, T.S. and Neema, S. (2013). Efficacy of different fungicides on incidence of alternaria leaf spot of sesame (*Sesamum indicum* L.). *J. Oilseeds Res.* **30** (1): 99-100.

Rajpurohit, T.S., Solanki, Z.S. and Ahuja, D.B. (2003). Screening of promising sesame genotype against Macromophomina stem and root rot under sick plot. In: ISOR 2003 Extended Summaries: *Nat. Seminar on stress management for oil seed for attaing self reliance in vegetable oils. Indian Soc. Oilseed Res.,* Hyderabad. pp. 351-352.

Rao, G.V.N. and Rao, M.A.R. (2001). Assessment of yield losses in sesame (*Sesamum indicum* L.) due to powdery mildew and its management. *Indian J. Plant Prot.* **29**: 165-167.

Rao, Y.P. (1962). Bacterial blight of sesame (*Sesamum indicum* L.) *Indian Phytopath.*, **15**: 297.

Rao, Y.P. and Durgapal, J.C. (1966). Seed transmission of bacterial disease of sesamum (*Sesamum orientale* L.) and eradication of seed infection. *Indian Phytopath.*, **19**: 402.

Rathaiah, Y. and Pavgi, M.S. (1975). Perpetuation of species of *Cercospora* and *Ramularia* parasitic on oil seed crops. *Ann. Phytopathol. Soc. Jpn.* **39**: 103.

Rathaih, Y. (1985). Phytophthora blight of sesamum, new to Assam. *J. Res. Assam Agric. Univ. Diphu*: 69-73.

Rivers, G.W. Martin, J.A. and Kinman, M. L. (1965). Reaction of sesame to Fusarium wilt in South Carolina. *Plant Dis. Reptr.* **49**: 383.

Sabet, K.A. and Dowson, W.J. 1960. Bacterial leaf blight of sesame (*Sesamum orientaale* L.) *Phytopath. Z.* **37**: 252.

Saharan, G. S., Mehta, N., and Sangwan, M. S. (2005). Diseases of oilseed crops. Indus Publishing Co. Pvt. Ltd. New Delhi.

Saksena, H.K. and Singh, D.V. (1975). Corynespora blight of sesame in India. *Indian J. Farm Sci.* **3**: 95.

Salazar-Huerta, E.J. and Qrtiz Enriquez, J.E. (1992). The effect of available soil water on the incidence of sesame (*Sesamum indicum* L.) root rot. *Revista Maxicana de Fitopatologia*, **10**: 44-48.

Samuel, G.S., Govindaswamy, C.V. and Vidhyasekaran, P. (1971). Studies of the Alternaria blight disease of gingelly. *Madras.Agric. J.*, **58**: 882.

Sehgal, S.P. (1963). Studies on the Phytophthora blight of sesame in Rajasthan. Ph. D. Thesis, Rajasthan University.

Sehgal, S.P. and Prasad, N. (1966). Variation in the pathogenicity of single zoospore isolates of seammum phytophthora: studies on the penetration and survival. *Indian Phytopath.* **19**: 154-158.

Sehgal, S.P. and Prasad, N. (1971). Instability of pathogenic characters in the isolates of sesamum phytophthora and effect of host passage on the virulence of isolates. *Indian Phytopath.* **24**: 295-298.

Shadakshari, V.G., Virupakshappa, K. and Siddaramaiah, A. L. (1989). Reactions of sesamum genotypes against powdery mildew and bacterial leaf spot. *Curr. Res.* **18**: 4-5.

Sharan, G.S. and Chand, J.N. (1988). Diseases of oilseed crops (In Hindi). Directorate of Publication Haryana Agricultural Univ., Hisar, India. 269p.

Shekharappa, G. and Patil, P.V. (2001). Chemical control of leaf blight of sesame caused by *Alternaria sesami*. *Karnataka J. Agric. Sci.* **14**: 1100-1102.

Shukla, B.N., Chand, J.N. and Kulkarni, S.N. (1972). Changes in sugar content of sesamum leaves infected with *Xanthomonas sesami*. *Indian Phytopath*. 25: 150.

Shukla, P., Yadav, R.N., Dwivedi, K. and Dwivedi, K. (1987). Studies of perpetuation of Corynespora blight of sesame. *Indian J. Plant Pathol*. 5: 10-13.

Siddaramaiah, A.L., Kulkarni, S., Desai, S.A. and Jayaramaiah, H. (1981). Studies on pod spot of sesamum caused by *Alternaria sesami* (Kawamura) Mohanty and Behera. *Agric. Sci. Digest* 1: 73.

Singh, B.P., Shukla, B.N. Nema, K.G. and Jain, A.C. (1977). Diseases of oilseed crops. Bull. JNKVV, Jabalpur (MP), India pp.104

Singh, R.N. (1970). Integrated control of bacterial disease in India. *Indian Phytopath*. 23: 155.

Singh, T. and Singh, D. (1980). Anatomy of penetration of *Macrophomina phaseoli* in seed of sesame In: *Recent Researches in Plant Sciences* (S.S. Bir. Ed). Pp. 603-606.

Singh, T. and Singh, D. (1990). Incidence of *Macrophomina phaseolina* (Tassi) Gold in sesame seeds of Rajasthan. *Indian J. Mycol. Pl. Pathol*. 20: 111-114.

Sirry, A. R., Amer, M. A., Elewa, I. S., Abtallah, S. M. and Abdel-El-Gawadm. (1980). Effect of P and K fertilizers on the growth and nutrients contents of sesame plant and their relation to root rot incidence. *Agric. Res. Rev*. 57: 29-38.

Srivasava, S.S.L., Rai, M. and Singh, D.V. (1997). Epidemiological study on bacterial blight of sesame. *Indian Phytopath*.50: 289-290.

Sundaraman, S. (1932). Administration report of the mycologists for the year 1930-31. pp. 20.

Sundaraman, S. (1933). Administration report of the mycologists for the year 1930-31. pp. 37.

Tripathi, N.N., Kaushik, C.T. and Yadava, T.P. (1977). Control of charcoal rot of sesamum caused by *R. bataticoala*. *Pesticides* 11: 35-37.

Tripathi, S. K., Bose, U. S. and Tewari, K. L. (1996). Economics of fungicidal control of Cercospora leaf spot of sesame. *Agric. Sci. diegest*. 16: 233-236.

Vajayat, R. and Chakravarti, B.P. 1978. Survival of *Pseudomonas sesami* and effect of antagonistic bacterium isolated from seeds on the control of the disease in the field. *Indian Phytopath*. 31: 286.

Vasudeva, R.S. (1961). Diseases of Sesamum. In: Sesamum monograph by A.B. Joshi Indian Central Oilseeds Committee, Hyderabad.

Verma, M.L. (2002a). Fungal and bacterial diseases of sesame and their managemen-challenges for the milliennium. (Prasad, D and Puri,S.N.)Jyoti Pub., New Delhi, pp.161-192.

Verma, M.L. (2002[b]). Effect of soil types on diseases and yield of sesame (*Sesamum indicum* L.) varieties sown on various dates in rainy season. *Asian Congr. Mycol. Pl. Pathol. and Symp. On Plant Health for Food Security*. Univ.of Mysore, pp.-53(Abstr.).

Verma, M.L. and Bajpai, R.P. (2001). Effect of bioinoculants on Phytophthora blight and Microphomina root/stem rot of sesame (*Sesamum indicum*). *Nat. Symp. Plant Prot. Strategies for Sustainable Horticulture, Soc. Plant Prot. Sci.* SHAUT, Jammu, pp.113(Abstr.).

Verma, M.L., Mehta, N. and Sangwan, M.S. (2005). Fungal and Bacterial Diseases of sesame. In: Diseases of Oil Seed Crops. (G.S. Saharan, N. Mehta and M.S. Sangwan, Eds.) pp. 269-318.

Vyas, S.C, Prasad, K.V.V. and Khare, M.N. (1983). Diseases of sesamum and niger and their control. Directorate of Research service, JNKVV, Jabalpur (MP.) 1983. pp. 1-16.

Vyas, S.C. (1981). Diseases of sesamum and niger in India and their control. *Pesticides.***15**: 10.

Vyas, S.C. Kotwal, Indra, Prasad, K.V.V. and Jain, A.C. (1984). Note on Seed borne fungi of sesamum and their control. *Seed.Res.* **12**: 93-94.

Wei, C.T. (1950). Notes on Cercospora. Mycological Paper No.34, CMI, Kew, England.

Whitcomb, R.F., Tully, J.G (Eds.) The Mycoplasmas. New York NY. Academic Press. pp. 545-560.

Yadav, R.N., Shukla, P., Dwivedi, K. and Dwivedi, K. (1988). Evaluation of some fungicides against Corynespora blight of sesamum (*Sesamum indicum* L.) *Farm Sci. J.*, **3**: 122-127.

Young, J.M., Dye, D.W., Brasbury, J.F., Panagopoulos, C.G and Robbs, C.F. (1978). A proposed nomenclature and classification of plant pathogenic bacteria. *N.Z Agric. Res.* **21**: 153.

Yu, S.H. and Park, J.S. (1980). *Macrophomina phaseolina* detected in seeds of *Sesamum indicum* and its pathogenicity. *Korean J. Plant Prot.***19**: 135-140.

6 Soybean Diseases and their Management

Sunil Kumar and H.K. Singh

Soybean (*Glycine max* L. Merrill) is a leguminous crop; it belongs to the family Leguminocae. It is rich in high quality protein (40-42 per cent), oil (18-20 per cent) and other nutrients like calcium, iron and glycines. It is a good source of isoflavones. Soybean helps in preventing heart diseases, cancer, HIV *etc.* (Kumar, 2007). Soybean protein is rich in the valuable amino acid lysine (5 per cent) in which most of the cereals is deficient. In addition, it contains good amount of minerals, salts and vitamins (thiamine and riboflavin). Its sprouting grains contain a considerable amount of vitamin C, minerals, salts and vitamins (thiamine and riboflavin) (Singh *et al.*, 2003). Soybean is the richest, cheapest and easiest source of best quality protein and fat. Hence, it is called as vegetarian meat and wonder crop. This crop is severely affected by a number of diseases and causes much yield losses.

Fungal Diseases

Collar Rot

The disease appeared in seedling stage of the crops and causes huge loss. Collar rot is the most important disease which causes heavier yield losses. Muthusamy and Mariappan (1991) reported 14–74 per cent yield loss due to this disease in Maharashtra. It is very difficult to manage this pathogen as they perpetuate in the soil as sclerotia and chlamydospores.

Symptoms

The infected plants gradually lose their colour and turn pale, followed by drooping. The affected roots, particularly the collar portion turn yellowish-brown. Affected plants can be easily pulled out from the soil. White to tan-brown mustard

Figure 6.1: Collar Rot.

seed like sclerotia are seen around the infected roots. The symptoms may be extended on stem, causing shrivelling of the stem (the fungus can also be seen naturally causing water-soaked spots on leaves) and finally result in the death of the plants.

Causal Organism

The causal organism of this disease is *Sclerotium rolfsii*. It is a well known polyphagous and most destructive soil borne fungus and was first reported by Rolfs (1892) as a cause of tomato blight in Florida. Later, Saccardo (1911) named the fungus as *Sclerotium rolfsii*. But, in India, Shaw and Ajrekar (1915) isolated the fungus from rotted potatoes and identified as *Rhizoctonia destruens* Tassi. However, later, studies showed that, the fungus involved was *S. rolfsii* (Ramakrishnan, 1930). Mundkur (1934) successfully isolated the perfect stage of *S. rolfsii*. Sclerotium is soil inhabitant basidiomycetes, produces abundant white fluffy, branched mycelium that forms numerous sclerotia but is usually sterile (does not produce spores) and cause serious diseases on many hosts by affecting the roots, stems, tubers, corns and other plant parts that develop in or on the ground. The perfect stage of the fungus is *Aethalium rolfsii*. The fungus produces two types of hyphae. Coarse, straight, large cells (2-9 um x 150-250 um) have two clamp connections at each septation, but may exhibit branching in place of one of the clamps. Branching is common in the slender hyphae (1.5-2.5um in diameter) which tend to grow irregularly and lack clamp connections. Slender hyphae are often observed penetrating the substrate. Sclerotia (0.5-2.0mm diameter) begin to develop after 4-7 days of mycelial growth. Sclerotia forming on a host tend to have a smooth texture, whereas those produced in culture may be pitted or folded. Serving as a protective structure, sclerotia contain viable hyphae and serve as primary inoculum for disease development.

Disease Cycle

The fungus overwinters mainly through sclerotia. Pathogen is spread by contaminated tools, infected transplant seedlings, moving water, infested soil, infected vegetables and fruits and in some hosts as sclerotia mixed with the seed. The fungus attacks tissues directly. However, the mass of mycelium produces secretes oxalic acid and also pectinolytic, cellulolytic and other enzymes and it kills and disintegrates tissues before it actually penetrates the host. Fungus once establishes in the plants, advances and produces mycelium and sclerotia quite rapidly, especially at high moisture and high temperature from 30 to 35°C.

Disease Management

Management of collar rot disease is difficult. Crop rotation provides only partial control. Cultural practices *i.e.* deep summer ploughing to bury the fungal sclerotia in surface debris, ammonia fertilizations, and calcium compounds application are effective in controlling the diseases. Soil solarisation and use of Pentachloronitrobenzene (PCNB) which is sold under the name of Brassicol, Quintozone or Terrachlor are very effective for controlling this disease. The control is attributed to the hydrolysis products of glucosinolates in to allyl and butenyl isothicyanates which are toxic to *Pythium aphanidermatum, Sclerotium rolfsii, Sclerotinia sclerotiorum* and *Phytophthora capsici* (Singh, 2009).

Cercospora Leaf Spot

Cercospora leaf spot is a common disease of soybean of many diseases which infect soybeans in late season. In India it is severely appeared Madhya Pradesh when soybean crop is about 30 to 40 days old. This disease causes a minor yield loss in India due to late appearance. As a seed disease, it's reduced the quality of seed.

Symptoms

Foliar symptoms usually are seen at the beginning of seed set and occur in the uppermost canopy on leaves exposed to the sun. Affected leaves are discoloured, with symptoms ranging from light purple, pinpoint spots to larger, irregularly shaped patches typically only on the upper leaf surface. As disease develops, affected leaves may become leathery and dark purple with bronze highlights. Symptoms may be confused with sunburn. Discoloration may extend to the upper stems, petioles and pods. Infection of petioles and severe symptoms may lead to defoliation of the uppermost leaves and give the appearance of a maturing crop. However, petioles of fallen leaves remain attached to the stem, and lower leaves of the plant remain green. Symptoms of purple seed stain are distinct pink to dark purple discoloration of seed. Discoloured areas vary in size from small spots to the entire surface of the seed coat; however, infected seeds may not show symptoms.

Causal Organism

Cercospora leaf spot is caused by *Cercospora kikuchii* (Teleomorph - *Mycosphaerella*). The fungus produces long, slender and colourless to dark, straight to slightly curved, multicellular conidia on short dark conidiophores. Conidiophores arise from the plant surface in the clusters through stomata and form conidia successively on new growing tips. Conidia are detached easily and are often blown long distances by the wind. The fungus is favoured by high temperatures and therefore is most destructive in the summer months and in warmer climates. Fungus produces non specific toxin cercosporin which acts as a photosensitizing agent in the plant cells *i.e.* it kills cells only in light. The pathogen remains over seasons in or on seed and as small black stomata in plant debris.

Disease Cycle

The fungus survives winter in infected crop residue and infected seed. Mostly early season infections do not cause symptoms but contribute to infection of foliage and pods later in the season. Warm and wet weather is favourable for infection. Foliar symptoms are the result of an interaction between a toxin produced by the fungus and sunlight. Weather conditions during flowering and plant maturity will affect the incidence of purple seed stain. Despite being caused by the same organism, there is no consistent relationship between the occurrence of *Cercospora* leaf blight and purple seed stain.

Management

Use the disease free seeds and resistant varieties to control this disease. Seed treatment is essential to eliminate the seed-borne inoculum. Disinfection of seeds by dip in 0.5 per cent copper sulphate solution for 30 minutes. Foliar application of fungicides namely hexaconazole @ (0.025 per cent), bavistin (0.025 per cent) and chlorothalonil (0.2 per cent) are economic and effective to control this disease. Applications made during pod-filling stages can reduce the incidence of purple seed stain, but may not affect soybean yield. Crop rotation with non-host crops such as alfalfa, corn and small grains and tillage to bury infested crop residue will reduce pathogen levels. If considering tillage, use proven conservation practices to maintain soil quality.

Downy Mildew

The visibility of this disease in only where, the humidity is about more than 80 per cent. The economic loss by this disease is very minute.

Symptoms

Seedlings that are infected from oospores on the seed can develop large chlorotic areas on the first and second pairs of true leaves. The disease is more common in late vegetative and reproductive growth stages. Lesions occur on upper surfaces of leaves as irregularly shaped, pale green to light yellow spots that enlarge into pale to bright yellow spots. Older lesions turn brown with yellow-green margins. Young leaves are more susceptible than older leaves, so disease is often found in the upper canopy. Lesion size varies with the age of the affected leaf. On the underside of the leaf, fuzzy, gray tufts may be seen growing from each lesion, particularly when humidity is high or leaves are wet, for example, early in the morning. Infected pods show no external symptoms, but the inside of the pod and seed may be covered with a dried, whitish fungal mass that appears crusty and contains spores. Infected seed can be smaller, appear dull white and have cracks in the seed coat.

Causal Organism

The causal organism of this disease is *Peronospora manshurica*. Downy mildew is a very common foliar disease of soybeans, but it seldom causes serious yield loss. The pathogen may also infect seed and reduce seed quality. Diseased plants are usually widespread within a field.

Disease Cycle

The pathogen is primarily soil borne through oospores lying in the diseased plant debris. *Peronospora manshurica* survives in leaves and on the surface of seed. Extended periods of leaf wetness are favourable for movement of the pathogen. High humidity and moderate temperatures favour infection. The increased resistance of older leaves and higher temperatures midseason usually stop disease development before extensive damage occurs.

Management

Use only resistant and certified seeds for sowing. However, many races of the pathogen have been identified, and varieties that are resistant to all known races have not yet been developed. Crop rotation and burial of infested crop residue using conservation tillage practices can reduce pathogen levels. Two to three foliar spray of fungicide such as sulphur fungicide should be done at the disease initiation and after that 15 days interval.

Frogeye Leaf Spot

Frog-eye leaf spot is a common disease of soybean especially in the crops sown during March-April (Maiti *et al.*, 1983). The reduction of yield in susceptible crops is up to 15 per cent (Sinclair and Backman, 1989).

Symptoms

Early season infections from infected seed result in stunted seedlings. On leaves, lesions are small, irregular to circular and gray with reddish-brown borders that most commonly occur on the upper leaf surface. Lesions start as dark, water-soaked spots that vary in size, and as lesions become old, the central area becomes gray to light brown with dark, red-brown margins. In severe cases, disease can cause premature leaf drop and may spread to stems and pods. Symptoms on stems are not as common or distinctive as foliar symptoms and appear as narrow, red brown lesions that turn light gray with dark margins when they mature. Lesions on pods are circular or oval and are initially red-brown and turn to light gray with a dark brown margin. Seed close to lesions on pods can be infected. Infected seeds have light to dark gray discoloured blotches that vary in size and cover the entire seed in severe cases. The seed coat often cracks.

Causal Organism

Frogeye leaf spot has become more prevalent in north hills zone of India. The causal organism of this disease is *Cercospora sojina*. It is especially problematic in continuous soybean growing fields. Diseased plants are usually widespread within a field.

Disease Cycle

The fungus survives in infected crop residue and infected seed. Early season infections contribute to infection of foliage and pods later in the season. Warm, humid weather promotes spore production, infection and disease development. Young leaves are more susceptible to infection than older leaves, but visible lesions are not seen on young, expanding leaves because the lesions take two weeks to develop after infection. It is common for disease to be layered within the canopy. During dry periods either no infection or very little but during wet or humid period infection are high.

Management

Resistant varieties should be used where disease is a potential problem. Several races of the pathogen have been identified, and varieties with resistance to all known races are available. Crop rotation and tillage will reduce survival of *Cercospora sojina*. Crops not susceptible to this pathogen are alfalfa, corn and small grains. If tillage is considered to promote decay of crop residue, great care should be taken to minimize soil erosion and maintain soil quality. Foliar fungicides applied during late flowering and early pod set to pod-filling stages can reduce the incidence of frogeye leaf spot and improve seed quality and yield.

Septoria Brown Spot

Septoria brown spot is also known as brown leaf spot was first time reported from United States in 1922. This disease has been also reported in most soybean growing areas of the world.

Symptoms

Symptoms are typically mild during vegetative growth stages of the crop and progress upward from lower leaves during grain fill. Infected young plants have purple lesions on the unifoliate leaves. Lesions on later leaves are small, irregularly shaped and dark brown, and are found on both leaf surfaces. Adjacent lesions can grow together and form larger, irregularly shaped blotches. Infected leaves quickly turn yellow and drop. Disease starts in the lower canopy and, if favourable conditions continue, will progress to the upper canopy. Lesions on stems, petioles and pods are not as common, but appear as brown, irregularly shaped spots ranging from small specks to 1/2 inch in diameter.

Causal Organism

Brown spot is the most common foliar disease of soybean. The pathogen of this disease is *Septoria glycines*. Disease develops soon after planting and is usually present throughout the growing season. Yield losses depend on how far up the canopy the disease progresses during grain fill. Diseased plants are usually widespread within a field.

Disease Cycle

Warm and wet weather favours the disease development. The fungus survives on infected leaf and stem residue. Disease usually stops developing during hot and dry weather, but may become active again near maturity or when conditions are more favourable.

Management

Use of resistant variety is good source of managing this disease but there are no known sources of resistance, but differences in susceptibility occur among soybean varieties. The host range of *Septoria glycines* includes other legume species and common weeds such as velvet leaf. Crop rotation with non-host crops such as alfalfa, corn and small grains and incorporation of infested crop residue into the soil will reduce the survival of *Septoria glycines*. If tillage is an option, use conservation tillage practices to maintain soil quality.

Rust

The occurrence of soybean rust is world-wide. However, in India it has been sporadic but occurred in epidemic form in certain states (Vyas *et al.*, 1997, Sarbhoy *et al.*, 1972, Sathe 1972). It was first time observed from Madhya Pradesh in 1994 and continued to be a major threat in soybean cultivation, resulting in 80 per cent loss in yield in susceptible varieties (Sharma and Mehta, 1996).

Symptoms

Soybean plants are susceptible at any stage of development, but symptoms are most common after flowering. Early symptoms of rust infection begin on lower leaves. Lesions begin to form on lower leaf surfaces, starting as small, gray spots and changing to tan or reddish-brown. Lesions are scattered within yellow areas that appear translucent if the affected leaves are held up to the sun. Mature lesions contain one to more small pustules that usually occur on lower leaf surfaces. These pustules produce uredospores and spore production may continue for weeks. Premature defoliation and early maturity occurs while an infection is severe.

Causal Organism

The causal organism of soybean rust is *Phakopsora pachyrhizi*. Soybean rust is an aggressive disease capable of causing defoliation and significant yield loss. Soybean rust is an endemic to India and found in most soybean growing areas of the world.

Disease Cycle

The rust pathogen can only survive on green tissue; thus, the pathogen is unable to survive in areas where killing frosts eliminate susceptible hosts. The movement of rust depends on rust spores increasing at sites where the pathogen has survived the winter, dispersal of the spores to new areas and establishment of the disease in

Figure 6.2: Soybean Rust.

those areas. These steps need to be repeated several times within a growing season in order for rust to cause an epidemic in the country. When spores land in new areas, infection takes place only when prolonged periods of leaf wetness (6 to 12 hours) and moderate temperatures occur in those areas. Cool, wet weather or high humidity favour soybean rust epidemics. Dense canopies also can provide ideal conditions that encourage disease development. Infection can spread rapidly to middle and upper leaves once the canopy closes.

Management

A limited number of resistant breeding lines have been identified; however, there is currently some commercially available soybean rust resistant variety (DSb 21) in India. Resistant varieties have been released in other countries, but none are resistant to all known races of the pathogen. Currently, foliar fungicides are the only viable option for managing soybean rust. To manage the disease effectively and profitably, fungicides need to be sprayed prior to infection or, at the latest, very soon after initial infection *i.e.* hexaconazole @ 0.1 per cent or propiconazole @ 0.1 per cent (Singh, 2009).

Charcoal Rot

Charcoal rot of soybean is a soil-borne fungal disease. This disease is favoured by hot, dry weather and poor soil fertility. Excessive seeding rates and lack of soil moisture also cause this disease. Upto 77 per cent yield losses have been recorded from major soybean growing areas.

Symptoms

Symptoms of charcoal rot usually appear after flowering. Initial symptoms are patches of stunted or wilted plants. Leaves remain attached after plant death. The lower stem and taproots of these plants are discoloured light gray or silver. When stems are split, black streaks are evident in the woody portion of the stem. In addition, the fungus produces numerous tiny, black fungal structures called microsclerotia that are scattered throughout the pith and on the surface of taproots and lower stems. These microsclerotia give the tissue a charcoal-like appearance. Infected seed either show no symptoms or having microsclerotia embedded in seed coat cracks or on the seed surface. Infected seed have lower germination, and if seed germinates, the seedlings usually die within a few days.

Figure 6.3: Charcoal Rot.

Causal Organism

Charcoal rot can be an important disease and is most yield-limiting when weather conditions are hot and dry. This disease is caused by *Macrophomina phaseolina*. This disease is more common is southern and North Eastern part of the India and causes huge losses.

Disease Cycle

The fungus survives in soil or soybean residue as microsclerotia. Microsclerotia infect roots of soybean plants, sometimes very early in the season. Many environmental factors like temperature, humidity, rainfall *etc.* affect microsclerotia survival, root infection and disease development. The fungus is more abundant in soil when pH is very acidic or alkaline. Charcoal rot is most prevalent during hot, dry weather, especially when it occurs during the flowering/pod formation stages.

Management

In summer crops, irrigation lowers soil temperature and increases soil moisture. These conditions are unfavourable for the disease. Most efforts on control of *M.*

phaseolina involve management of populations of microsclerotia. Growing small grains, such as wheat or barley, can reduce microsclerotia numbers. Corn is also a host of *M. phaseolina* so it will not reduce levels of the fungus when planted in rotation with soybean. The fungus is less damaging to corn than to soybean. Fields with minimal or no tillage may have fewer symptoms because of lower soil temperature and greater water-holding capacity. Avoid excessive seeding rates so that plants do not compete for moisture, which increases disease risk during a dry season.

Fusarium Wilt and Root Rot

Fusarium wilt and root rot of soybean is a common and widespread disease in major soybean growing regions. This disease is caused by more than 10 different species of Fusarium. The yield loss due to this disease is up to 15 per cent.

Symptoms

Symptoms of Fusarium wilt are more noticeable under reduced moisture and hot conditions and are often misdiagnosed as those of Phytophthora root rot. Infected plants have brown vascular tissue in the roots and stems and show wilting of the stem tips. However, external decay or stem lesions are not seen above the soil line. Foliar symptoms include scorching of the upper leaves, while middle and lower canopy leaves can turn chlorotic and later wither and drop from the plant. Young plants are at the greatest risk to root rots caused by *Fusarium* species. Infected plants may exhibit poor or slow emergence, and seedlings are often stunted and weak. Seedlings with root rot have reddish-brown to dark brown discoloured roots. Infected plants may have poor root systems and poor nodulation, which may cause the plants to wilt and finally die.

Causal Organism

Fusarium is a very common soil fungus, and more than 10 different species are known to infect soybean roots and cause root rot. The species *Fusarium oxysporum* is responsible for causing Fusarium wilt. Although Fusarium root rot is a widespread disease in the country, the economic impact on yield is not well documented.

Disease Cycle

The fungus survives in the soil either as spores or as mycelium in plant residue. Certain weeds may serve as hosts to some pathogenic *Fusarium* species. The fungi can infect plants at any stage of soybean development but infection is particularly favoured when plants are weakened. Stresses such as herbicide injury, high soil pH, iron chlorosis, nematode feeding and nutritional disorders can all predispose plants to infection. After infection, damage to plants can be worsened if soil moisture is limited because of the compromised root systems.

Management

Varieties have varying levels of susceptibility, but no resistant varieties have been described. Reducing or eliminating stress factors, such as use of herbicides that cause injury to soybeans, wet soils and soybean cyst nematode, can help reduce root rot problems. Growing of tolerant varieties to iron deficiency chlorosis should be considered if the root rot seems associated with iron deficiency chlorosis. If *Fusarium* is a problem in a field, seed treatments with bavistin @ 2g/kg seed may protect seedlings in subsequent years.

Powdery Mildew

Powdery mildew of soybean was first time reported in 1947 from North Carolina. In India this is not a common disease of soybean. In other countries the estimated yield losses have been recorded up to 13 per cent.

Symptoms

The most common and characteristic sign of powdery mildew is white, powdery fungal growth appeared on aboveground plant parts, particularly the upper surface of leaves. Powdery mildew usually does not appear until mid- to late reproductive stages. Initially, small fungal colonies form and grow together as they enlarge. Eventually, entire surfaces of infected plant parts are covered with white fungal growth. Advanced symptoms include yellowing of plant tissues and premature defoliation.

Causal Organism

The powdery of soybean is caused by *Microsphaera diffusa*. The disease is more prevalent in cooler than normal seasons.

Disease Cycle

Microsphaera diffusa is a biotrophic parasite. The fungus survives in infested crop residue. That the pathogen survives between crop seasons through cleistothecia in soil. The favourable conditions for the disease development are cool, cloudy weather and low humidity. Powdery mildew of soybean is severely affect when crop sown in late season.

Management

Planting of resistant varieties minimise the disease and early sowing escape the disease. Chemicals such as sulphur fungicide effectively manage the powdery mildew; however, there are limited situations where fungicide use will be profitable. Efficacy of some plant extracts and plant products against the pathogen has been experimentally demonstrated Nemadole (a neem product) and *Allium cepa* (onion), *Allium sativum* (garlic), rhizome of ginger and neem leaves (*Azadiracta indica*) are non phytotoxic but fungicidal and at par with Karathane in the suspension of powdery mildew of pea. Several fungi such as *Ampelomyces*, *Tilletiopsis* and *Verticillium* and insects (*Thrips tabaci*) are natural biocontrol agents of the powdery mildew.

Anthracnose Stem Blight

Colletotrichum truncatum causes the anthracnose stem blight of soybean. Anthracnose is generally a late season disease that is prevalent on maturing soybean stems throughout the world. Soybean, however, is susceptible to infection throughout the growing season. Diseased plants are usually widespread within a field.

Symptoms

Infected seed may or may not show symptoms. When seed symptoms do occur, they appear as brown discoloration or small gray areas with black specks. Foliar symptoms include reddish veins, leaf rolling and premature defoliation. On stems and petioles, symptoms typically appear as irregularly shaped red to dark brown blotches during early reproductive stages. Damping off may occur if infected seed is planted. Leaves, pods and stems may also be infected without showing symptoms. Petiole infection may result in a shepherd's crook. Early infection of leaf petioles may cause premature defoliation and yield loss. Infection of young pods results in seedless pods at maturity while pods infected later contain seeds that are infected. Near maturity, black fungal bodies that produce small, black spines and spores are evident on infected stems, petioles and pods.

Disease Cycle

The fungus overwinters as mycelium in crop residue or infected seed. Although plant stand may be affected by early season infection, most infection occurs during the reproductive stage of the crop. Spores produced by the fungus are sensitive to drying; thus, free moisture for 12 hours or longer is necessary for successful infection. Warm, wet weather favours infection and disease development. The most important factors affecting the infection are temperature and moisture. Moderate temperatures between 13° and 26°C favour infection. No infection occurs at temperatures above 27° and at 13°C also the disease is considerably reduced. A relative humidity of above 92 per cent is necessary for infection, the optimum being close to 100 pre cent. A 10 hour wet period is reported to be necessary for conidial infection and new lesions usually appear in 3 -7 days depending on prevailing temperature.

Management

There are no known sources of resistance to anthracnose, but soybean varieties differ in susceptibility. The seed must be disease free hence it should be collected from only healthy pods. Usually, seed produced in dry areas or free from infection. Crop rotation and tillage will reduce survival of *Colletotrichum* species. Non-legume crops such as corn are not susceptible to this pathogen. If tillage is considered, great care should be taken to minimize soil erosion and maintain soil quality. Foliar fungicides labelled for anthracnose are available. Benlate, Ziram, Vitavax, Ferbam and lime sulphur, in order listed had been recommended for foliar sprays. Bavistin, Vitavax and Agroson GN were recommended for seed treatment. Applications should be made during the early to mid-reproductive

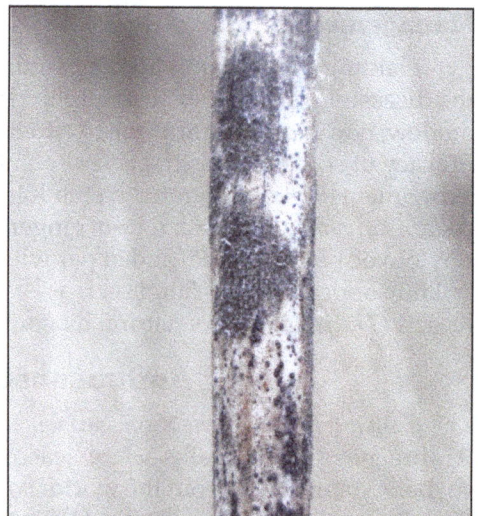

Figure 6.4: Anthracnose Stem Blight.

growth stages of the crop, although there are limited situations where fungicide use will be profitable. There are many reports of biological control of the anthracnose of bean through seed bacterization and through inoculation with avirulent strains of the pathogen (Sticher *et al.*, 1997; Van Loon *et al.*, 1998).

Phytophthora Root and Stem Rot

Phytophthora root and stem rot is a soil borne disease and causes severe economic losses. It was first time reported in 1951 from Ohio (Allington and Chamberlain, 1948) and has been spread throughout most of the soybean growing areas of United States.

Symptoms

The most characteristic symptom of Phytophthora root rot, however, is a dark brown lesion on the lower stem that extends up from the taproot of the plant. The disease can infect soybeans at any growth stage from seed to maturity. Early season symptoms include seed rot and pre- and post-emergence damping off. Stems of infected seedlings appear water-soaked, while leaves may become chlorotic and plants may wilt and die. On older plants, symptoms vary depending on the variety. For susceptible plants, leaves become chlorotic between the veins and plants wilt and die, with the withered leaves remaining attached. Varieties that are not fully susceptible may appear stunted, but plants are typically not killed. The lesion often reaches as high as several nodes and will girdle the stem and stunt or kill the plant.

Causal Organism

Phytophthora sojae causes the Phytophthora root and stem rot of soybean. Phytophthora root and stem rot is an economically important disease of soybeans that is most severe in poorly drained soils. Diseased plants often occur singly or in patches in low-lying areas of the field that are prone to flooding.

Disease Cycle

Phytophthora sojae survives on crop residue or in the soil as oospores. Optimum soil moisture is 15 to 20 per cent is needed for oospores germination to produce structures that release swimming spores, called zoospores, under saturated soil conditions. The zoospores are attracted to soybean roots. Infection occurs via the roots, and from there the pathogen colonizes the roots and stems. Disease is most common in poorly drained soils, but may occur in other soils as well.

Management

Management of Phytophthora root rot is by planting resistant varieties. Many race-specific resistance genes (called *Rps* genes) to *Phytophthora sojae* have been identified in soybean breeding lines. Some of these genes have been incorporated in commercial soybean varieties; thus, there are soybean varieties available that have complete resistance to a specific race of *Phytophthora sojae*. There are numerous races (now called pathotypes) of *Phytophthora sojae*, and many pathotypes can exist in a single field. Furthermore, new pathotypes can develop that can infect varieties with specific *Rps* genes. Partial resistance is available to *Phytophthora sojae*. Partial resistance is effective against all races of *Phytophthora sojae*; however, it is only expressed after the first true leaves emerge, not in very young seedlings. Continuous soybean production may increase disease severity. But rotation to non-hosts may reduce disease severity because oospores can survive in soil for long periods of time. Disease is more severe in no-till fields because these fields can be wetter. If tillage is considered to improve drainage, use proven conservation tillage practices to maintain soil quality. Where *Phytophthora sojae* is a serious problem, seed treatments with metalaxyl as an active ingredient can provide some protection. Seed treatments are especially helpful with poor quality seed and in fields with a history of this problem.

Pythium Root Rot

Pythium root rot causes huge yield losses of the soybean crop every year. The root rot pathogenic fungi are major threat for this crop as these fungi attack on the root of the plant and destroy the proper functioning of the plant to take water and other nutrients upward. The root incidence ranged 10-17 per cent and losses ranged 6.75-15.5 per cent (Haq *et al.,* 2012).

Symptoms

Pythium species cause pre- or post-emergence damping off. Infected seed appear rotted and soil sticks to them. The water-soaked lesions appear on the hypocotyl or cotyledons that are latterly converted into a brown soft rot. Diseased plants are easily pulled from the soil because of rotted roots. Older plants become resistant to soft rot, but root rot may cause plants to become yellow, stunted or wilted if infection is severe.

Causal Organism

Several species of *Pythium* are reported to cause this disease. Early planting dates increase the risk of disease in the major soybean growing areas. Diseased plants often occur singly or in small patches in low-lying areas of the field that are prone to flooding.

Disease Cycle

The pathogen survives either in plant residue or in soil as oospores. Severity of disease depends on the amount of the pathogen in the soil, plant age and environmental conditions at the time of infection. Saturated soil is critical for infection for all *Pythium* species. *Phytophthora, Pythium* produces zoospores that swim in free water and infect the roots of plants. In general, *Pythium* species that are prevalent in the north infect plants at lower temperatures (10 to 15°C), and *Pythium* species in the south infect plants at warmer temperatures (30 to 35°F), although there are exceptions.

Management

Planting in cold, wet soils should be avoided to reduce infection by *Pythium* species that infect at low temperatures. Where *Pythium* is a problem, seed treatments with Apron, metalaxyl or strobilurins as active ingredients can provide some protection. Resistance to metalaxyl/mefenoxam has been accepted; however, they are generally considered more effective than strobilurins. Soil applications of metalaxyl at transplanting time followed by weekly sprays of potassium phosphonate (1g/l) plus acibenzolar-S-methyl (0.025 g/l) also significantly reduce the root rot infection. No-till soils often have higher soil moisture and lower soil temperatures, factors that increase the risk of *Pythium* infection. If tillage is considered to improve drainage, use conservation tillage practices to maintain soil quality.

Rhizoctonia Root Rot

This disease commonly appears in high humid and rainfall areas. In India it is first time reported by Singh *et al.* (1974). Under favourable conditions it's causes substantial losses and may become destructive at all the growth stages of the crop. Pratt and Wrather (1998) estimated soybean yield loss of 9.29 per cent, 9.14 per cent and 8.67 per cent in 1994, 1995 and 1996 respectively and the greatest economic loss over the three year period was due to soybean root rot caused by *R. solani* and *F. solani*.

Symptoms

Rhizoctonia root rot is one of the most common soil borne diseases of soybeans. Diseased plants usually occur singly or in patches in the field. Disease is typically more common on the slopes of fields. *Rhizoctonia* infects young seedlings, causing pre- and post-emergence damping off. Infected seedlings have reddish-brown lesions on the hypocotyls at the soil line. These lesions are sunken, remain firm and dry and are limited to the outer layer of tissue. If seedlings survive the damping off phase, infections may expand to the root system, causing a root rot. The root rot phase may persist into late vegetative to early reproductive growth stages. Older infected plants may be stunted, yellow and have poor root systems.

Causal Organism

The causal organism of Rhizoctonia root rot is soil borne fungus *Rhizoctonia solani*. This pathogen is survives in the soil as a sclerotia. The yield loss is estimated up to 48 per cent in the United States.

Disease Cycle

The fungus survives on plant residue or in soils as sclerotia. When soils warm, the fungus becomes active and infection may occur soon after seed is planted. The fungus grows better in aerated soils; thus, disease is more severe on light and sandy soils. Symptoms may disappear if infected plants grow out of the root rot problems although plants may remain stunted. Many strains of *Rhizoctonia* can infect corn, alfalfa, dry bean and some cereal crops.

Management

Resistance has been reported in some varieties; however, there are no varieties being developed for resistance to Rhizoctonia root rot. Eliminating stress factors, such as use of herbicides that cause injury to soybean roots, can help to reduce root rot problems. Seed treatments with Bavistin or Benlate (2 g/kg) are effective against *Rhizoctonia* and same fungicide can be used as foliar sprays 2 – 3 times for good control of disease.

Bacterial Disease

Bacterial Pustule

Bacterial pustule is one of the most important disease of soybean in India and causes more than 20 per cent yield loss. The percentage yield loss in the soybean crop due to this disease has been estimated by 15, 21, 38 and 53 respectively at 10.1 – 25 per cent, 25.1 – 50 per cent, 50.1 – 75 per cent and more than 75 per cent infection levels (Shukla, 1994).

Symptoms

Lesions are found on outer leaves in the mid- to upper canopy. Lesions start as small, pale green specks with elevated centres and develop into large, irregularly shaped infected areas. Unlike bacterial blight, no water soaking is associated with lesions, but each lesion is surrounded by a greenish-yellow halo. A pustule may form in the centre of some lesions, usually on the lower leaf surfaces. Pustules crack open and release bacteria. Bacterial pustule will not cause leaves to tatter like bacterial blight.

Causal Organism

This disease caused by *Xanthomonas axonopodis* pv. *glycines*. Bacterial pustule occurs mid- to late season when temperatures are warmer and more favourable

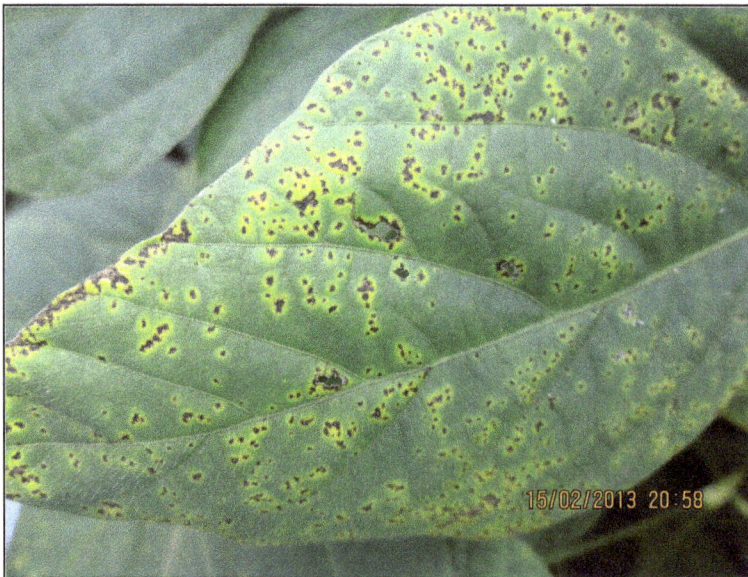

Figure 6.5: Bacterial Pustule.

for disease development. Symptoms may be mistaken for bacterial blight, Septoria brown spot or soybean rust. Diseased plants are usually widespread within a field.

Disease Cycle

Bacteria survive winter in crop residue and seeds and are spread by rain and wind. Infection occurs through leaf stomata or wounds. Rainy weather favours disease development. Unlike bacterial blight, high temperatures do not slow disease development.

Management

Avoid planting extremely susceptible varieties. Some varieties are marketed as resistant to this disease. Rotation and tillage reduce survival of *Xanthomonas axonopodis* pv. *glycines*. Other legume crops may be hosts; non-hosts include alfalfa, corn and small grains. If tillage is considered, use proven conservation tillage practices to maintain soil quality.

Viral Disease

Soybean Yellow Mosaic

This viral disease is the most destructive disease of soybean in India. It was first reported in 1960 and is now known to occur throughout the country. The loss of yield depends upon the stage at which the crop is infected. If the infection is early in the season there may be total loss of seed yield.

Symptoms

Disease appeared in the field when the crop is about one month old. Two types of symptoms appeared depending upon the host response. The general pattern of development of both symptoms is the same. The first visible sign of the disease is the appearance of yellow spots scattered on the lamina surface. They are mostly round in the shape. In yellow mottle, the spots are diffuse and expand rapidly. The leaves show yellow patches alternating with green areas that also turn yellow. Such completely yellow leaves gradually change to a whitish shade and ultimately become necrotic. These colour changes of affected plants are so conspicuous that the disease can be spotted in the field from a distance. In case of necrotic mottle, the centre of yellow spots develops necrosis which is demarcated by finer veins. The virus becomes systemic in the plant and all newly formed leaves show signs of mottle from the very beginning. Number of size of spots per plant and seeds per pod are greatly reduced.

Causal Organism

Four viruses causing yellow mosaic disease of legumes across the South Asia have been identified as bipartite begomoviruses (genus *Begomovirus*, family *Gemini viridae*). The soybean strain of MYMV occurring in north India is distinct from the strain occurring in southern and western India (Usharani *et al.,* 2004). A strain of MYMIV, designated as MYMIV-Cp causes golden mosaic of cowpea. It has restricted host range and transmission by *Bemisia tabaci*. These viruses have evolved

Figure 6.6: Soybean Yellow Mosaic.

independently of the begomoviruses in plant species of other families. The paired particles of the virus measure 30 × 18 nm. The particle contains two circular ssDNA molecules which account for 20 per cent of the particle weight. The coat protein contains one polypeptide with MW of 28.5 kDa.

Management

Certified and healthy seeds use for sowing. Cultivar PK 21-22 of soybean is tolerant to the disease. Control of the disease through prevention of population build up of the vector has also been recommended. Sprays of 0.1 per cent metasystox, starting when the crop is about a month old or as soon as a single diseased plant is seen in the field, can give relief from severe incidence of the disease. Anthio is effective at 0.2 per cent when used as spray 3 times.

Nematode Disease

Soybean Cyst Nematode

In India the most important pathogen of soybean is soybean cyst nematode. In high-yielding production fields or during years when soil moisture is plentiful, damage from soybean cyst nematode may not be obvious. However, yield losses up to 40 percent on susceptible varieties are still possible. When symptoms are associated with damage, infected plants usually occur in patches within a field.

Symptoms

Obvious symptoms may not develop, even though yield loss occurs. Noticeable symptoms of soybean cyst nematode include stunting, low or no canopy closure and chlorotic foliage. Infected plants have poorly developed root systems. Soybean cyst nematode infection also may reduce the number of nodules formed by the beneficial nitrogen-fixing bacteria necessary for optimum soybean growth. Signs of soybean cyst nematode include white females that are most readily seen in the field starting about six weeks after crop emergence. To see them, roots must be dug and soil carefully removed. However, the only way to get a reliable diagnosis as to the amount of soybean cyst nematode in the soil is through analysis of a properly collected soil sample by a diagnostic laboratory. Plant damage is not just limited to direct and indirect effects of feeding by the nematodes. Wounds caused by infecting nematodes and by maturing females serve as entry points for other soil borne pathogens. Diseases such as brown stem rot, Rhizoctonia root rot, sudden death syndrome and charcoal rot are more severe in the presence of soybean cyst nematode.

Causal Organism

The soybean cyst nematode that causes the disease is known as *Heterodera glycines*. The body of females is swollen, pearly white and lemon shaped and usually varies between 0.6 to 0.8 mm in length and 0.3 to 0.5 mm in diameter. The male is wormlike about 1.3 mm long and 30 – 40 μm in diameter. The males remain in the root for a few days during which they may or may not fertilize the females and then they move into the soil and soon die. Cysts are typically lemon shaped. Mature cysts of Indian population's measures 470 – 1010 × 370 – 730 μm. Each females produces 300 to 600 eggs most of which remain inside her body when the females die. Eggs in the gelatinous matrix may hatch immediately and the emerging second stage juveniles may cause new infestation.

Disease Cycle

Soybean cyst nematode survives in the soil as eggs within dead females called cysts. These eggs can survive several years in the absence of a soybean crop. The second-stage juvenile (J_2) hatches from the eggs and infects soybean plants. After infection, these juveniles migrate to the vascular system before setting up specialized feeding cells within the root. As they feed, the nematodes become immobile. The juveniles moult three more times before maturing into adults, with females becoming so large they burst through the outer surface of the roots. A female will produce 200 to 300 eggs that are deposited in an external egg mass or are retained within her body. Soybean cyst nematode can complete four or more generations during the growing season, depending on planting date, soil temperature and length of the growing season, host suitability, geographic location and maturity group of the soybeans. Conditions that favour soybean growth are also favourable for soybean cyst nematode development. High soil pH may be used to predict where soybean cyst nematode is more problematic. Areas of fields with soil pH levels of 7.0 to 8.0 typically have more soybean cyst nematode compared to areas with soil pH 5.9 to 6.5.

Management

The number of soybean cyst nematode in a field can be greatly reduced through proper management, but it is impossible to eliminate soybean cyst nematode from a field once it is established. Soil tests are recommended prior to every third or fourth soybean crop to monitor soybean cyst nematode population densities (numbers). Resistant varieties are available to manage soybean cyst nematode. The three most common sources of resistance are PI 88788 (most common), PI 548402 (Peking) and PI 437654 (also referred to as Hart wig or PUSCN-14). Resistant varieties are not resistant to all soybean cyst nematode populations. Most resistant varieties contain only one source of genetic resistance. Rotating sources of soybean cyst nematode resistance may help prevent the development of more damaging soybean cyst nematode populations. Soybean cyst nematode -resistant varieties, even high-yielding varieties, can vary considerably in how well they control nematode population densities. Greater soybean cyst nematode reproduction will result in a higher soybean cyst nematode egg population in the soil at the end of the growing season, and consequently, higher numbers of soybean cyst nematode in subsequent seasons. Thus, growers must consider how soybean cyst nematode-resistant soybean varieties affect soybean cyst nematode population densities, in addition to how well the varieties yield, to maintain the long-term productivity of the land for soybean production. If soybean cyst nematode is a problem, rotation should include non-host crops (usually corn) and resistant soybean varieties. Years of non-host crops may decrease soybean cyst nematode numbers by as much as 90 percent in south, but only 10 to 40 percent in the north. Maintaining adequate soil fertility, breaking hardpans, irrigation and controlling weeds, diseases and insects improves soybean plant health. These practices help plants compensate for damage by soybean cyst nematode, but do not decrease soybean cyst nematode numbers. Zero tillage practices may slow soybean cyst nematode movement and lower population densities. Soil that remains on tillage and harvest equipment can move soybean cyst nematode and should be removed before equipment is relocated from an infested to a non-infested field. Seed treatments labelled for use on soybean cyst nematode may provide early season protection. A limited number of nematicides labelled for use on SCN can be applied at planting.

References

Allington, W. B. and Chamberlain, D. W. (1948). Brown stem rot of soybean. *Phytopathology*. 38: 793 – 802.

Cruzi, M. (1931). Aleumi csidi 'Canorena Pedale' da sclerotium observation in Italia. *Atti Academia Nazionale des* Lincei. Rendiconti, **14**: 233-236.

Haq, M. Inam-Ul, Mehmood, S., Rehman, H. M., Ali, Z. and Tahir, M.I. (2012). Incidence of root rot diseases of soybean in Multan Pakistan and its management by the use of plant growth promoting rhizobacteria. *Pak. J. Bot.*, **44**(6): 2077-2080.

Higgins, B. B. (1927). Physiology and parasitism of *Sclerotium rolfsii* Sacc. *Phytopath*, **17**(7): 417-448.

Kolte, S. J. (1985). *Diseases of annual edible oilseed crops*. Vol. 3. Rapeseed-Mustard and Sesame Diseases. CRC Press, Inc. Boca Raton, Florida. USA, p. 97.

Kumar. 2007. A study of consumer attitudes and acceptability of soy food in Ludhiana. MBA research project report, Department of business management, Punjub Agricultural University, Ludhiana, Punjab.

Maiti, S., Kumar, S., Verma, R. N. and Vishwadhar (1983). Current status of soybean disease in North East India. *Soybean Rust Newsletter*, **6** (1): 14-21.

Mundkur, B. B. (1934). Perfect stage of *Sclerotium rolfsii* Sacc. in pure culture. *J. Agric. Sci.*, **4:** 779-781.

Muthusmy, S. and Mariappan, V. (1991). Disintegration of sclerotia of *Macrophomina phaseolina* (Soybean isolate) by oil cake extract. *Indian Phytopath.*, 44: 271-273.

Pratt, P. W. and Wrather. J. A. (1998). Soybean disease loss estimates for the Southern United States, 1994 to 1996. *Plant Disease*, **82:** 114-116.

Ramakrishnan, T. S. (1930). A wilt of zinnia caused by *Sclerotium rolfsii*. *Madras Agric. J.*, **16:** 511-519.

Rolfs, P.H. (1892). Tomato blight some hints. *Bulletin of Florida Agricultural Experimental Station*, p. 18.

Saccardo, P. A. (1911). Notae mycologicae. *Annales Mycologici*, **9:** 249-257.

Sarbhoy, A. K., Thapliyal, P. N. and Payak, M. M. (1972). *Phakopsora pachyrhizi* on soybean in India. *Current Science*, **41:** 264-265.

Sathe, A. V. (1972). Identification and nomenclature of soybean rust from India. *Current Science*, **41:** 59.

Sharma, N. D. and Mehta, S. K. (1996). Soybean rust in Madhya Pradesh. *Acta Botanica Indica*, **24:** 115 - 116.

Shukla, A. K. (1994). Pilot estimation studies of soybean yield losses by various levels of bacterial pustule (*Xanthomonas campestris* pv. *glycine*) infection. International *Journal Pest Management*. 40(3): 249-251.

Sinclair, J. B. and Backman, P. A. (1989). *Compendium of Soybean Diseases*, edn. 3, pp. 19-21. American Phytopathological Society, Minnesota 55121, the USA.

Singh Chhidda, Singh Prem, Singh Rajbir. 2003. *Modern Techniques of Raising Field crops, Oxford* and IBH Publishing Co. Pvt. Ltd. pp. 273.

Singh R., Shukla, T.N., Dwivedi, R.P., Shukla, H.P. and Singh,P.N. (1974). Studies on soybean blight of *Rhizoctonia solani*. *Indian J. Mycol. Plant Pathol.* 4: 101-102.

Singh, R. S. (2009). Plant Diseases. Oxford and IBH Publishing Co. Pvt. Ltd. New Delhi.

Sticher, L. B. Mauch-Mani and Metroux, (1997). Systemic acquired resistance. *Annual Review Phytopathology*, **35:** 235.

Townsend, B.B., and H.J. Willetts. 1954. The development of sclerotia of certain fungi. *Ann. Bot.* 21: 153-166.

Usharani, K. S.; Surendranath, B.; Haq, Q. M. R. and Malathi, V. G. (2004). Yellow mosaic virus infecting soybean in northern India is distinct from the species infecting soybean in southern and western India. *Current Science*, **86**(6): 845.

Van Loon, L. C., Bakker, P. A. H. M. and Pieterse, C. M. J. (1988). Systemic resistance induced by rhizosphere bacteria. *Annual Review of Phytopathology*, **36**: 453.

Vyas, S. C., Rajasekaran, G. and Geeta, M. (1997). Soybean rust. *Soybean Processor of India Digest* (May), pp. 33 -41.

7 Major Diseases of Sunflower and their Management

Shailesh Godika, R.P. Ghasolia, Shashikant Goyal and Jitendra Sharma

Sunflower (*Helianthus annuus* var. *macrocarpus* (DC) Cockerell) is an important edible oil seed crop. It belongs to the family Asteraceae. The sunflower head is composed of about 1000-2000 individual flowers. The fertile disc florets bear the seed which is white, black or striped grey and black. The seeds contain 40-50 per cent oil and 50-55 per cent meal, which contain high protein (35 per cent), calcium, phosphorus, iron, potassium and vitamin E. The sunflower is a native of North America, where it is used in dyes, food preparations and medicines.

A large array of pathogenic fungi in or on the seed is associated with sunflower which induces pre and post-emergence mortalities in field. Several pathogens *viz., Alternaria, Fusarium, Macrophomina* and *Plasmopara* are seed transmitted and can cause loss of seedlings. The developing plants of sunflower are prone to infection by *Alternaria* blight, downy mildew, rust and a large array of foliage damaging pathogens like *Phoma exiqua, A. alternata, A. zinniae, Cercospora helianthi, Septoria helianthi* etc. Amongst stem and root diseases, charcoal rot or dry root rot (*Macrophomina phaseolina*) and stem or collar rot (*Sclerotium rolfsii*) are damaging in patches in many areas. Head rot (*Rhizopus* sp.) has been commonly observed in Punjab, Haryana and U.P. on spring sown crop. Currently, sunflower is known to be infected by *Pseudomonas solanacearum*, Cladosporium blight, Phoma blight, Fusarium wilt, sunflower necrosis and a mosaic virus. These are sporadic at present but could be serious in future. Although exact estimate of yield losses due to diseases of sunflower have not been investigated but it has been reported that diseases account to the tune of 30 per cent reduction during *Kharif* and about 5-10 per cent during post *Kharif* seasons.

Downy Mildew

The disease was unknown in India till 1984. In 1985, it has been reported to occur in a serious form in Maharashtra. The causal fungus *Plasmopara helianthi* is considered to be of North American origin. It has been distributed rapidly by seed trade. Preliminary observations in Maharashtra indicate the possibility of large scale reduction in yield due to attack of this disease.

Symptoms

Infection at the seedling stage can occur under optimal environmental conditions, resulting in a systemic disease. Damping-off or seedling blight occurs when seedlings are exposed to high concentrations of inoculum in the soil combined with cool (12-13°C) and very wet soils. Downy mildew infections can be subdivided into two major categories based on symptoms. Systemic symptoms are produced when young seedlings are infected through the root system, often resulting in plant death. If seedlings survive, they are severely stunted with chlorotic and puckered leaves. Chlorosis is an indication of the portion of leaves being colonized by the pathogen. Chlorosis often remain associated with the leaf veins, but can also cover entire areas of the leaf. With extended periods of cool (15-18°C) weather coupled with high humidity or leaf wetness from rain or dew, a dense cottony, white growth appears on leaves, generally on the underside. If systemically infected plants reach maturity, they produce few viable seeds, and heads characteristically face upwards. Head size reduced, along with number and weight of seeds per head. Rarely, systemic symptoms may also appear on leaves in the middle or upper leaves of the plant, with no downy mildew symptoms being observed on the lower leaves. Secondary infections occurs when wind blown zoospores from adjacent plants land

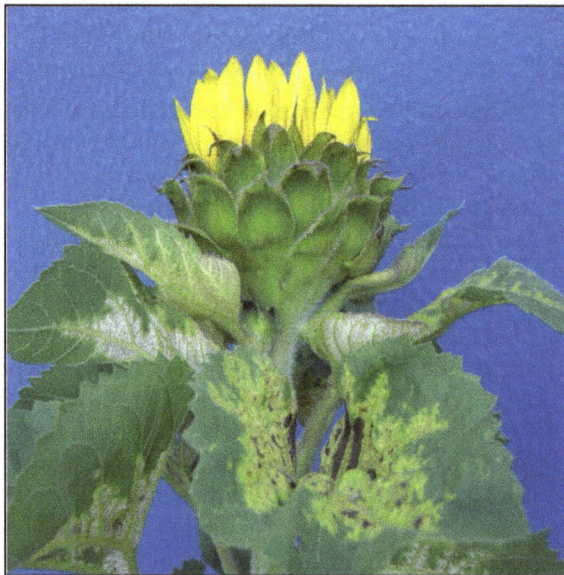

Figure 7.1: Downy Mildew of Sunflower.

on leaves of uninfected plants, provided that sufficient leaf moisture is available, but do not become systemic. These lesions are generally chlorotic and angular in shape, being limited by leaf veins. Found directly beneath yellow lesions on the underside of leaves is the white cottony sporulation of the fungus, which is diagnostic. Plants are susceptible to secondary infection for longer periods of time than with the systemic root infections, yet rarely result in yield loss. Both types of infection have been observed on wild sunflower species as well.

Causal Organism

Plasmopara helianthi.

Systemic Position (Acc. To Krik *et al.*, 2008)

The pathogen belongs to Phylum- Oomycota, Class- Oomycetes, Order- Pernosporales, Family- Pernosporaceae, Genus- *Plasmopara*, Species- *helianthi*

Etiology

Mycelium of Plasmopara are aseptate (coenocytic).The mycelium grow inter-cellularly and establishes systemic infection. Globular haustoria and enter the host cell and draw nutrients. Cell wall made of cellulose. Zoospores biflagellate (bear tinsel and whiplash flagella). The zoospores infect the host and cause primary infection. In downy mildew characteristic sporangiophores of definite growth, emerge singly or in tufts through stomata, on the lower surface of leaves. The term downy mildews etymologically means hairy fungi. The sporangiophores look like hairs, emerging from the stomata. The branches of sporangiophores arise at right angles and are irregularly spaced.

Disease Cycle and Epidemiology

Basal gall symptoms occur independently of systemic infection. The root infection may result in formation of galls on primary roots at the base of plants. Such plants are less vigorous and subject to lodging. Oospores of the fungus are observed in all affected tissues, but more oospores are found in affected roots than in any other organ. The pathogen survives through oospores in the residue of the preceding sunflower crop or through oospores on seeds in systemically infected plants. Seed transmission has been clearly demonstrated. Germinating oospores produce zoospores mostly under wet conditions. Zoospores cause infection. Seed borne infection usually results in low percentage of systemically infected plants. The outbreak of the disease in severe form is to a large extent due to soil borne inoculum. Rain in the early phases of growth and age of seedlings at the time of infection are important factors in this context. Heavy soils with poor drainage are more conducive to infection.

Management

Cultural Practices

Use of resistant varieties offers the best control. Choice of planting sites and disposal of infected crop residues, also give a fairly good control over the disease.

Since wild and volunteer sunflowers are hosts for the pathogen, eliminating these plants help to reduce overall inoculum build-up in the fields.

Chemical Control

Seed treatment with metalaxyl @ 3 g per kg of seed has been found to give effective control.

Genetic Resistance

Using resistant cultivars, where available, is the most efficient control measure for downy mildew. However the appearance of new physiological races makes this technique inconsistent, and possibly ineffective in all production systems.

A new sunflower downy mildew resistant hybrid LSFH-171 was released from Maharashtra state for commercial cultivation for the benefit of the farmers in major sunflower growing states of India.

Cultivars namely as NSH-23, Sunbred-2077, NSSH-303, K-678 and MLSFH-93 can also be exploited for commercial use (Shrishikar, 2008., 2013).

Biological Control

Nagaraju *et al.* (2012) reported that seed priming with conidial suspension of plant growth promoting fungi (PGPF) at 1×10^8 cfu ml^{-1} significantly increased seed germination and seedling vigor as compared to non primed control. Plants raised with primed seeds demonstrated a significant reduction in downy mildew disease severity and provided a maximum of 61 per cent protection under green house conditions when compared to the untreated control. The tested PGPF, promoted growth and induced systemic resistance (ISR) in sunflower plants against the downy mildew disease caused by *P. halstedii*, due to their growth promoting and bio-control abilities. There is a strong correlation between root colonization and resistance induction in PGPF treated plants.

Rust

Rust disease is prevalent throughout the sunflower growing areas in India. It is more prominent in the *rabi* season and in the *kharif* season the appearance is usually late. This is one of the most destructive disease of sunflower. This disease in conjunction with *Alternaria* blight may cause yield loss up to 40 per cent.

Symptoms

Uredo pustules appear first on the lower leaves. They are small, circular 0.5 to 1 mm in diameter, powdery, orange to black. Uredo pustules appear on the younger leaves and later spread over the entire vegetative surface covering stems, petioles, floral bracts and petals. Uredia often coalesce to cover large areas on the affected plant parts. Towards maturity, the uredo- sori are replaced by telia and the black rust stage appears.

Causal Organism

Puccinia helianthi

Figure 7.2: Rust.

Systemic Position (Acc. To Krik *et al.*, 2008)

The pathogen belongs to Phylum *Basidio mycota,* Sub phylum *Puccinio mycotina,* Class *Puccinio mycetes,* Order *Pucciniales,* Family *Pucciniaceae,* Genus *Puccinia* Species *helianthi.*

Etiology

Mycelium of *puccinia helianthi* are septate. The mycelium is intercellular with haustoria and limited to parts of leaves or other aerial parts and causing local infection. Karyogamy occurs through spermatia,. Absence of basidioma. *Puccinia helianthi* is obligate parasite in nature. Explosive discharge of basidiospores. Absence of clamp connections and presence of dolipore septa. Formation of bi-celled teliospore. Teliospores are resting spores of *Puccinia helianthi.*

Disease Cycle and Epidemiology

Puccinia helianthi is a macrocyclic, homothallic and autoecious rust producing all the stages of spore forms on a sunflower. It is restricted in its host range to annual and perennial *Helianthus* species. *Puccinia helianthi* mainly survives through teliospores on leaves left in the field or on the soil surface. Possibility of survival of the fungus as teliospores or uredospores on the seed surface exists though as yet no positive information on seed transmission is available. Uredia, sporidia, pycnia, aecia may appear on volunteer seedlings among plant debris of the previous year's crop and survival of the pathogen through such volunteer plants is also a great possibility. Primary infection may result from sporidia, from germinating teliospores, aeciospores or from uredospores on volunteer plants at high altitudes and carried through air currents. Secondary infection occurs through repeatedly produced uredospores in a crop season. Day temperature of 25.5° to 30.5°C with relative humidity of 86 to 92 per cent enhances intensity of rust attack.

Chemical Control

Two to three foliar sprays of dithiocarbamate fungicides like maneb, mancozeb or zineb at 0.2 to 0.3 per cent have been found effective in control of the disease. Sulphur fungicides like lime-sulphur, wettable sulphur (0.2 per cent), sulphur dust (15 kg per ha) also give good control. Nickel compounds, Bordeaux mixture, Miltox and systemic fungicides, benodanil and oxycarboxin have also been reported to be effective. Resistant varieties offer the best means of control. Tall and late forms are more resistant than dwarf and early ones.

Alternaria Blight

The disease is considered as a destructive one. It is widely distributed wherever the crop is grown. A reduction in seed and oil yield by 27 to 80 and 17 to 33 per cent respectively has been reported. A negative correlation has been established between an increase in disease intensity and yield components and oil content. The most affected components due to disease are number of seeds per head, followed by the seed yield per plant. The disease also affects the quality of sunflower seeds by affecting germination and initial vigour of the seedlings.

Symptoms

The disease is characterised by the development of dark brown to black, circular to oval spots varying from 0.2 to 5mm in diameter. The spots are surrounded by necrotic chlorotic zone with grey white necrotic centre marked with concentric rings. Spots first appear on lower leaves, later spread to middle and upper leaves. At later stages, spots may be formed on petioles, stem and ray florets. Spots first appear on lower leaves, later spread to middle and upper leaves. At later stages, spots may be formed on petioles, stem and ray florets.

Causal Organism

Alternaria helianthi

Systemic Position (Acc. To Anisworth 1973)

The pathogen belongs to Phylum- Deuteromycota, Class- Hyphomycetes, Order- Moniliales, Family- Dematiceae, Genus- *Alternaria*, Species- *helianthi*.

Etiology

Mycelium of *Alternaria helianthi* is septate. The genus stands distinct from other genera by its transverse and longitudinal-septate (muriform) conidia. The conidia bear distinct beak which may be very long or short. The conidia are formed in chains (acropetal). The beak is light in colour, which may serve as conidiophore. The conidiophore is dark than the hyphae and is sympodula, bearing geniculations marked with scars of detached conidia.

Colonies usually effuse velvety, light powdery olivaceous brown. Conidiophores arise singly or in groups of 2-3, simple or branched; 38.40-138.24 × 5.76-11.52 µ in size, conidia is geniculate with 1-2 scars, 1-5 septate solitary, non-catenate, acropleurogenous cylindrical to obclavate, pale light brown, smooth, thin- walled, beakless, basal scar dark and distinct, constricted at the septa 23.04-138.6× 15.4-38.4 µ in size.

Disease Cycle

Pathogen is reported to survive on seed, host debris and weed host. Successive crops of sunflower may also help to multiply the pathogen. Hot weather and frequent rain during the milk and wax stages of development favour infection. The fungus is reported to produce a toxic metabolite, a phenolic substance in culture and leaf tissues.

Control

Occurrence and severity of the disease depend on the season and planting dates. Mid-September planting of sunflower remains free from most of the major diseases including *Alternaria* infection which only shows traces. Mancozeb (0.3 per cent) sprayed four times at an interval of seven to 10 days proved very effective in controlling the disease with increase in yield. Other fungicides, namely, zineb, ziram and captafol are also effective. Resistant sources identified are EC 132846, EC 132847, EC 132361 and EC 126184.

Charcoal Rot

Charcoal rot is of economic importance particularly in the arid areas, where a reduction in seed weight of 30 to 46 per cent has been reported. The most common symptom of the disease, under field conditions is the sudden wilting of plants, usually after pollination. In India the disease has been mainly reported in Maharashtra and Karnataka.

Symptoms

Early symptoms are not visible on infected plants, but they become weak, mature early and when dry, show a presence of black ashy discolouration of the stem. Black micro-sclerotia are formed in huge number on the affected portion. Sometimes the disease causes seedling blight, damping off, root rot or basal stem rot. Symptoms begin as chlorosis of lower leaves, progress acropetally, finally resulting in plant death.

Causal Organism

Macrophomina phaseolina

Systemic Position (Acc. To Anisworth 1973)

The pathogen belongs to Kingdom- Fungi, Phylum- Deuteromycota, Class-Hyphomycetes, Order- Aganomycetales, Family- Aganomycetaceae, Genus-*Macrophomina*, Species- *phaseolina*.

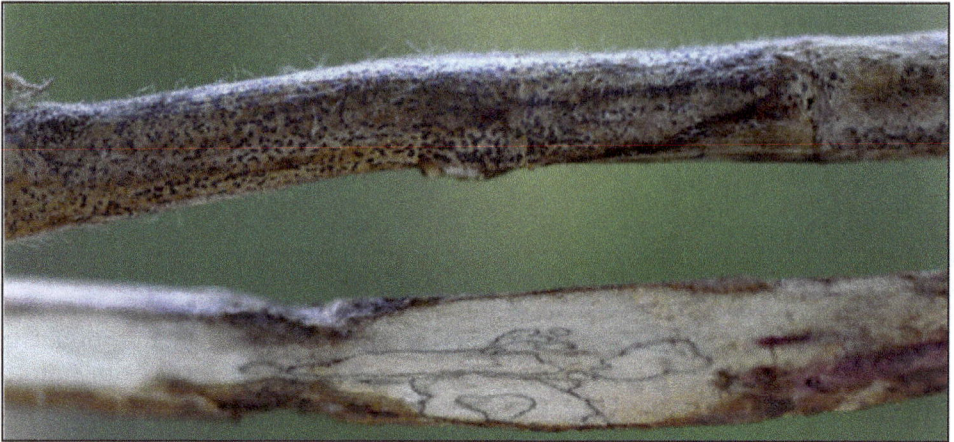

Figure 7.3: Outer and Inner Stem Symptoms of Charcoal Rot.

Macrophomina phaseolina is not an aggressive pathogen capable of attacking vigorous plant tissue but can overcome senescing tissue at grain filling stage under soil moisture stress and high temperature causing a serious damage.

Epidemiology

Pathogen survives as sclerotia in soil and crop residues. It is also reported to be seed borne in sunflower. The age of the plants appears to be a very important factor influencing the infection. Moisture stress and higher temperature favour development of the disease.

Control

Practically no field control is available by using chemicals. Seed borne inoculum can, however, be minimised by treatment of seeds with thiram (3 to 4 g per kg seed).

Powdery Mildew

Powdery mildew of sunflower caused by *Erysiphe cichoracearum* is an important disease found throughout the year reducing the yield to considerable extent (Kolte, 1985). The development of powdery mildew is directly related to atmospheric temperature therefore, the time of sowing has a subsequent epidemic. It is a well established fact that occurrence and disease severity depend upon the season and sowing period.

Symptoms

Symptoms of the disease appear as white to grey floury patches on the upper surfaces of older leaves, but still green. Occasionally the symptoms appear on stem and bracts. White to grey areas enlarge, coalesce and cover most plant parts. As the season advances cleistochecia become visible as black pinpoints over the

Figure 7.4: Powdery Mildew.

white mildew areas. Pathogen survives through cleistothecia. The disease has not been observed to be of any economic importance. This mildew appears as small, circular, whitish spots on both surfaces of the leaves. These enlarge to form larger spots covering the entire surface as dirty white growth. Symptoms start from lower leaves and gradually younger leaves are attacked. With advance in age, mildew turns grey and numerous, minute black bodies of cleistothecia appear scattered over the lower surface of the leaves. Pathogen survives through cleistothecia. Development of cleistothecial stage of this fungus under Indian conditions appears to be governed by the presence or absence of the hyper-parasite *Cicinnobolus cesatii*. Dinesh *et al.* (2010) from Karnataka revealed that the variation of disease incidence in the various locations was mainly attributed to climatic factors, such as temperature, relative humidity and distribution and amount of rain, followed by cultural practices like sanitation and other suitable management practices. Maximum disease severity was recorded where sunflower was grown under irrigated conditions.

Causal Organism

Erysiphe cichoracearum f sp. *helianthi.*, *Erysiphe cichoracearum*

Systemic Position (Acc. To Krik *et al.*, 2008)

The pathogen belongs to Phylum- Ascomycota, Sub-phylum- Pezizomycotina, Class- *Leotiomycetes*, Order- *Erysiphales*, Family- *Erysiphaceae*, Genus- *Erysiphe*, Species- *cichoracearum*

Morphology

Erysiphe cichoracearum is obligate ectoparasite. Mycelium of *Erysiphe cichoracearum* is septate, superficial and hyaline. The ascomata are minute cleistothecia with simple external appendages and contains globose, ovoid asci, arranged in basal layers. There are no paraphyses. The superficial mycelium produced enormous number of conidia usually on the leaf surface which appear like a mass of white powder. Conidia are the main source of spread of disease.

Control

Remove and destroy plant debris at the end of the season. Grow crop in an area that receives full sun during most of the day. Space plants for good air circulation. Spraying wettable sulfur @ 3gm per ltr or carboxin @ 1 ml per ltr.

Sclerotium Wilt or Sclerotinia Rot

Sclerotinia sclerotiorum (Lib.) de bary is a widespread and destructive pathogen of sunflower of causing wilt and stem rot. Sackston (1961) from Canada was first to report the pathogen causing head rot. The disease is initiated by *Sclerotia* which germinate myceliogenically and infect the sunflower roots (Yang and Morris, 1927; Zimmer and Hoes, 1978; Huang and Dueck, 1980). Infection appears in sunflower fields in the late bud or bloom stage and spreads during the remainder of the season. This disease cause much yield reduction in sunflower field. Hua *et al.* (1987) showed that the disease caused 45 per cent yield reduction. 1000 seed weight was reduced by 31 per cent oil content by 2.7 per cent and germination by 20-64 per cent.

Symptoms

Initial symptoms of the disease are noticed 40 days after sowing. Sickly appearance of plants can be rotted from a distance and a row effect can be observed in heavily infested soil. Later the entire plant withers and dies. White cottony mycelium and mustard-seed-type sclerotial bodies are conspicuous on the affected stem near soil level. About 10 to 11 per cent plants have been reported to be affected, amounting to 10 to 11 per cent loss in yield, if the sunflower crop is planted in July or August or in February or March in India.

Causal Organism

Sclerotinia sclerotiorum

Systemic Position (Acc. To Anisworth 1973)

The pathogen belongs to Phylum *Ascomycota*, Class- *Discomycetes*, Order- *Helotiales*, Family- *Sclerotinaceae*, Genus- *Sclerotinia*, Species- *sclerotiorum*.

Sclerotinia sclerotiorum (Lib.) de Bary is a wide spread and destructive pathogen of sunflower causing wilt, stem rot and head rot disease. Sackston (1961) from Canada was first to report the pathogen causing head rot. The disease is initiated by sclerotia which germinate myceliogenically and infect the sunflower roots. Under natural conditions, the diseased plants exhibits 3 kinds of symptoms (1)

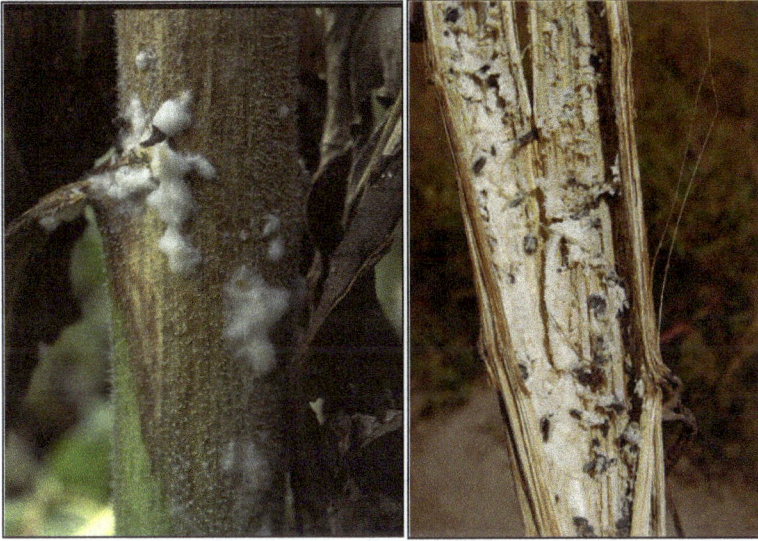

Figure 7.5: *Sclerotium* **Wilt or Sclerotinia Rot.**

wilting without any visible symptoms of rotting (2) stem rot and (3) head rot. The presence of black irregular sclerotia in the diseased plant parts was considered to be the diagnostic criteria for identification of affected plants. Infection appears in sunflower fields in the late bud or boom stage and spreads during the remainder season.

Disease cause much yield reduction in sunflower fields. Hua *et al.* (1987) showed that the disease caused 45 per cent yield reduction in the highest yielding plots. 100 seed weight was reduced by 31 per cent, oil content by 2.7 per cent and germination by 20- 64 per cent.

Control

The disease can be controlled by collection and destruction of plant debris. Seed treatment with captan or carboxin 3g/kg and drenching the base of the plant with cheshunt compound 3g per litre of water.

Rhizopus Head Rot

Rhizopus stolonifer from U.S.A. and India (Yang *et al.*, 1979; Narsimhan and Praksah. 1983) and *Rhizopus microspores* from India and South Africa (Raut, 1975; Swart, 1988) respectively have been reported to cause Rhizopus head rot of sunflower. Arnan *et al.* (1970) reported that the infection occurred only following the injury to the heads after flowering possibly from birds which were implicated in the spread of the disease.

Symptoms

The disease first appears as dark spots on the back of ripening heads, followed by a watery, soft rot that later turns dark brown. As disease progresses, heads dry prematurely, shrivel, and tissues appear to shred. The fungus can be observed inside shredded heads as coarse, threadlike mycelial strands that are later followed by the appearance of fungal reproductive structures (fruiting structures) called sporangia that look like small black dots about the size of pinheads. The disease also can be recognized on the flower side of the head by the appearance of a grayish, fuzzy substance covered with sporangia. These dark fruiting bodies are sacks filled with spores that are easily broken, thereby spreading- the spores to neighboring plants by wind. Symptoms can often seem to appear simultaneously on opposite sides of the head. The fungus also causes a systemic effect in the plant on the side where the head became infected. If the peduncle becomes infected, it can allow the head to fall completely off, increasing the potential for severe yield loss.

Causal Organism

Rhizopus sp.

Systemic Position (Acc. To Krik *et al.*, 2008)

The pathogen belongs to Phylum *Zygomycota*, Sub phylum *Mucoromycotina*, Order *Mucorales*, Family *Mucoraceae*, Genus- *Rhizopus* spp.

Morphology

Mycelium of Rhizopus is aseptate (coenocytic). The mycelium is distinguishable into aerial hyphae called stolons and root like rhizoids, which remain submerged in the substratum. Stolons creeping, attached to seed surface by means of brown rhizoids; sporangiophores 1-4 mm long; sporangia hemispherical; columella broad, hemispherical; sporangiophores oval, irregular, round, unequal in size, 9-12 µ long ×7.5 -8 µ diameter and grey in colour.

Disease Cycle and Damage

Even though *Rhizopus* spp. spores are found everywhere in the environment, mechanical injury on heads is a prerequisite for infection and disease development. Afterwards, the disease becomes more severe in warm, humid environments, especially under irrigation. Thus, as spores are carried to sunflower plants through air currents, and under conditions of high humidity, infection is initiated through wounds created by hail, birds, or insects.

Damage and economic losses depend on the creation of wounds, the environmental conditions present, and the time of season that wounding and infection occur. Research has demonstrated that head rot rarely occurs before flowering, so it appears that mature tissues are required to support the growth of *Rhizopus* spp. Yields are negatively affected because seeds in infected heads fail to fill properly and have reduced weights. This can obviously affect profits for confectionary seed growers since payment is based on seed size. Yield also can be affected even if infection occurs late in the season due to the loss of seeds that have fallen to the ground from affected heads. Oil seed growers also may be adversely affected- by head rot due to bitter or poor quality oils obtained from infected plants. Yildirim (2010) reported that the mechanical or physical damage on back of sunflower head results in infection of the head rot disease and significantly reduced the seed yield.

Favourable Conditions

The disease is of importance in wet weather and causes loss in yield. Injury to the flower head is necessary for infection. Injury before flowering or during the early stage of head development is unlikely to favour infection even though the inoculum may be present. Susceptibility of the flower head is increased as its age advances. Maximum rotting is noticed at the soft dough stage. Seed development is severely impaired depending on the stage of maturation at the time of *Rhizopus* infection and rot development. Larvae of *Heliothis armigera* have been reported to pre- dispose heads to infection. Spread of the disease is positively correlated with birds visiting the beads in search of seeds.

Control

To have effective control of the disease, simultaneous application of compatible insecticide and fungicide beginning with the completion of flowering stage is suggested. Injury to the head should be avoided as far as practicable.Avoid mechanical damage after flowering. Control head moth infestation before or at flowering. Select varieties with head types that turn down after flowering. Avoid planting sunflowers near water that consistently harbors many birds.

Viral Diseases of Sunflower

Several virus and virus like diseases have been reported by various workers on sunflower, both in India and abroad. Sunflower mosaic, chlorotic mosaic, yellow ring mosaic, yellow mosaic, yellow spot, chlorotic leaf/mosaic, greening, cucumber mosaic, curl mosaic and Mycoplasma like organisms (strain of tomato big bud, aster yellows and phyllody) have been reported.

Battu and Pathak (1965) reported viral disease of sunflower caused by sunflower mosaic virus from India. They found mosaic symptoms on infected plants with ring and chlorotic spots on leaves. Martens *et al.* (1970) from Kenya and Fivawo (1987) from Tanzania have reported another viral disease of sunflower caused by rugose mosaic virus in which some parts of plants become wrinkled and lighter in colour.

(A) Sunflower Mosaic Disease

Several kinds of symptoms have been described from India. Mosaic symptoms accompanied by ring spots or chlorotic spots which had a tendency to coalesce have been frequently reported. Another mosaic virus described as mosaic and chlorotic rings that were more common on young leaves which make the plants stunted, producing malformed heads and shrivelled seeds. Symptoms as small circular spots on leaves which coalesced to form typical mosaic pattern, cupping and malformation of leaves have also been reported. Sunflower mosaic virus is reported to be mechanically sap transmitted and also by several aphid vectors. The important vectors are *Aphis gossypii, A.craccivora, A.malvae, Rhopalosiphum maidis* in a non-persistant manner, both under laboratory and natural field conditions. The host

Figure 7.6: Sunflower Mosaic Disease.

range of this virus ranged from narrow, infecting only the cultivars of sunflower to as many as more than 25 plant species belonging to different families.

(B) Aphid Transmitted Virus

This virus has been reported from Cambridge. The chief symptoms produced are chlorosis and thickening of lower leaves. The interveinal areas of these leaves become bright yellow near the tip and margins, many irregular necrotic patches develop later on and these symptoms are reported to be similar to the sugarbeet yellowing virus. This virus is transmitted by an aphid vector, *Myzus persicae*.

(C) Yellow Ring Mosaic Virus

This virus has been reported from India. It produces mosaic and yellow spot symptoms. Severe mosaic accompanied by stunting and malformation of young leaves in the form of yellow rings has also been reported. This virus is transmitted both by mechanical sap and also by leaf and cleft grafting and has an additional host, *Chenopodium amaranticolor*.

(D) Tobacco Streak Virus

This virus has been reported from Argentina. It produces mosaic symptoms followed by necrosis and vein swelling severe necrosis and chlorosis on the leaves and curling of glumes are the additional symptoms. It is transmitted through mechanical sap and also by dodder. *Nicotiana clevelandi, N.rustica, Chenopodium amaranticolor, Gompherena globosa* are the additional hosts of this virus.

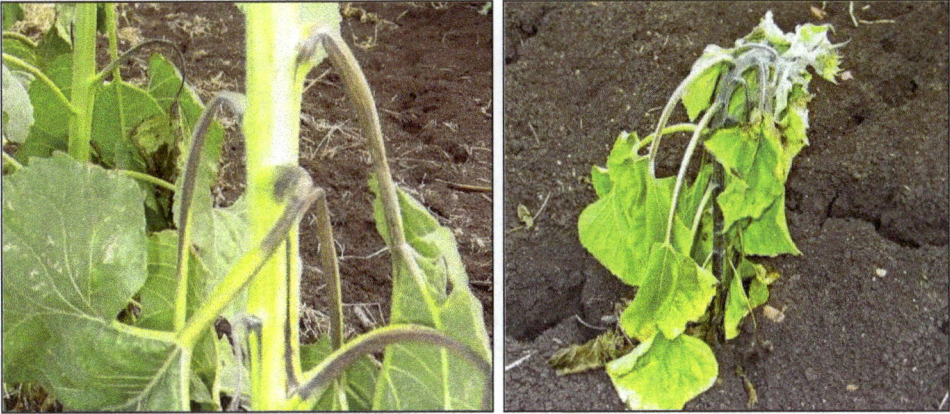

Figure 7.7: Tobacco Streak Virus.

(E) Yellow Blotch Disease

Distinct bright yellow blotches occurred on leaves crinkling was absent. Also, short, irregular yellow vein bands appear in the beginning and later coalesced to form "Y" or concentric rings with either yellow or a green centre. Some yellow bands coalesced to form brilliant yellow blotches measuring 1-3 cm in diameter. In mildly affected plants, only a yellow vein-net involving some or the entire leaf surface was observed. Under field conditions, some plants show more severe symptoms after yellow blotch appearance. The youngest leaves were twisted, reduced in size and often curled downwards.

(F) Leaf Crinkle Disease

The symptom of this disease always develops first as yellow blotch, but later leaf crinkling symptoms dominate. The electron microscopic studies of the group of symptoms of this disease revealed the presence of spherical particles measuring 26.8+0.15 mm in diameter.

(G) Sunflower Necrosis Disease (SND)

This disease first noted in severe form in parts of Karnatka in India during summer 1997. The disease incidence ranged from 10-80 per cent in both open pollinated and hybrid varieties.

Symptoms

The characteristic symptom of this disease include sudden necrosis and or/ necrosis of the lamina, petiole, stem floral calyx and corolla. Early infected plants remain stunted with malformed heads having chaffy or deformed seeds. In some cases, the heads become partially twisted and show sterility.

Causal Organism

Tobacco streak virus belongs to genus *Ilavirus*. Transmitted by thrips.

Management of Virus Diseases

Like in any other virus diseases of crop plants, there are no direct methods of controlling the viruses of sunflower. Since the sunflower mosaic is transmitted both by mechanical means and also by aphid vectors in a non-persistant manner, it is very difficult to protect the plants from infection of the virus through this type of vectors. However, the following methods helps much in reducing the incidence and further spread of the virus diseases in the field and thereby reducing the yield losses. Clean cultivation by removing the weeds both inside and neighbouring which helps to destroy the virus source and avoid the primary inoculum to the crop. Prophylactic sprays with suitable insecticides to control the insect vectors which come from outside the field and also harbouring inside the field, helps to avoid the primary sources and also further spread in the field. Careful destruction of the infected plants as soon as they are noticed in the field will also help to avoid further spread of the viruses in the field by destroying the source of inoculum within the field.

Phytoplasma Diseases of Sunflower

Phyllody Disease

Causal Organism

Phytoplasma (Group: *Candidatus* Phytoplasma *aurantifolia*)

Symptoms

The brief symptoms of phyllody of sunflower are stunting, leaf deformation, hypertrophy, flora malformations, sterility and virescence and leaves show the presence of many zonal necrotic areas. Such plants produce abnormal heads and floral parts were transformed into green leaf like structures. The flower head may be transformed into twisted elongated stalks. Internodes were reduced and plant became stunted. Such plants remain green for longer time. The capitulum instead of producing normal ray and disk florets, showed wedge shaped sectors with hypertrophied flowers. The ovary, calyx, corolla, anther tubes and branches of stylets show green colorued, short rudimentary leaves. It has been reported to be transmitted by leaf hopper, *Orocious albicinctus*, both under laboratory and in field conditions.

Integrated Disease Management in Sunflower

IDM as practiced in the developed countries is a sophisticated technology achieving computerised on line service. Practical IDM methods have been suggested for less developed countries or for traditional agro-ecosystem. It is based on efficacy, ecology and economics employing routine and supervisory practices with minimum use of agro-chemicals being backed by epidemiology and management components. Education and extension should also form an integral part of IDM. The current IDM, therefore, could be based on the following principles of management choosing any resistant or tolerant hybrid cultivar for cultivation.

Compulsory seed treatment with available effective fungicides prior to sowing, deep ploughing, followed by double harrowing and removal of volunteer plants as land preparation components. Addition of FYM prior to addition of recommended inorganic fertilizers, compulsory crop rotation of at least three four years. In rainfed situation with known history of disease outbreak, intercropping with suitable crop. Since there are no known threshold limits of diseases, actual survey and monitoring of diseases by way of diagnostic symptoms and signs. Rainfall prediction through national satellite network during kharif, and then link the spraying technology with rain pattern. Alternaria epidemic is ensured when sunflower head stage is caught in heavy rain. Two sprays of mancozeb at 0.25 per cent with appropriate stickers before head emergence can prevent damage by the disease. Crop rotation if coupled with green manuring legume crop can be of great use in reducing soil or stubble inhabiting spore forms. During post-rainy seasons optimum use of irrigation water, avoidance of water stagnation, avoidance of direct contact of water with stem, wider spacings between plants and planting on ridges has been found useful in curtailing disease outbreaks. There are some areas where research gaps exists *e.g.*, resistance to Alternaria, development of biocontrol system, forecasting models for disease epidemics, role of rhizobacteria and VAM *etc.* Integrated disease management is possible provided the components of management are suitably amended in IPM or in the total crop management system.

Legislative Approach

Failure to check the downy mildew pathogen of sunflower from introduction into India is a glaring example of how relaxed quarantines could be harmful. Since the aggressive race 1 has entered the country, it is still important to have a strict vigilance on the entry of downy mildew for virulent races through quarantine means. Presently downy mildew is restricted to Central and South India, domestic quarantine measures can be adopted to prevent it from spreading to North India where it could pose a threat during *rabi* and spring seasons.

Regulatory measures are, thus, an important component of management of diseases as it has been known to prevent entry of downy mildew pathogen into Australia.

Host Resistance Approach

The best option for disease management is host resistance. In sunflower development programme in India, it has become amply clear that incorporation of resistance of downy mildew and rust diseases into high yielding hybrids and cultivars is possible. Several sources of resistance to rust are known and the R_1 and R_2 genes have been widely utilised in development of rust resistant hybrids. The rust resistance has been originally derived from accession 953-88 (sunrise x Texas wild annual 0 carrying R_2 gene). In India, several sources of resistance have been reported. These include 24 RHA lines 12 CMS, 20 hybrids, 5 in-breeds, 4 wild species and 86 open pollinated populations. The race flora of the rust pathogen has not been understood clearly in India. It is, therefore, necessary to understand the genetics of the pathogen monitor the races and programme the resistance breeding accordingly. Downy mildew resistance breeding is well advanced in the USA, Russia and European countries. Thus, when the disease was first detected in India, a well suited three-tier screening system for downy mildew was developed and several sources of resistance to race 1 has been identified. Based on the breeding programme, two downy mildew resistant hybrids for endemic pocket of Maharashtra have been released. These hybrids LSH-1, CMS 338A x MRHA- 2) And LSHO3 (CMS-207A x MRHA-1) have very good yield potential but posed some problem of seed production which are currently being refined. Nine high yielding hybrids such as KBSH- 1, ICI-302, NSH-22, PKVSH-38, ITC-601, MSFH-17, SPIC-105, Jwalamukhi, NARF-114, KBSH-31 *etc.*, have fairly high degree of resistance mainly because the majority of them have either one or both the parental lines derived from resistant sources. Alternaria blight is by far the most serious and damaging disease during rainy season. The outbreak of the disease is closely linked with rainfall and it is also difficult to control when weather conditions are favourable. Attempts have been made to identify sources of resistance to the disease. However, high degree of resistance is not found in the available germplasm lines. Field tolerance has been observed in few lines and fortunately the hybrids often exhibited greater tolerance than cultivars. There are no specific efforts as yet in the country to evolve cultivars resistant to head rot, stem rot, dry root rot or other disease problems. Techniques of screening for disease resistance are to be evolved before breeding suitable practices that suppresses the growth pathogens but selectively favour growth of the host crop.

Cultural Approach

Ploughing, harrowing, weeding, planting time, organic amendments, fertiliser, use, intercropping, planting method, micronutrient balance, water use, *etc.*, have impact on disease development of sunflower. The oldest known cultural practice of crop is basically a classical approach for reduction of diseases. A six year rotation in sunflower is found useful in the control of downy mildew in Russia and European countries. Crop rotation also reduces Alternaria blight, rust and soil borne diseases. It is very clear that outbreak of sunflower downy mildew is ensured by water stagnation, deep sowing, continuous sunflower growing and those practices that delay emergence of seedling. The influence of several practices on different diseases has been examined at number of locations in India. It has been observed that early planting of sunflower during kharif reduces damage by Alternaria. Also intercropping, spacing, manuring has great effect on diseases which could be exploited fully for the advantage of disease management summarizes the beneficial effects of some major cultural practices on sunflower diseases. It is, therefore, suggested that emphasis on crop rotation, intercropping, timely planting, proper water use and field. Populations of sunflower be given a priority in management of disease.

Biological Approach

The soil borne pathogens particularly *Sclerotium*, *Macrophomina* and *Fusarium* can be efficiently reduced by the biocontrol agents like *Trichoderma*. There is a great potential for application through seed as these systems have been widely used in other crops.

Chemical Approach

The use of fungicides in management of sunflower diseases has been extensively documented. However, the utilisation of this strategy is restricted mainly because of cost, availability and suitable time of application. Chemical control in rainfed situation need to be restricted to seed treatment. This is an assured, safe and cost effective strategy and demonstrated throughout the country. A large number of pathogens in or on the seed can be effectively controlled by a range of available fungicides such as, Captan, Carbendazim, Carboxin, Thiram *etc*. In case of irrigated crop management, limited use of spray fungicides particularly Mancozeb (0.25 per cent) or Carboxin (0.05 per cent) plus Mancozeb (0.2 per cent) effectively reduces Alternaria blight and rust. However, the benefits of seed treatment have been found to be nearly equal to three sprays particularly in resistant hybrids. The response of chemical control of rust in terms of yield of seed is higher in susceptible cultivars like Modern and EC 68414 but the chemical control response was negligible in resistant hybrids like MSFH-8, VSF-1 and VSF-2. However, since the losses in oil yield are not proportional to the degree of severity of rust in sunflower cultivars, the resistant hybrids exhibited high response to oil increase by chemical control of the disease. Thus, chemical sprays could be used efficiently provided the loss potential, cost benefit and some knowledge of likely outbreak of disease are known. There is no work on the epidemiology and forecasting of Alternaria blight, rust and downy mildew which will go a long way in optimising the chemical control strategy for sunflower diseases.

References

Agrios G.N. (2012). Plant Pathology. Fifth Edition. Academic Press. *An imprint of Elsevier*. U.K.

Anisworth, G.C. (1973). Introduction and keys to higher taxa. In: The fungi. An Advanced Treatise (G.C.Anisworth,F.K.Sparrow and A.S. Sussman eds.) Vol. 4 A. 1-7. Academic press, New York, London.

Arnan, M.,Pinthus, M.J. and Kenneth, R.G. 1970. Epidemiology and control of a sunflower head rot in Israel caused by *Rhizopus arrhizus*. *Canadian J. Pl. Sci.* 50 (3): 283-286.

Battu, A. N. and Pathak, H.C. (1965). Observations on a mosaic disease of sunflower.

Dinesh, B.M.; Shripad K; Harlapur, S. I.; Benagi, V. I.; Mallapur, C. P. (2010). Prevalence of powdery mildew in sunflower growing areas in northern Karnataka; Karnataka *J. Agric. Scie* 23 (3): 521-523.

Dube H.C. (2013). 4th Edition, scientific publishers (India).

Fivawo, N. C. (1987). Diseases and pests of sunflower in Tanzania. TARO. Newsletter 2(3): 6-8

Hawksworth, D.L. (1995). Anisworth and Bisby's Dictionary of the fungi (8th ed.) C.A.B. International, Oxon, U.K.

Hua Z.F., Zhang, X.W., Ren, X.Y. and Wang, C.B. (1987). Occurance and damage by *Sclerotinia sclerotiorum*. In sunflower on crops of China No. 2: 51-54.

Huang, H.C. and Dueck, J. 1980. Wilt of sunflower from infection by mycelia-germinating sclerotia of *Sclerotinia sclerotiorum*. *Canadian J. Pl. Pathol.* 2: 47-52.

Kolte, S. J. 198. Diseases of annual edible oil seed crops. Vol. II CRC Press Inc Boca. Raton, Florida 47 pp.

Krik, P.M. (2008). Anisworth and Bisby's Dictionary of the fungi (10th ed.) C.A.B. International,Oxon, U.K.

Martens, J.W., Ravagan, G. and Mc. Donald W.C. (1970). Diseases of sunflower in Kenya. E. *Afr. Agric. For. J.* **35** (4): 389-395.

Nagaraju, A.; Murali, M.; Sudisha, J.; Amruthesh, K. N.; Murthy, S. M.(2012). Beneficial microbes promote plant growth and induce systemic resistance in sunflower against downy mildew disease caused by *Plasmopara halstedii;p; Current Botany*, 3 (5): 12-18.

Narasimhan, V. And Prakasam, N. 1983. Studies on head rot of sunflower and its control. In processing of the national seminar on management of diseases of oil seed crops, Madurai, TNAU, India. *Phytopathology.* 18: 317

Raut, J.G. 1975. Fungi associated with sunflower seeds. *Curr. Sci.*, 44: 787-790.

Sackston, W.E. (1961). *Botrytis cinerea* and *Sclerotinia sclerotiorum* in seeds of sunflower. *Plant Disease Reoporter.* 44B: 664-668.

Shirshikar, S. P. (2010).Sunflower necrosis disease management with thiomethoxam; Helia, vol. 33 No. 53 pp. 63-68.

Shirshikar, S. P. (2008). Integrated management of sunflower necrosis disease, Helia; Vol. 31 No. 49 pp. 27-34.

Shirshikar, S. P. (2008). Response of newly developed sunflower hybrids and varieties to downy mildew disease; *Helia,* **31** (49): 19-26.

Singh R.S. (2009). Plant disease. 9th Edition Oxford and IBH publishing company Ltd.

Swart,S.H. 1988. Head rot of sunflower in South Africa caused by *Rhizopus microspores*. *Phytopathologica* 20 (4): 391-392.

Yang, P.A. and Morris, H.E. (1927). *Sclerotinia* wilt of sunflowers. *Montana Agric. Exp. Stn. Bull.* 208. 32 pp.

Yang,S.M., Morris, J.B., Unger, P.W. and Thompson, T.E.(1979). *Rhizopus* head rot of cultivated sunflower in texas. *Pl. Dis. Reptr.* 63 (10): 833-835.

Yıldırım, I.; Turhan, H.; Ozgen, B. (2010). The effects of head rot disease (*Rhizopus stolonifer*) on sunflower genotypes at two different growth stages; *Turkish J. Field Crops,* **15** (1): 94-98.

Zimmer, D. E. And Hoes, J.A. 1978. Diseases, pp. 225-265. In: Sunflower Science and technology. A SA Monograph 19. J.F. Carter ed. American Society of Agronomy, Madison,WI. 505.

8 Diseases of Rapeseed-Mustard and their Management

H.K. Singh, R.B. Singh, Mahesh Singh and Ramesh Singh

Rapeseed-mustard is group of crops, contributes 32 per cent of the total oilseed production in India, and it is the second largest indigenous oilseed crop. Out of 75.55 m tones of rapeseed-mustard in the world from over 30.51 m ha area, India shares 7.80 m tones of rapeseed-mustard production from the area of 6.50 m ha with productivity of 1208 kg/ha. (Anonymous, 2013). Despite considerable increase in the productivity and production under Technology Mission, huge amount of money is spent on the import of edible oil. A wide gap exist between the potential yield and the yield realized at the farmer's field, which is largely because of number of biotic and abiotic stresses to which the rapeseed-mustard crop is exposed. The disease occurring on these crops are described here as under:

Alternaria Blight

Alternaria blight is one of the most widespread and destructive disease of rapeseed and mustard in India, particularly in the Indo-Gangetic plains (Vasudeva, 1958). The disease is also called the dark spot (Louvet, 1958) or grey spot (McDonald, 1959) based on symptoms produced on the host. In India, it was probably first noted on *Sarson* (*Brassica campestris* var. *sarson*) in 1901 at Tirhoot (Trihut) near Pusa, Bihar (Butler, 1918 and Bhowmik, 2003).

Geographical Distribution

Three species of *Alternaria*, *viz.*, *A. brassicae* (Berk.) Sacc., *A. brassicicola* (Schw.) Wilts. and *A. raphani* Groves and Skolko have been found to affect the rapeseed and mustard crop quite commonly throughout the world. The disease caused by *A. brassicae* is more destructive and occurs more frequently than the one caused by other species. It is reported to occur in China, Canada, England, Japan, France, Germany, Holland, India, Poland, Sri Lanka, Spain, Sweden, and Trinidad (Kolte, 1985; Verma and Saharan, 1993, Singh *et al.*, 2013, Singh *et al.*, 2017).

Economic Importance

Disease by infecting the leaf reduced the photosynthetic area, which results in ultimate low production. Siliquae are also directly affected by the disease and result in severe yield losses. Besides quantitative loss in yield, the quality of the seed, *i.e.* seed size, seed colour and germination capability, are also adversely affected due to the disease. In northern parts of India, at different locations, the loss in yield due to the disease ranged between 10 to 70 per cent, maximum being in *B. campestris* var. yellow *sarson* or *B. campestris* var. brown *sarson*. The reduction in oil content also varies with the crop species and variety involved. Reduction is higher in rapeseed than in Indian mustard, and varied between 4.7 to 36 per cent, and 2 to 29 per cent, respectively (Ansari *et al.*, 1988[a]; Kaushik *et al.*, 1984, Singh *et al.*, 2013, Singh *et al.*, 2017).

Symptoms

Alternaria blight attacks all the green aerial parts of the plant reducing its photosynthetic area and vigour. The disease is usually seen during mid-December to early January as minute dark-brown or black spots on lower leaves of young plants. At older age of leaves, the spots turn into circular, dark-brown, sunken necrotic lesions surrounded by light yellow halo and bear conidiophores and conidia in concentric rings at the grayish-white centre giving them a target board appearance (Figure 8.1A). Under congenial weather condition, the lesions enlarge; coalesce with each other resulting in chlorosis and defoliation of leaves. Gradually, the disease progress to the upper leaves, stems and pods (Figures 8.1B-D). The spots on young stems and green pods appear as black specks. On stem, they turn into dark-brown elongated lesions with pointed ends while on pod; they become round to elliptical and sunken, rarely encircle the stem or pod. Normally, lesions are seen only on the surface of the stem or pod of that mostly faces the air currents. The grayish-white centre of mature lesions becomes laden with dark spore mass to disseminate the disease within and between the fields. Infected pods show pre-mature ripening; they shrink and shatter easily. Seeds, particularly those directly below the pod-surface lesions show dark-brown spots on seed coat, become under sized, deformed, discoloured and infected. Such seeds, when peeled show dark-brown speaks on their cotyledons (Bhowmik, 2003, Singh *et al.*, 2013 and Singh *et al.*, 2017).

Habitat and Host Range

Alternaria species are either parasites on living plants or saprophytes on organic substrates. The host range of pathogenic *Alternaria* is very broad. Mostly *A. brassicae* and *brassicola* have been found to infect number of crucifers, *viz.*, *B. campestris* var. *toria* and *B. juncea*, *Brassica rapa* var. yellow *sarson*, *B. campestris* var. brown *sarson*, *B. alba*, *B. nigra*, *B. carinata*, *B. napus*, *B. tournefortii*, *B. chinensis*, *B. perkinensis*, *B. rugosa*, *B. oleracea* var. *botrytis*, *B. oleracea* var. *capitata*, *Raphanus sativus* and *Eruca sativa* under field as well as glasshouse conditions. In addition, it has been observed

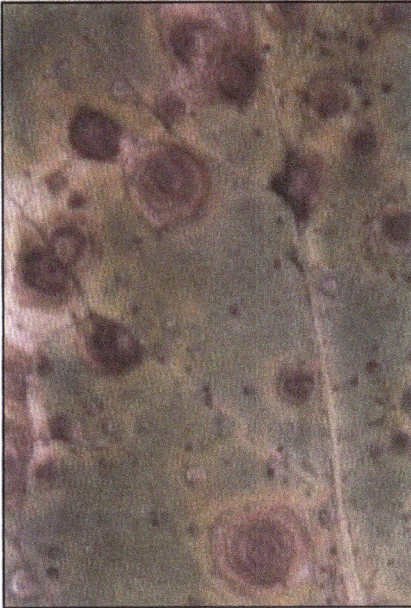

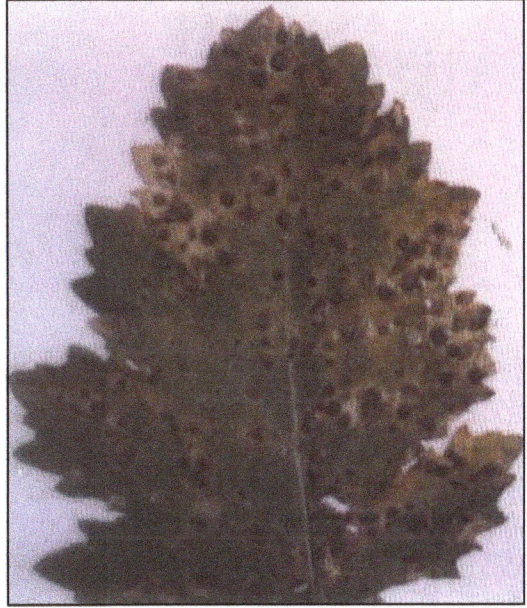

A: Concentric rings on the spots.   **B: Alternaria spots on leaf.**

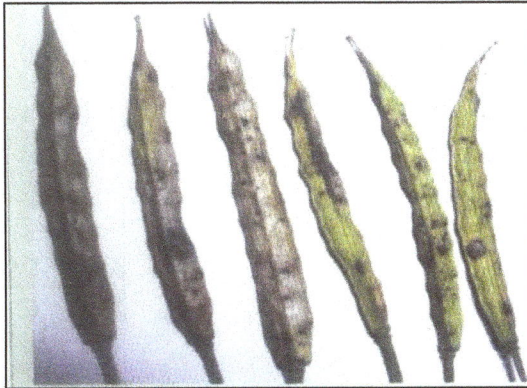

C: Alternaria spots on siliquae.   **D: Alternaria infection on stem.**

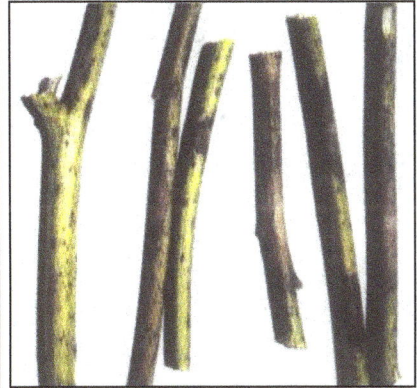

Figures 8.1A-D: Alternaria Blight Symptoms on different Parts of *Brassica juncea* (L.) Czern. and Coss (Courtesy: Shukla *et al.*, 2003).

on weed host like *Camelina sativa, Crambe abyssinica, Anagalis arvensis* (hirankhuri), *Crambe maritime, Convolvulus arvensis* and *Chenopodium album* (bathua) usually found in *brassica* field (Saharan and Kaidan 1983a; Tripathi and Kaushik, 1984; Verma and Saharan, 1994; Mehta *et al.*, 2002; Saharan and Mehta, 2002; Singh *et al.*, 2017). These hosts can play a major role in disease perpetuation.

Etiology

The three Alternaria species can be readily distinguished by microscopic observation. (Mridha, 1983, Vishwanath *et al.*, 1999). Studies on pathogenic

Table 8.1: Comparative Morphological Characteristics of *A. brassicae, A. brassicicola* and *A. raphani*

Fungal Structures	A. brassicae	A. brassicicola	A. raphani
Mycelium	Septate, brown to brownish-gray	Septate, grayish black to olive gray	Cottony whitish to greenish gray, or dark olive
Conidiophores	Dark, septate, arise in fascicles, measuring 14-74 μ x 4-8 μ	Olivaceous, septate, branched measuring 35-45 μ x 5-8 μ	Septate, olive-brown, simple or branched, measuring 29-160 μ x 4-8 μ
Conidia	Brownish black, obclavate, muriform with long beak, longer and wider and more septations than that of *A. brassicicola* and *A. raphani*, produced singly or sparingly in chains of 2 to 3 (Figure 8.2B)	Dark, cylindrical to oblong, muriform without beak, few septations and smaller than those of *A. brassicae* and *A. raphani*, produced in long chains of 8 to 10 spores (Figure 8.2A)	Olive-brown to dark, obclavate, muriform with poorly developed or no beak, wider than those of *A. brassicae*, less uniform in shape than either of the other two species, more or less pointed at each end, appear singly or in chains of upto 6 spores (Figure 8.2C)
Spore body length (μ)	96-114	45-55	45-58
Spore beak length (μ)	45-65	None	10-25
Spore body width (μ)	17-24	11-16	13-21
Overall spore length (μ)	148-184	45-55	60-83
Transverse septation	10-11	5-8	6-9
Longitudinal septation	0-6	0-4	3-6
Rate of growth and sporulation on media	Rudiment growth, grow slowly	Produce well-developed black sooty colony with distinct zonations, grow faster, sporulates abundantly	Cottny mycelia colony distinguishes this form the other species, less abundant sporulation
Formation of chlamydospores in culture or on host	Formed less frequently in culture	Not known	Usually olive brown chlamydospores are formed in culture as on the partially decayed affected plant parts
Infection	Penetrates leaf only through stomata	Penetrates leaf directly or through stomata	Direct penetration

Source: Meena *et al.,* 2010.

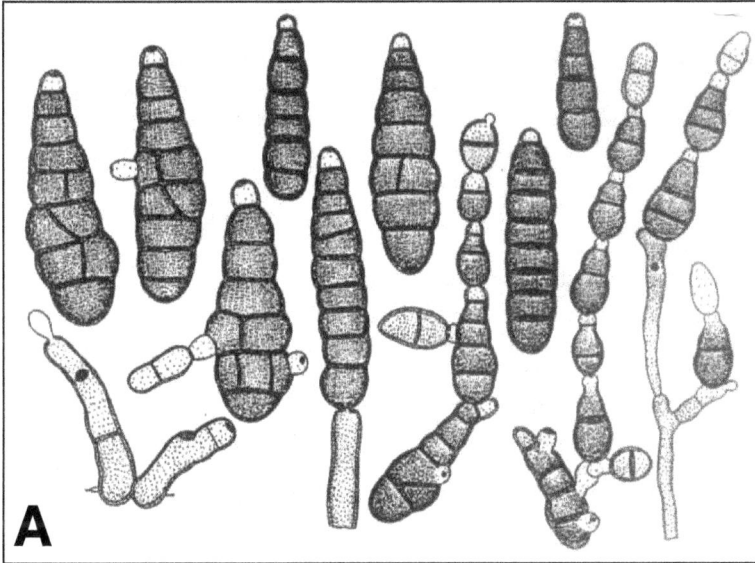

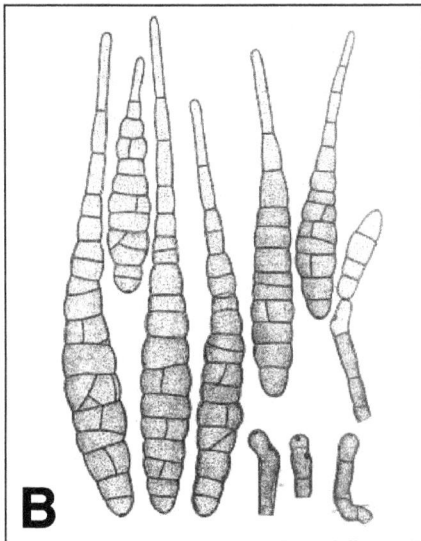

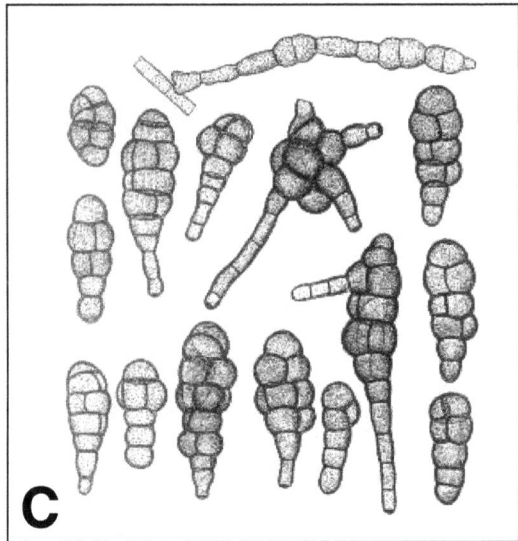

Figure 8.2A-C: A: *Alternaria brassicicola* **(x 650), B:** *A. brassicae* **(x 650), C:** *A. raphani* **(x 650) (Courtesy: Ellis, 1971).**

variability have to be the foundation for development of pre-breeding populations as strategic defence mechanism. Three distinct *A. brassicae* isolates, A (highly virulent), C (moderately virulent) and D (avirulent) are prevalent in India (Vishwanath and Kolte, 1997). Variations among the Indian isolates of *A. brassicae* has also been noted (Meena *et al.*, 2005; Patni, *et al.*, 2006; Singh *et al.*, 2009, Singh *et al.*, 2017). Khan *et al.* (2007[a]) reported variation in conidial length, breadth and septation in eight isolates collected from different places of district Aligarh.

Disease Cycle

The pathogen perpetuates through seeds (Figure 8.3), plant debris, soil and weed hosts. Recently it has been proved that in north India, seed borne inoculum does not cause infection in the coming season due to high temperature conditions during summer months in storage, however, the possibility of its survival through seed in hills cannot be ruled out. *A. brassicae* survives on plant debris buried in the field at 7.5cm depth in the form of microsclerotia and chlamydospore formed in the infected leaves. Local dispersal of the inoculums takes place through air, rain splashes and occasional driving rain drops. The primary infection appears on the cotyledonary leaves where sporulation takes place. These spores act as the source of secondary infection for the entire crop. *Alternaria brassicae* has been found to infect number of crucifers, *viz.*, *Brassica rapa* var. yellow *sarson*, *B. campestris* var. brown *sarson*, *B. campestris* var. *toria* and *B. juncea*, *B. nigra*, *B. alba*, *B.carinata*, *B. napus*, *B. chinensis*, *B. perkinensis*, *B. tournefortii*, *B. rugosa*, *B. oleracea* var. *capitata B. oleracea* var. *botrytis*, *Raphanus sativus* and *Eruca sativa* under field as well as controlled conditions. In Addition, it has been observed on weed host like *Camelina sativa*, *Crambe abyssinica*, *Crambe maritime*, *Anagalis arvensis* (hirankhuri), *Convolvulus arvensis* and *Chenopodium album* (bathua) usually found in *brassica* field. These hosts can play a major role in disease perpetuation. Air-born spores of *A. brassicae* form the primary source of inoculums of this polycyclic disease (Kolte, 1985, Bhowmik, 2003, Singh *et al.*, 2013, Singh *et al.*, 2017). The primary infection on cotyledonary leaves or true leaves on initially infected few plants, which serve as a source of secondary infection for the whole crop (Figures 8.3 and 8.4).

Figure 8.3: *Alternaria* **Perpetuates through Mustard Seeds.**

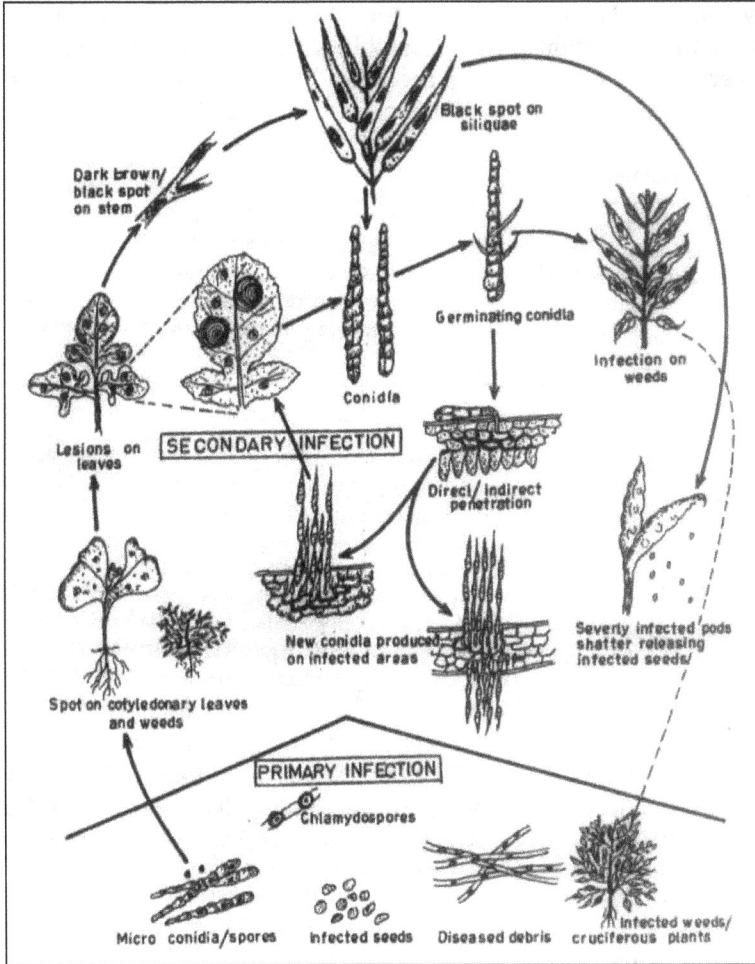

Figure 8.4: Disease Cycle of Alternaria Blight of *Brassicas* (Courtesy: Saharan and Mehta, 2002).

Epidemiology

Alternaria blight initiation starts with primary infection of cotyledonary and lower leaves which remain moist with dew during early hours of the day and these in turn act as the source of inoculums for its secondary infection. The relative humidity during this period of crop growth showed range between 46-96 per cent during the day and 73-92 per cent during the night.

Sporulation in *A. brassicae* and *A. brassicicola* on naturally infected leaf discs of oilseed rape requires RH ≥ 91.5 per cent and 87 per cent, respectively. The optimum temperature for sporulation is 18-24°C for *A. brassicae* and 20-30°C for *A. brassicicola* at which both the pathogens produce spores in 12-14 hours. Above 24°C, sporulation in *A. brassicae* is inhibited. Interrupting a 16 h wet period at 20°C with a dry period of 2 h at 70 or 80 per cent RH does not affect sporulation in either fungs but a dry interruption of 3 to 4 h inhibits sporulaion in both. Exposure of both fungi to alternating wet (18 h at 100 per cent RH, 20°C) and dry periods (6 or 30 h at 55-65 per cent RH, 20°C) does not affect the concentration of spores produced in each wet period. Sporulation times are not affected by either the host type or the age of the host tissue (Humpherson-Jones and Phelps, 1989). Alternate light and darkness is better for growth of fungus than continuous light or complete darkness (Ansari *et al.*, 1989). Susceptibility of oilseed *Brassicas* to *A. brassicae* increases with the age of the plant, prevailing temperature and duration of host surface wetness (Bhowmik, 2003).

Disease incidence is also influenced by the concentration of inoculums. The incidence of the disease on pods increased as the inoculum concentration was increased from 80-104 spores/ml. The incubation period decreased as temperature increased from 6-20°C. The incubation period also decreased as wetness period increased from 2-12 h, as inoculums concentration increased from 80 to 2×10^3 spores/ml and as leaf age increased from 4-10 days (Bhowmik, 2003, Singh *et al.*, 2013).

There is a positive correlation between weather factors and disease progress. Severe blight is associated with low temperature (minimum 2-12°C, maximum 16-26°C), high RH (80-96 per cent), average rain fall of 30 mm and wind velocity of 2-6 km/h (Sinha *et al.*, 1992; Dang *et al.*, 1995; Gadre *et al.*, 2002; Singh and Singh, 2011). Gupta *et al.* (2003) correlated the weather factor and Alternaria blight development and reported that 85 days old plant showed highest disease severity with negative correlation with maximum and minimum temperature and positive correlation with relative humidity. Yadav and Barar (2003) observed susceptibility of the disease during rosette to flowering stage of the crop when relative humidity ranges 81-94 per cent with total rainy days of 4-11 days.

Resistance

Among the cultivated oleiferous *Brassicas*, the digenomic species, especially *B. carinata* exhibits high resistance to diseases incited by fungi. Even though a high level of resistance to Alternaria blight is rare in cultivated species. The existences of wide inter and intraspecific variation in reaction to the disease is a common feature. Lesion type and sporulation capacity seems to indicate the levels of resistance in *Brassica* spp. The presence of epicuticular wax in some cultivars of *B. napus* and *B. oleracea* var. *alboglabra* confers a physical type of resistance to *A. brassicae*. Wax layer being hydrophobic, may affect conidial germination by impeding movement of foliar exudates to surface, by preventing water droplets to stay on the leaf surface besides providing a sort of mechanical barrier to the invading pathogen but it has apparently no effect on the pathogen after its entry into the host (Bhowmik, 2003).

Young *Brassica* plants show a high degree of resistance to *A. brassicae* but they become susceptible with age. Resistance to *A. brassicae* in some *Brassica* genotypes has been associated with quantitative changes in certain biochemical compound following infection. The resistant strain showed higher accumulation of anthocyanins of the halo area surrounding the disease lesion and had scanty sporulation. Similar changes in the levels of total phenols, sugars, nitrogen and ascorbic acids, *etc.* in rapeseed and mustard cultivars following infection with *A. brassicae* have been also correlated with their resistance or susceptibility to the disease by other workers (Chattopadhyay, 1989; Begum *et al.*, 1993; Gupta *et al.*, 1984; 1995, Bhowmik, 2003, Singh *et al.*, 2013, Singh *et al.*, 2017, Singh *et al.*, 2018).

Screening of Germplasm for Resistance

Screening of germplasm/promising lines for resistance to *A. brassicae* is usually made on the basis of disease rating scale in the field from natural infection/and with supplementary artificial inoculation. Resistant genotypes further evaluated under artificial inoculation condition.

1. Disease Rating Scale

Brassica cultivars are rated for disease intensity by assigning numerical scores on 0-5 rating scale (Hussain and Thakur, 1963; Rai *et al.*, 1976; Conn *et al.*, 1990; AICRP R-M, 2004) as (Table 8.2) given below:

Table 8.2: Reaction of *Brassica* sps. to Alternaria Blight Caused by *Alternaria* spp.

Rating Scale	Description of Scale	Host Reaction
0	No visible symptoms	Free/Immune
1	1-10 per cent leaf or pod area covered with small pin-head spots on the leaves and superficial pin-head spot on pods	Resistant
2	11-25 per cent leaf or pod area covered with small (generally less than 3mm) spot on the leaves and superficial pin-head spot on pods	Moderately resistant
3	26-50 per cent leaf area covered with bigger (more than 3mm) spot with initiation of coalescence on the leaves and deep lesion on pods	Moderately susceptible
4	51-75 per cent leaf or pod area covered with bigger commonly coalescing spot on the leaves and deep lesion on pods	susceptible
5	76-100 per cent leaf or pod area covered giving blighting appearance	Highly susceptible

The disease index is calculated by the formula (Townsend and Heuberger, 1943) as under:

$$Disease\ index = \frac{Sum\ of\ all\ numerical\ ratings}{No.of\ leaves\ examined \times Maximum\ rating\ score} \times 100$$

The numerical scale is based on the extent of leaf area damaged, which is primarily related to the quantity of spores deposited on its surface at a particular point of time. This type of visual scale is useful only at the initial stage of screening in discarding the highly susceptible materials, but classification of cultivars into various disease reaction categories on the basis of such scale is not comparable and does not give a reliable estimate of their reaction to the disease (Bhowmik and

Munde, 1987). Hong and Fitt (1995) also noted increasing number of lesions/cm² (severity) of dark leaf spot with the increase in inoculums concentration.

Presently modified scale of Conn *et al.*, 1990 is being used as 0 to 9 point scale by the scientists working under All India Coordinated Research Project on Rapeseed-Mustard (AICRP R-M, 2015) system in the country, which is described as 0=No lesion (Immune); 1= Non sporulating pinpoint size or small brown necrotic spots, less than 5 per cent leaf area covered by the lesions (Highly resistant); 3= small, slightly sporulating larger brown necrotic spot, about 1-2 mm in diameter with a distinct margin or yellow halo, 5-10 per cent leaf area covered by lesions (Resistant); 5= moderate sporulation, non-coalescing larger brown spots, about 2-4mm in diameter with a distinct margin or yellow halo, 11-25 per cent leaf area covered by the lesions (Moderately resistant); 7=moderately sporulating, coalescing, larger brown spots about 4-5 mm in diameter, 26-50 per cent leaf area covered by the lesions (Susceptible); 9 = profusely sporulating, rapidly coalescing, brown to black spots measuring more than 6mm in diameter without margins covering more than 50 per cent leaf area (Highly susceptible).

Management

(A) Use of Resistant Varieties

Improved commercial varieties of rapeseed-mustard in India are resistant to this disease and genotypes reported highly tolerant to *A. brassicae* from other species under artificial conditions are: *Brassica (sinapis) alba; B. hirta; B. napus* lines Lethbridge, Westar, WRG-1, Midas, HNS-3, UP 75-BN-5403, Target-84, Exotic-131 and *B. napus; B. carinata* lines PC-3 and -5, BCON-1, CE-8 and -9, and PPSC-1; *B. tournefortii; B. nigra; B. oleracea var. alboglabra; B. fructiculosa; B. juncea* lines BC-115, CSR-448, DIRA-251, PHR-2, Zem-1, RC-781, RH-8312, RS-104, RSK-21, RW-5453, B-2, K-41732; and *B. campestris* lines Tobin and PYS-3 (Hussain and Thakur, 1963; Rai *et al.*, 1976; Tripathi *et al.*, 1980; Saharan *et al.*, 1981; Munde and Bhowmik, 1985; Bhowmik and Munde, 1987; AICRP (R and M). 1990; 1996; Sharma and Singh, 1992ᵃ; Chahal and Jaura, 1994, Bhowmik, 2003, Singh *et al.*, 2017).

The cultivars Gulivar, Tower, GSL-1 and EC-126743 of *B. napus*; PC-5 and HC-1 of *B. carinata*; TM-7, TMV-2, YRT-3, KRV-Tall, PHR-1, Vardan, Saurabh, RC-1401, CSR-43, -142, -142-2, -343, -622 and -741, RH-8113 and -8114 and B-108(3) of *B. juncea*; and Torch, T-8 and -22, DYS-1 and Jata *sarson* of *B. campestris* have consistently been tolerant to *A. brassicae* under field conditions over years across the country (Kolte, 1987; Saharan and Chand, 1988; Dikshit and Srivastava, 1991; Bhowmik, 2003, Kumar and Chauhan, 2005, Mehta, 2014). The different genotypes of *Brassica* spp. namely BAUSM-92-1-1, RC-376, PBC-9221, EC-399299, JMM-99-01, PBN-2002, PBN-2001, RH-9912, JMM-992, PBC-2002-2, NPC-14, PBC-2004-1, NPC-15, GSL-1, HNS-004, ONK-1, NUDB-26-11, OCN-13, CAN-133, RGN-55, NRCR-837, NPN-1, PBC-2002, PHR-2, EC-399299 remained consistently field tolerant in mid-eastern of India (Singh and Singh, 2005ᵃ; Singh *et al.*, 2006; Singh *et al.*, 2007; Singh *et al.*, 2008; Kumar and Singh 2008; Kumar *et al.*, 2009; 2014; 2015c; AICRP R-M, 2016, Singh *et al.*, 2017).

(B) Cultural Management

A number of cultural practices or their modifications that are usually followed by farmers, a variety of chemical pesticides and combinations of agro-practices and chemicals may provide effective control of this disease.

The field and its neighbourhood area should be cleaned of weeds before taking up sowing operations. Sound and bold seeds of recommended cultivar that are free from discoloured ones should only be sown. Usually, rapeseed and mustard sown by mid-October escapes the severe attack of Alternaria blight in north India (Singh *et al.*, 1998; Yadav *et al.*, 2002; Prasad *et al.*, 2003; Bhowmik, 2003, Kumar and Kumar, 2006; Singh and Singh 2006). In the eastern part, delay in sowing of Indian mustard by one month beyond 22nd October under West Bengal conditions increased the incidence of *Alternaria brassicae* and *A. brassicicola* on leaves by 37 per cent and pods by 31 per cent (Dasgupta, 1991).

Use of recommended seed rate at sowing is also essential, as crop density per unit area directly affects the disease incidence and yield. High fertilizer doses, especially of nitrogen increase the severity of Alternaria blight in oilseed rape (Stankova, 1972; Kumar and Kumar, 2006) but Meena *et al.* (2002) reported decrease in severity with increase level of nitrogen. Significant reduction in disease severity and highest yield of mustard were obtained under field conditions at Faizabad, UP when NPK were used at 100: 40: 40 kg/ha along with three sprays of Indofil M-45 (Singh and Chauhan, 1997). In general, blanced fertilizer dose (N: P: K: 80: 40: 40) coupled with prophylactic sprays of fungicide (Mancozeb) provide maximum increase in yield. Recommended dose of NPK (90: 60: 40Kg/ha) along with 40 Kg sulphur/ha gave maximum control of Alternaria blight (Kumar and Kumar, 2006, Singh *et al.*, 2013, Singh *et al.*, 2017).

(C) Biological Management

Indiscriminate use of pesticide has leaded to the problems like pest resistance, pest resurgence and environment pollution. The integrated pest management programme in the management of pest is one of the very potent components of crop management. Till date chemical management was the only option against the problem. In many European countries, organic agriculture has rapidly been transformed from a farmers' movement to an institutionalized part of agricultural policy. In certification, compliance with published organic standards is verified through annual inspections on farms (Laura and Juha, 2004). However, some reports indicate possibility of biological management of the disease. Phyllosphere residents *Aureobasidium pullulans* and *Epicoccum nigrum* reduced the infection by *A. brassicicola*, especially when they were inoculated 14 h before the pathogen (Pace and Campbell, 1974). Spray of soil isolates of *Trichoderma viride* at 45 and 75 days after sowing could manage Alternaria blight of Indian mustard (*Brassica juncea*) as effectively as mancozeb (Meena *et al.*, 2004). Botanicals *viz.*, bulb extract of *Allium sativum* has been reported to effectively manage Alternaria blight of Indian mustard (Meena *et al.*, 2004; Patni and Kolte, 2006; Nigam, *et al.*, 2011; Singh *et al.*, 2013). Application of aqueous bulb of garlic either alone or in combination with *Trichoderma* spp. and fungicides like mancozeb were found effective in controlling the disease severity at different places (Singh and Singh, 2007; Singh *et al.*, 2008 and Meena *et al.*, 2008, Singh *et al.*, 2013, Singh *et al.*, 2017, Singh *et al.*, 2018).

(D) Chemical Management

Seed treatment with brassicol or captafol @ 4 g/kg seeds (Chahal and Sekhon, 1980; Chahal, 1982; Saini, 1982), iprodione @ 1.25g a.i./kg seeds (Stovold *et al.*, 1987), bavistin (carbendazim) @ 2.5 g/kg seeds or dithane-M 45 (mancozeb) @ 2 g/kg seeds eradicate the seed borne infection of *A. brassicae*. At least 24 h storage period after their treatment is needed for effective eradication of infection (Kumar and Singh, 1986, Bhowmik, 2003, Pandey *et al.*, 2018).

Among the more popular chemicals, spraying with difolatan (captafol); dithane-M 45, dithane- Z 78 and cuman L (dithiocarbamates); bavistin (carbendazim); panolil (guazatin); baycor (bittertanol); syllit (dodecyl guanidine acetate); daconil (chlorothalonil); wettable sulphur; dueter (fentin hydroxxide); fentin acetate; Topsin-M (thiophanate-methyl) and TPTH (triphenyl tin-hydroxide) @ 0.2-0.25 per cent provide effective control of the disease and result significant increase in crop yield (Chahal and Sekhon, 1980; Kaushik *et al.*, 1983; Singh and Bhowmik, 1985; Chahal, 1986[b]; Kolte *et al.*, 1987; Tripathi *et al.*, 1987; Dasgupta, 1991; Chattopadhyay and Bagchi, 1994; Mridula *et al.*, 1994; Gupta *et al.*, 1997; Godika *et al.*, 2001; Bhowmik, 2003; Singh and Singh 2005[a]; Singh and Singh 2006; Khan *et al.*, 2007[b]; Singh *et al.*, 2011, Singh *et al.*, 2017, Pandey *et al.*, 2018; Singh *et al.*, 2018).

Iprodione is also highly effective and caused minimum defoliation (Kumar, 1996; Shivpuri *et al.*, 1988) and increased the crop yield and oil content as compared to untreated crop (Chattopadhyay and Bhunia, 2003, Murkherjee *et al.*, 2003; Bhowmik, 2003; Singh and Bhajan, 2004 and Singh and Singh, 2005[b]). The amount of residues of iprodione in edible commodity (seeds) was less than the recommended MRL (0.5 mg/kg) safety level [AICRP (R and M), 1995, 1996; Chatterjee *et al.*, 1997]. Singh and Maheshwari (2003) reported Cantaf as most effective fungicide for the control of disease on leaves and pod, while Singh and Singh, (2007) found Mancozeb treatment was best against Alternaria blight of *Brassica compestris* var. yellow *sarson*. Iprobenfos @ 0.1 per cent and Propiconazole @ 0.05-0.1 per cent were also found effective against Alternaria blight in case of Indian mustard under field condition at Kumarganj, Faizabad (Kumar *et al.*, 2009). Spray operations with fungicides should be taken up with the first appearance of disease spots and then repeat two to three times at 10 to 15 days intervals depending upon the crop variety, its growth stage and the severity of infection.

White Rust or White Blister

White rust occur on aerial parts of a large number of cultivated and wild species of cruciferous plants having economic importance such as turnip (*Brassica rapa* sub sp. *rapa*), cauliflower (*Brassica oleracea* var. *botrytis*), cabbage (*Brassica oleracea* var. *capitata*), Chinese cabbage (*Brassica oleracea* sub sp. *pekinensis*), radish (*Raphanus sativus*), mustard or sarson (*B. juncea* var. *juncea*), rape or *lahi* (*B. napus* var. *napus*), black mustard (*B. nigra*), Brussels sprouts (*B. oleracea* var. *gemmifera*), cress and *taramira* (*Eruca sativa*). Non-crucifer crops affected by white rust include spinach

(*Spinacia aleracea*) and sweet potato (*Ipomoeae panduratae*). In India, this disease has been reported to occur on turnip, radish, *Eurca sativa*, different species of oil bearing *Brassica* and some weeds such as *Cleome viscosa*. Among the cultivated crops the oilseed crops like *Eruca* and *Brassica* suffer heavy losses due to distortion of floral organs. Combined infection of leaves and inflorescences may cause up to 60 per cent yield loss in *Brassica juncea* (Indian mustard). In late sown crops the yield loss due to floral infection may vary from 23 to 54.5 per cent. The disease is more prevalent in colder area as compared to temperate region (Bhowmik, 2003, Singh and Singh, 2013).

Distribution

This disease is wide spread in countries like Germany (Klemm, 1938), Palestine (Rayss, 1938), Fiji (Parham, 1942), Brazil (Viegas and Teixeira, 1943), India (Chowdhary, 1944), France (Darpoux, 1945), Romania (Savulescu, 1946), Turkey (Bremer *et al.*, 1947), Japan (Hirata, 1954), United Kingdom and USA (Walker, 1957), Canada (Greelmn, 1963; Conners, 1967 and Petrie, 1975), Pakistan (Perwaiz *et al.*, 1969), Newzealand (Hammett, 1969) and China (Zhang *et al.*, 1984). White rust causing pathogen is frequently associated with the downy mildew pathogen *Pernospora parasitica* (Pers. ex. Fr.). It may occur individually or together on cultivated oilseed *Brassica* and other cruciferous species throughout the world (Channon, 1981). According to Butler (1918) there is very frequent coexistence of white rust and downy mildew, and it is not easy to separate their effects, but white rust produces the greatest deformities in the stem and flowers (Awasthi *et al.*, 1995, 1997). In mid eastern India such as Uttar Pradesh, Bihar and Jharkhand white rust develop in the month of December and continued till February. Staghead production, though observed sporadically, was economically important in *Diara* land areas of Uttar Pradesh (Singh and Singh, 2005; Singh and Bhajan, 2005, Singh and Singh, 2013).

Avoidable Yield Losses

There are number of reports where the yield loss due to white rust alone has been calculated. Berkenkamp (1972) reported yield losses due to white rust in north-central Alberta upto 1-2 per cent in 1971, while in Manitoba, losses varied from 30-60 per cent (Bernier, 1972). Petrie and Vanterpool (1974) estimated the yield losses of 60 per cent due to *A. candida* staghead infection in rapeseed in Saskatchewan (Canada). Petrie (1975) reported 6.0 per cent reduction in average seed yield in turnip rape. Bains and Jhooty (1979) estimated 17-30 per cent losses in yield of rapeseed-mustard crop in the country. Barbetti (1981) recorded the annual yield loss of 5 to 10 per cent due to stagheads in rapeseed in Western Australia. Saharan *et al.* (1984) observed 23.0 to 54.5 per cent yield loss in late sown Indian mustard (*B. juncea*). Combined infection (leaf and staghead) led to 89.8 per cent losses in yield (Lakra and Saharan, 1989). Singh and Bhajan (2005) reported highest avoidable seed yield loss due to disease upto 28.2 per cent in cv. Varuna under late sown condition. Singh and Singh (2007) recorded losses in *B. campestris* var. *toria* due to hypertrophied host tissues called staghead appear to be about 34 per cent when average receme length is 10 cm. Combined infection with white rust and downy mildew pathogens lead to 37-47 per cent and 17-32 per cent reduction in siliquae number and seed yield, respectively.

Occurrence of white rust in mid-eastern India was always reported with appearance of Alternaria blight in Indian mustard. The losses due to combined effect of these diseases were recorded 7.09 per cent to 40.87 per cent (Singh and Singh, 2006 and Kumar *et al.*, 2009, Singh and Singh, 2013).

Symptomatology

The disease may occur on all the parts of the host except the roots. There are two types of infection caused by *A. candida* on *Brassica* spp. general or systemic and local (Butler, 1918; Walker, 1957). Local infection is manifested by white or creamy yellow, raised, scattered, zoo-sporangial pustule (1 or 2 mm in diameter) on the under surface of leaves (Figures 8.5A and B). The upper surface, corresponding to the pustule on the lower surface becomes tan yellow and prevalence of the disease is easily recognized from the upper surface of the affected leaves. After the complete development of the pustule, it ruptures and releases a chalky dust of spores (sporangia) (Figure 8.5B). Necrosis of the leaf tissue is not observed immediately. When these pustule ages and affected leaves are about to become senescent, the necrosis around or in the pustule may be seen.

When the plants are infected through the seedling or at the flowering stage, it is reported that the fungus becomes systemic in plant tissue and causes distortion, hypertrophy, hyperplasia and sterility of inflorescence which is known as staghead phase (Petrie, 1973; Verma and Petrie, 1980). Affected flowers show malformation. The petals may become green like sepals and the stamens may be transformed into leaf like or carpelloid structures (Figure 8.5C). The petals and stamens persist in the flowers instead of falling early as in the case of normal plants (Figure 8.5C). The stamens are sometimes changed into thick, club-shaped sterile bodies. The ovules and pollen grains are usually atrophied, resulting in complete sterility (Sharma, 2006, Sangwan and Mehta, 2001; Bhowmik, 2003, Singh and Bhajan, 2005; Singh and Singh, 2007, Singh and Singh, 2013).

In nature completely systemic infected plant is seen rarely. Such plants are characterized by stunted, thickened growth without any primary or secondary branches and it bears only the white rust pustules on its surface.

Causal Organism

The disease is caused by fungus *Albugo candida* (Pers. ex. Lev.) Kuntze, belongs to division Eumycota, subdivision Mastigomycotina, class Oomycetes, order Peronosporales and family Albuginaceae. It is an obligate parasite. The mycelium is intercellular producing small globose knob-shaped haustoria in the host cell. Often a large number of haustoria are seen in single cell. The haustoria remain functional for a short time after which loses viability. The sporangiophores are formed by the vertical growth of broad, short, clavate stalks beneath the epidermis, which is raised up due to pressure of these structures and is separated from underlying cells. The sporangiophores are hyaline, clavate and free from each other laterally and are thick walled, especially towards the base, 30-45 x 15-18 µm in size. The sporangia formed in basipetal succession in chains on sporangiophores (Figure 8.5D and E). The sporangia formed in the initial stage are thick walled and closely pressed against in

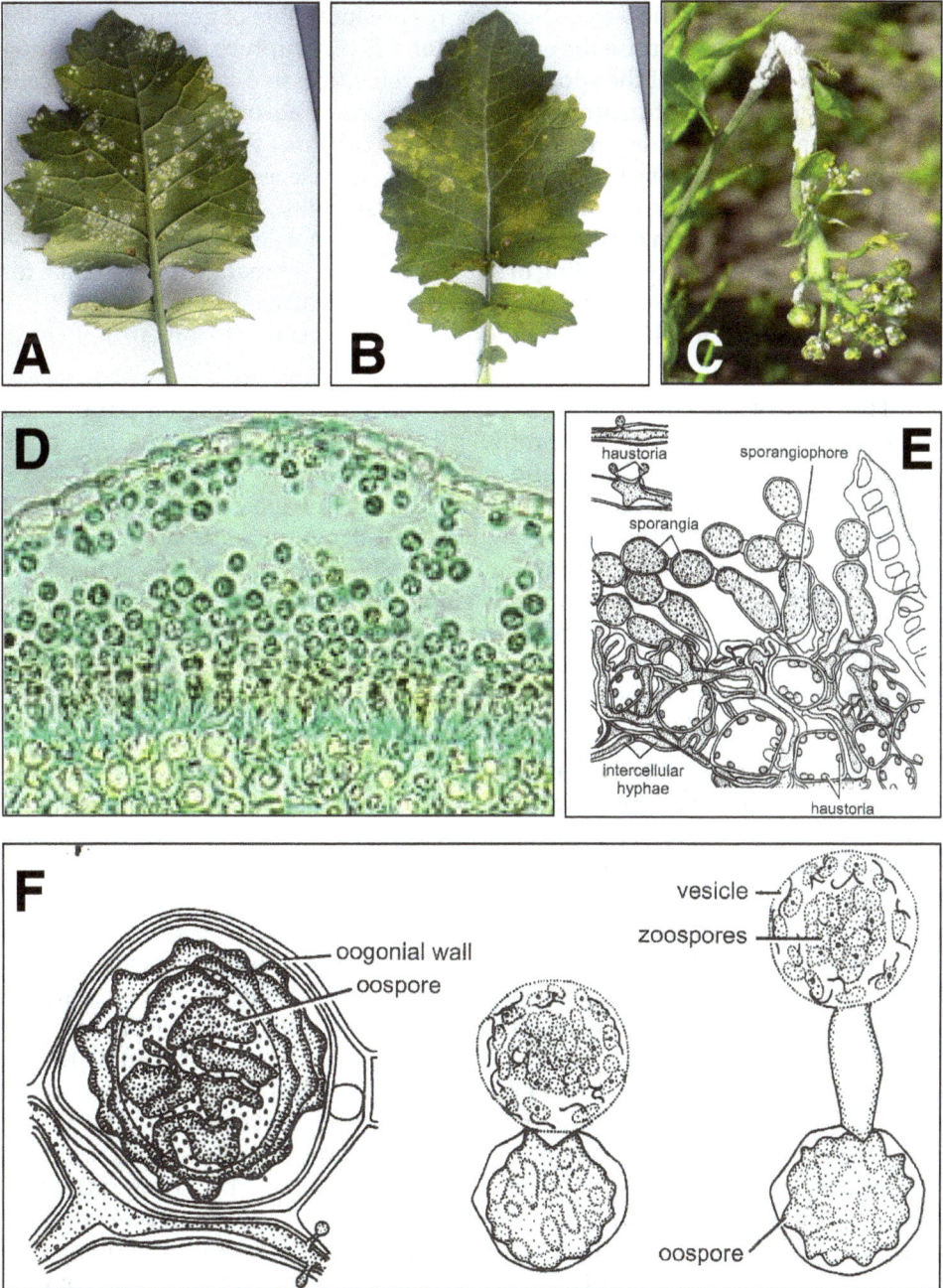

Figures 8.5A-F: A: White Rust Affected Leaf (Lower Surface); B: White Rust Affected Leaf (Upper Surface); C: White Pustules on Swollen Floral Parts (Staghead); D: *Albugo candida* Producing Chains of Unicellular Sexual Sporangia; E: Intercellular Hyphae, Haustoria, Sporangiophores and Sporangia; F: Fertilised Oogonium Enclosing Oospore within its Persistent Wall and Germinating Oospore to Release Zoospores (Courtesy: Mehta, 2014).

the lower unruptured epidiermis. Such spores are incapable of germination. Their only function appears to raise the epidermis and finally rupture it, thus facilitating the dispersal of the rest of the sporangia. Between successive sporangia small cells (pads) are present that function as disjuncture. After absorbing moisture these pads swell and disintegrate, freeing the sporangia from the sporangial chains. The pustules/sori or blisters eventually rupture and release the powdery mass of sporangia. The sporangia are hyaline, nearly spherical to oval with uniform thin wall and 14-16 x 16-20 µm in size. The sporangia germinate most readily at the time when the pustules rupture. Chilling of sporangia at low temperature is essential for the planospore production. Reduction in water content of sporangia to about 30 per cent is also essential for their germination. The sporangia can germinate only upto 6 weeks after their formation (Bhowmik, 2003 and Sharma, 2006).

The apex of the sporangium is drawn out to form a rounded papilla and few large vacuoles appear in the protoplasm. After sometime these become quite spherical and discharge outside and then the entire mass segments into 4 to 8 polyhedral cells each forming planospores. Each planospore is provided with a contractile vacuole. The planospores, which are 4-8 in number separate from each other and escape from the sporangium one or sometimes in groups, adhered to each other. Sometime the whole mass is emptied into a sort of bladder formed by the swelling of the papilla. The planospores are biflagellate. After a few minutes of activity, the planospores loose their flagella, come to rest and then germinate by a germ tube, which penetrates the epidermis of the host directly or through the stomata.

The oogonia and antheridia are formed from the mycelium in the systemically invaded tissues in the intercellular spaces particularly in the hypertrophied part. The antheridium is clavate, paragynous and contains 6-12 nuclei. The oogonium is globose, terminal or intercalary and multinucleate (about 100 nuclei). The contents of the oogonium are clearly defined into a periplasm and a single central oosphere. At the contact point of antheridium with the oogonium, the wall becomes very thin and a papilla of the oogonium protrudes into the antheridium passes through the thin spot into the oosphere. The nuclei in the periplasm disorganize and the nuclei in the antheridium and the oosphere undergo two mitotic divisions. A granular body known as the coenocentrum appears in the oosphere and the single female functional nucleus is attached to a point near it. The fertilization tube penetrates into the coenocentrum, ruptures and discharges a single male nucleus, which fuses with the female nucleus. The fertilization tube collapses and the coenocentrum disappears. Some author suggested heterothallism in *Albugo*. The oospore is formed by the development of a wall around the oosphere. Mature oospores are globose, chocolate brown, 30-35 µm in diameter (generally 45 µm with oogonial wall sometimes upto 60 µm) (Figure 8.5F) epispore thick, verrucose to tuberculate or with low blunt ridges which are often confluent, irregularly branched and sometimes seemingly smooth. From the fusion nucleus, about 30 nuclei are formed by further division. One of these divisions is the reduction division ((Bhowmik, 2003, Sharma, 2006, Singh and Singh, 2013, Mehta, 2014).

Perpetuation

The disease causing fungus *Albugo candida* perpetuates through oospores lying in the soil in diseased plant debris or moving with diseased pieces along with the seed. Perennial weed host may also serve as source of primary inoculum. Some cruciferous weeds are known to carry systemic, persistent and non-symptomatic infection. According to Butler (1918), early disease attacks in annuals plants are due to infection by zoospores arising from germinating oospores left in the soil. Other workers like Chupp (1925) and Takeshita and Linn (1953) considered the survival of the pathogen through soil borne oospores. While zoospores from germinating oospores of *A. candida*, may act as primary inoculum for the young host plant (Verma and Petrie, 1980), no experimental evidence is available about the direct role of soil-borne oospores in initiating infection to the next season crop especially under our sub-tropical climate. Interestingly, report from Verma and Bhowmik (1988) shows that oospores of *A. candida* do not remain viable in field soil through summer months under north Indian conditions. They could not detect oospores in washings from soil samples collected from white rust infested mustard fields (cv. Pusa bold) in the flowering season nor could they recover viable oospores from hypertrophied host organs buried 6-inch below the soil surface after six months. This aspect needs confirmation (Figure 8.6).

However, the most important means of perpetuation of the pathogen is through oospores carried with seeds, and sowing of such contaminated seeds results in both locally and systemically infected plants in the field (Barbetti, 1981; Verma and Petrie, 1980). In India, seed samples from six cultivars of *B. juncea* and one from *B. campestris* showed heavy contamination with *A. candida* oospores whose number per gram of seeds ranged from 12.0 to 81.0 (Verma and Bhowmik, 1988). Lakra and Saharan (1989) estimated 8.75×10^5 oospores in one gram of hypertrophied cup-shaped leaves and 21.85×10^5 oospores in one gram of hypertrophied staghead position. Oospores can remain viable for 21 years under dry storage conditions (Bhowmik, 2003 Singh *et al.*, 2013 and Mehta, 2014).

Possibly, the pathogen also survives in cooler places on various host species such as turnip, cabbage, cauliflower, radish, horse radish, cress, *Brassica juncea, B. campestris, B. campestris* var. glauca, *B. napus, B. nigra, Capsella bursa-pastoris, Cleome viscosa, (Polanisia icosandara), Coronous didyama,* Eruca sativa, *Merremia marginata, Nasturtium palustra, Saussurea lappu, Symbrium irio, Cardamine subumbelata,* (Butler, 1918; Damle, 1944; Singh *et al.*, 1993). The sporangiospores from these hosts act as primary inoculum to initiate the disease in rapeseed and mustard during the crop season (Singh and Singh, 2013).

Germination of Oospores

Oospores of *A. candida* are difficult to germinate. deBary (1866) first noticed oospores of *A. candida* to germinate by the production of sessile vesicle from which zoopores were released (Figure 8.6). Vanterpool (1959) described the sound method of germination, they discharge tube with terminal vesicle type but the germination was irregular and poor (< 4 per cent). Later, Petrie and Verma (1974), and Verma and Petrie (1975) reported a third type of germination in which oospores

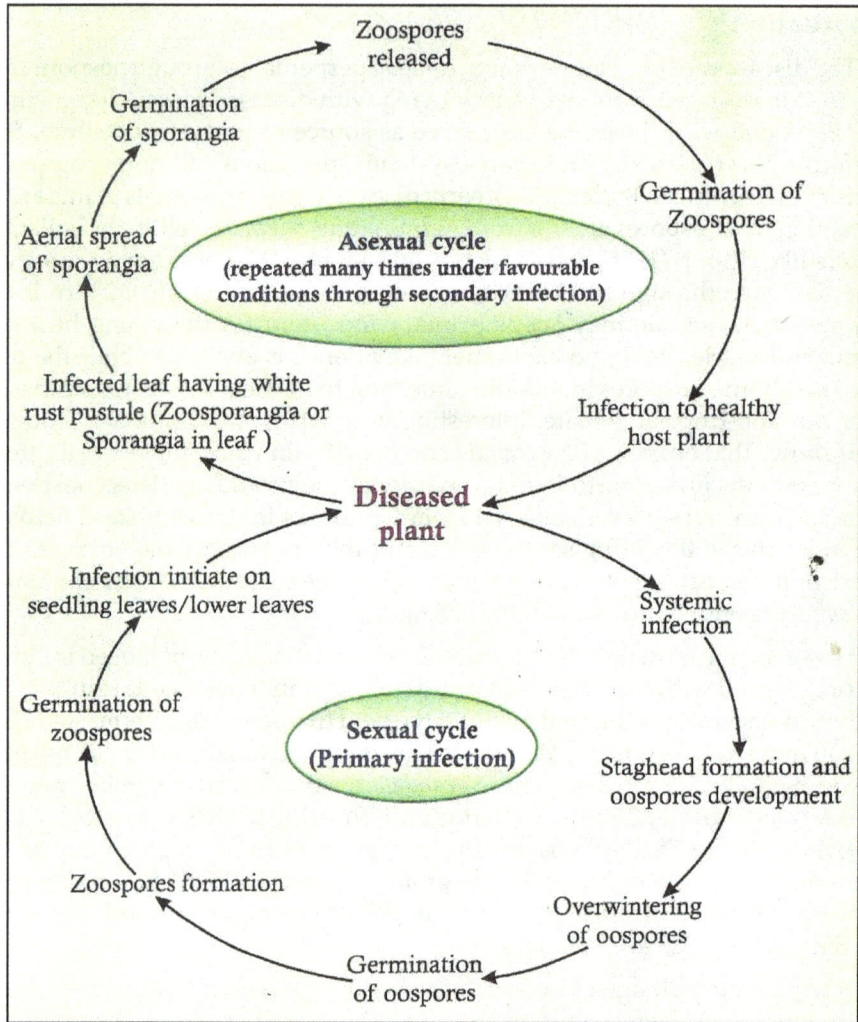

Figure 8.6: Perpetuation/Disease Cycle of White Rust (Courtesy: Shukla *et al.*, 2003).

germinated by the production of one or two simple or branched germ tube giving rise directly to mycelium, though germination by the production of sessile vesicle was the most common type. These workers used three methods to induce high percentage of oospore germination: the first was that of Hiura (1930) in which oospores were incubated on moist filter paper for about 21 days at 10 and 15°C, and this gave maximum germination of 65 and 88 per cent, respectively at the two temperatures. High germination also resulted from slow leaching of oospores for 15 days or more on sintered glass filter at 13°C. In the third method, oospores were subjected to washing in sterile water on a rotary shaker at 200 rpm for 3-4 days at 18-20°C, followed by a day in still-culture at 13°C. This was the most rapid method of securing high germination. Dormancy is not a factor since oospores germinated two weeks after their collection (Petrie and Verma, 1974). In India germination of oospores was obtained by subjecting them to continuous leaching with water for 15 days at 15°C. Germination was poor (av. 21.36 per cent), and took place by both sessile vesicle and germ-tube, the latter type being rare (Verma and Bhowmik, 1988).

A further improvement in percentage (≤ 80 per cent) germination of oospores was obtained by agitating them for upto 24 h in sterile distilled water containing a mixture of B-glucuronidase and aryl sulfastase followed by 3 days of washing on a rotary shaker at room temperature and 15 h of chilling at 13°C. Effectiveness of enzymes on stimulation of germination depend on the stage of oospore maturity. Both aging and chilling of oospores stimulate their germination (Liu and Rimmer, 1993 and Singh and Singh, 2013, Mehta, 2014).

Epidemiology

The incidence and severity of white rust on *Brassica* species are profoundly influenced by the prevailing weather conditions of the region. Prevalence of low temperature coinciding with high humidity – the effect of the latter being more dominant, cloudiness and less sunshine hours during flowering stage of the crop, are the conducive conditions for epidemic development of white rust disease. According to Sempio (1940), temperature most favourable for *A. candida* infection to red radish was 16 to18°C, though the disease developed most readily at 12 to 21°C, relative humidity at 60 to 80 per cent produced more severe infection than at 98 to 100 per cent, and 1 to 3 days darkness following infection stimulated the disease but continued darkness reduced it (Bhowmik, 2003).

Verma and Bhowmik (1989[b]) studied the epidemiology of white rust on Indian mustard (cv. Pusa bold) and reported that mean temperature (MT) less than 16°C and mean relative humidity (MRH) more than 60 per cent are essential for development of the disease. Infection rate was highest (0.574) when MT and MRH was 14.7°C and 73.25 per cent, respectively and rainfall around 230 mm. Infection rate declined with the increase in MT and decrease in MRH. Long sunshine hours for prolonged periods coincided with poor rust development. At Hissar, Haryana, dew fall more than 100 mm, MT of 10-18°C, MRH more than 65 per cent, potential evapotranspiration <60 per cent, cloudy weather and wind velocity of 2.7-3.4 km/h favoured faster increase in the number and size of rust pustules and epidemic development of the disease (Saharan *et al.*, 1988; Lakra and Saharan, 1991). At Pantnagar, Uttarakhand, high incidence of stagheads due to white rust and downy mildew (*Peronospora parasitica*) infection, was obtained with only 2-6 h sunshine per day concomitant with mean maximum and minimum temperature of 21-25°C and 6-10°C, respectively, MRH of 68-73 per cent and rainfall upto 161 mm (Kolte *et al.*, 1985, Bhowmik, 2003, Singh and Singh, 2013).

In South India at Dharwad, Karnataka, prevalence of moderate MT (23-24°C), high MRH (83.8-86.3 per cent) and high mean rainfall (3.1-9.3 mm) invited heavy incidence of both leaf infection (75-86 per cent) and floral malformation (41.8-86.3 per cent) on mustard crops (cv. Varadan) sown from the first week of June to third week of August 1988. On the other hand, crops sown during the first to third week of either September or October 1988, showed considerable reduction in both leaf infection and floral malformation. This was mainly attributed to the prevalence of low MRH (73.2-76.9 per cent) and low rainfall (0.0-0.6 mm) during the period there was not much change in MT (22.8-23.9°C) (Hegde and Anahosur, 1994 and Bhowmik, 2003,).

Plant age also influence disease development, older leaves are more susceptible than younger ones (Kumar *et al.*, 1995). Susceptibility of mustard plants (cv. Pusa bold) to white rust increases with age till they are about 60 days old and then declines (Verma, 1987). Medium aged turnip rape (*B. campestris*) show maximum infection than younger and older ones. Leaves develop more rust pustules, when detached from plants at the end of a dark period than at the end of a light period (Verma *et al.*, 1983). Incubation period was longer but rust intensity was higher on inoculated mustard plants maintained at 8-26°C than at 10-31°C (Verma and Bhowmik, 1986). In north-eastern zone of the country, generally white rust incidence escapes with normal sowing (October). The disease increased with delay in sowing time; however, some cultivars may be equally susceptible at all date of sowing (Gupta *et al.*, 2004; Singh and Singh, 2006). Kumar *et al.* (1995) also reported that the older leaves are more susceptible than younger leaves and symptoms appear earlier in inoculation of lower surface than their upper surface. Sangeetha and Siddaramaih (2007) reported maximum temperature positively correlated with disease index of white rust. Maximum temperature from 26-29°C and mean relative humidity more than 65 per cent favoured the development of disease (Bhowmik, 2003 and Singh and Singh, 2013).

Screening of Germplasm for Resistance

For preparation of disease sick plots, the dried staghead may be ground to powder form and this powder containing oospores may be incorporated into the soil. After preparation of the sick plot, the test material may be planted in the sick plots. Under Indian conditions late planting (beyond the second fortnight of October) has found to be quite favourable for severe occurrence of white rust. After every 2 to 5 rows of test material, susceptible check may be planted. If possible the known resistant check may also be planted; this help in monitoring of the disease and easy determination of the reaction of the test material in comparison with the susceptible and resistant checks. If the disease appearance is low during the course of experiments, then under such situation the inoculum may be sprayed. The rating for measurement of degree of disease should be done at leaf and siliqua formation stage. Generally, a five point (0-5) scale is used.

Disease Rating Scale

Use of visual white rust rating scale is important in selecting for components of partial resistance (Fox and Williams, 1984). They suggested scale of 9 point having five defined categories of severity as follows:

0= No symptoms; 1= small, pinpoint to larger brown necrotic flecks under inoculation point on upper surface occasionally necrosis extending to lower epidermis, no sporulation; 3= very sparse, one to few minute, scattered pustules on lower surface; 5= few to many scattered pustules on upper surface, none to few scattered pustules on lower surface; 7=many to few scattered pustules on upper surface, many scattered, small to large pustules on lower surface; 9= very few to no pustules on upper surface, many large coalescing pustules on lower surface. This technique was actually suggested for recording infection on cotyledons.

Kolte *et al.* (1986) suggested a 5 point scale for measuring white rust on leaf as follows:

0= No symptoms; 1= 1-10 per cent leaf area infected; 2= 11-25 per cent leaf area infected; 3= 25-50 per cent leaf area infected; 4= 51-75 per cent leaf area infected; 5= 75 per cent or more leaf area covered by the symptoms.

At present 0-9 scale is being followed by the All India Coordinated Research Project on Rapeseed-mustard system (Anonymous, 2010), which is approved by AICRP-RM Plant Pathologists, as under:

0= No lesions (Immune); 1 = Non-sporulating pinpoint size white pustules, less than 5 per cent leaf area covered by pustules (Highly Resistant); 3 = Small roundish slightly sporulating larger white curdy pustules, about 1-2 mm in diameter with a distinct margin or yellow halo, 5-10 per cent leaf area covered by pustules/spot (Resistant); 5 = Moderately sporulating, non-coalescing larger pustules, about 2-4 mm in diam. with a distinct margin or yellow halo, 11-25 per cent leaf area covered by the spots (Moderately resistant); 7 = Moderately sporulating, coalescing larger brown spots about 4-5 mm in diam. 26-50 per cent leaf area covered by the pustules (Susceptible); 9 = Profusely sporulating, rapidly coalescing larger pustules measuring more than 6 mm in diam. without margins covering more than 50 per cent leaf area (Highly susceptible).

Management

(A) Cultural

Date of Sowing

Sowing date, being dependent upon the weather conditions, affects crop growth as well as the incidence and severity of the rust disease. In northern region of country, the most favourable time for sowing rapeseed and mustard is around the middle of October, when the mean daily temperature ranges between 25-26°C (Gangasaran and Giri, 1984). Early sown crops maturing by the third week of March, generally escape the severity of both white rust (Bhowmik, 1985[b]). Late sown crops on the other hand, suffer most from white rust infection (Kuamr *et al.*, 1995). At Pantnagar, Uttarakhand, Indian mustard had the least incidence (0-20 per cent) and severity (0-9 per cent) of the staghead phase of white rust when sown in the first week of October whereas, sowing in late October or in November invited high incidence (10-30 per cent) and severity (23-40 per cent) of rust infection (Kolte *et al.*, 1986). In Uttar Pradesh also, disease escape in mid-October sown crops and severity increased with delayed sowing (Singh and Singh, 2006).

Field Sanitation

Collect and burn the infected plant debris particularly the staghead and other malformed parts, lying in the field after harvest, to reduce the primary source of inoculums. Destroy volunteer *Brassica* species and other volunteer weeds from the field and surrounding areas.

Deep Summer Ploughing

After harvesting the crops, plough down the left over inoculums deep into the soil. Summer ploughing exposes the pathogenic materials lying beneath the soil surface to hot sun and reduce the sources of inoculums.

Cropping System

Rotate the crop with rabi cereals, pulses, non cruciferous vegetables and oilseeds. Avoid growing of *Brassica* crops continuously in or around the same field. Practice varietal rotation to avoid intensification of disease. Long term rotations eliminate soil inoculums more effectively.

Seed

Use certified seed of varieties recently released for the zones. While using own produced seed, use bold, shining, true to the variety seeds obtained from a disease free crop. In case of any doubt, sieve the foreign material and under size seed out and treat the seed with metalaxyl (Apron 35 SD) @ 6 g/kg seed. Choose at least two or more varieties and change them every two year, so that pathogen gets diverse genotype and ultimately disease spread would be checked.

Nutrition

Apply farmyard manure or compost to balance the nutrient supply. Practice green manuring, using sunhemp or dhaincha (*Sesbania* spp.) after harvesting mustard. These, in addition to nutrients, add organic matter to soil, which serve as substrate for the native biocontrol micro-organism. It increases the water holding and nutrient retention capacity of the soil also. Add NPK fertilizers, where necessary, to make up for a 120: 60: 60 kg dose per hectare. Avoid use of excess nitrogen, as it predisposes the crop to most of the diseases.

Thinning and Weeding

Weeds serve as a carrier of disease and high plant density favour the disease development. Hence, thin down plants with in the rows after 15 to 20 days of sowing, to maintain plant-to-plant distance of 15 cm. Remove the weeds twice in season. Rogue out the mixture and stray diseased plants, as these serve as foci of infection.

Irrigation

Excess irrigation must be avoided. Provide only need-based light irrigation, generally two, as high humidity and frequent irrigation favour diseases development. Drain out excess water, if any, from the field promptly.

(B) Host Resistance

Host resistance is the most important means of disease control and has been used in two ways to study variation in resistance of the host itself and to study variation in pathogenicity of the pathogen.

Sources of Resistance

The supply of gene (s) for disease resistance is the first concern in an ongoing resistance breeding programme. Considerable exertions have been made to evaluate cultivars resistance to *Albugo* in oilseed *Brassica* crops. Williams and Pound (1963) reported that resistance in radish cvs. China Rose Winter (CRW) and Round Black Spanish (RBS) is governed by a single dominant gene. Humaydan and Williams (1976) identified a single dominant gene, Ac-1 in radish cv. Caudatus. Bonnet (1981) reported that radish cv. Rubiso-2 contains a single dominant gene.

Fan *et al.* (1983) reported that the resistance of *B. napus* cv. Regent is governed by three genes, Ac 7-1, Ac 7-2 and Ac 7-3. In *B. napus* a number of resistance sources to *A. candida* in India have been reported, *e.g.* Gulivar (Gupta and Singh, 1994; Saharan *et al.*, 1988), GSL-1 (Saharan, 1996, 1997), GLS-1501 (Gupta and Singh, 1994; Saharan, 1996, 1997), HNS-1 (Bhardwaj and Sud, 1993; Saharan *et al.*, 1988), HNS 3, Tower (Bhardwaj and Sud, 1993; Saharan, 1996), HNS 4 (Jain *et al.*, 1998; Saharan, 1996, 1997), Midas (Chahal, 1991-92; Jain *et al.*, 1998; Saharan, 1996, 1997; Saharan *et al.*, 1988), Regent (Bhardwaj and Sud, 1993; Saharan *et al.*, 1988), GSB 7006, Norin 14 (Gupta and Singh, 1994), H 715, HNS 1, Norin, Tower 1, 2, 3, 4 (Saharan *et al.*, 1988), EC 174243, GSB 101, GSL 706, HNS 8, Tower 60 (Jain *et al.*, 1998), Altex, EC 129126, EC 129127, EC 131625, EC 131626, EC 132121, GSA, GSB, HNS 1, Karat, Mary, Nikolas, VR-OLGA, VR-WW-1313 (Bhardwaj and Sud, 1993 and Bhowmik, 2003).

In *Brassica juncea* a number of resistant sources to *A. candida* have been reported from different places namely CSR 142, Domo-4, EC 126741, EC 126745, EC 126746, EC 129126-1, KOS-1, RC 781, PHR-1, Scimitar, SSK-1, T-4, YRT-1, Zem-1, Domo, Lethbridge, Sikkim Sarson-1, YSIK-741, DIR 519, DIR 1507, EC 126743-2, DIRA 313-7, RN 246, DOMO, YS-7b, Zem-2, EC 126743, EC 126743-1, EC 129121-1, RH 8541-46, RW 81-59, Blaze, Metapolka, Newton, Purbiraya, Stoke, RC 1001, RC 1405, RC 1408, RC 1424, RC 1425, RH 8545, WRR 3-1, SV 7739035 x RH 30-12-15, SV 7739035 x RH 30-10-3, SV 7739035 x RH 30-16-6, SV 7739035 x RH 30-2-17, Chamba-1, CSR 741, RC 295, RC 398, RC 1424, RC 1499, ZEM-1, DIR-1501, DIR-1571, RAURD-9710, PHR-2, PBR-2, PBR-92, RGN-3, RGN-13, RK-2002, RAURD-9503, RAURD-9505, JYM-10, NDRE-4, WRR-9801, EC 126126, EC 126747, RC 1401, RC 1499 (Parul and Bandyopadhyaya, 1973, Kaushik and Saharan, 1980, Kolte, 1987, Srivastva and Verma, 1987, Saharan *et al.*, 1988, Wood and Petrie, 1989, Lakra and Saharan, 1989, Yadav and Singh, 1992, Jat, 1992, Bhatia, 1994, Jain, 1995, Gupta and Singh, 1994, Saharan, 1996, Jain *et al.*, 1998, Dang *et al.*, 2000, Saharan 2000, Bhowmik, 2003, Li *et al.*, 2007, Kumar *et al.*, 2009, Singh and Singh, 2013).

In *B. campestris* different genotypes such as Prain, YST-6, Tobin, NDYS-2, YSB-9, YSWB-8925 and *B. campestris* var. toria genotypes PT-303, YSK-8502, BS 15, BSH-1 were resistant to white rust infection (Kolte and Tiwari, 1980, Singh and Kolte, 1981, Kolte *et al.*, 1985, Kolte, 1987, Jain, 1995, Jain *et. al.* 1998, Bhowmik, 2003, Kumar *et. al.* 2009, Singh and Singh, 2013).

In *B. carinata* HC 1, HC 2, HC 3, HC 4, HC 5, HC 7, HC 9001, HC-9605, PC-2, PC-3, PC 5, DIR 1510, DIR 1522 and PBC 9921 were reported resistant to this disease (Bhardwaj and Sud, 1993., Saharan *et. al.* 1988, Jain, 1995., Jain *et. al.* 1998, Gupta and Singh, 1994, Saharan, 1996, 1997, Dang *et al.*, 2000; Saharan, 2000; Bhowmik, 2003, Kumar and Kalha, 2005; Kumar *et al.*, 2009).

In *Brassica napus* genotypes Tower, GS-7027, HNS-4, HNS-10, Midass, Norin (Dang *et al.*, 2000); EC-3389, PBN-2001, EC-339000, PB-2002, DGS-1, HNS-9605 and GSL-1 (Kumar and Kalha, 2005; Kumar *et. al.* 2009, Singh and Singh, 2013) were rated as resistant against this disease.

(C) Botanical/Biological

Field and laboratory experiment were conducted by using botanicals *Azadiracta indica, Ocimum sanctum, Datura stramonium, Alium sativum* and commercial product Ovis and Zetron. *Azadiracta indica* was most effective amongst all (Meena and Jain, 2002). Meena *et al.* (2003) evaluated the efficacy of plant extracts (Garlic and *Acacia nilotica*) and isolates of *Trichoderma viride* against white rust in field experiment and reported application of garlic extract at par with mancozeb in reducing white rust incidence and number of staghead per plot.

Five botanical *i.e. Eucalyptus globosus, Parthenium hysterophorus, Datura stramonium, Azadiracta indica* and *Calotropis procera* were tested against white rust and compared with mancozeb. All the plant extract were effective to control the disease and increase the seed yield. Amongst three *E. globosus, A. indica* were at par with mancozeb which was the most effective (Godika and Pathak, 2009). Yadav (2009) also evaluated extracts of *Alium sativum, A. cepa, A. indica, C. procera* and *E. globosus* and reported *A. sativum* as most effective in controlling the severity of diseases.

(D) Chemical Control

In the absence of resistance to white rust in commercial cultivars of rapeseed and mustard in India, the use of chemicals constitutes the most effective means of its control. At least, two to three spray applications of protectant fungicides like mancozeb (diathane M-45), zineb (diathane Z-78), polyram, copper oxychloride (miltox, blitox-50), thiram, captan, captafol (difolatan), benomyl, thiovit (sulphur), cumin, thiophenate-methyl (topsin-M) and chlorothalonil, *etc.*, starting from the first appearance of the disease and then followed by one or two applications during the flowering stage of the crop provide effective control of both foliar and floral infection, and increase crop yield (Bhowmik, 1985; Dubey and Mishra, 1994; Gupta *et al.*, 1977; Gupta and Sharma, 1978; Kapoor and Sugha, 1995; Lakra and Saharan, 1988[b]; Perwaiz *et al.*, 1969; Verma, 1987). In Sikkim, India, blitox-50 was found to be the most effective amongst 9 fungicides, both in controlling the white rust disease and in increasing the yield of mustard (Srivastava and Verma, 1989). Similarly, minimum rust intensity and floral infection, and highest seed yield of Indian mustard were obtained following mixed application of topsin-M (0.25 per cent) + blitox-50 (0.10 per cent) in a field trial at Bihar (Dubey, 1996). Both blitox and zineb are effective against mixed infection of white rust and downy mildew (*Peronospora parasitica*) on Indian mustard (Bains and Jhooty, 1979).

Among the systemic chemicals, metalaxyl (ridomil) is an excellent product possessing both protective and eradicative activity against *A. candida* (Dueck and Stone, 1979; Stone *et al.*, 1987). Pre-sowing treatments of seeds with metalaxyl-35 (apron SD-35) at 5-8 g a.i./kg provides protection to seedlings against primary infection from germinating oospores (Mathur *et al.,* 1991; Saharan, 1984; Sharma and Kolte, 1985; Stone *et al.,* 1987), but it was ineffective in providing protection to plants against white rust which appeared after 70 days of sowing in the field (Bhowmik and Singh, 1984). As soil drench, metalaxyl @ 0.25-1.0 kg a.i./ha prevents foliar infection of seedlings though it may show slight phytotoxic effect (Stone *et al.,* 1987).

Spray application of metalaxyl at 0.2 per cent conc. controls both foliage infection and staghead formation, reduction in the latter type of infection becomes more evident when plants are sprayed at the flowering stage (Stone *et al.,* 1987; Lakra and Saharan, 1988[b]). Both metalaxyl (ridomil) and readymix (redomil + mancozeb 75 WP) at 0.1 per cent conc. provide effective control of white rust, the two are highly persistent on Indian mustard (cv. Pusa bold) and show both upward and downward movement in plants including their absorption through roots. Out of two other systemic chemicals namely, aliette (fosetyl) and pervicur (promocarb), the former is comparatively less effective than metalaxyl, shows only upward movement and absorption through roots while the latter is ineffective (Saharan *et al.,* 1984; Verma, 1987, Singh and Singh, 2013).

At the IARI, New Delhi, two spray applications with either 0.2 per cent dithane M-45 or 0.1 per cent readymix on mustard cv. Pusa bold at fortnightly intervals starting from mid-January onwards were highly effective against white rust and in improving seed yield/plant, and readymix performed better in respect to both disease control and plant yield (Verma, 1987). In the rainy season at Karnataka under high pressure of white rust disease, seed treatment plus two sprays with metalaxyl MZ at 2 g a.i., gave maximum disease control (94 per cent reduction of leaf infection and 96 per cent reduction of floral infection) and contributed to a 81.3 per cent increase in seed yield of mustard (Hegde and Anahosur, 1993).

Metalaxyl-25 (ridomil-25) shows complete compatibility with captafol and metasystox, the chemicals, respectively effective against Alternaria blight and aphid pest (*Lipaphis erysimi*) of rapeseed and mustard. Application of metalaxyl-25 either or in conjunction with captafol or with captafol + metasystox provided significant reduction in intensity, respectively of white rust, white rust + Alternaria blight, and of these two diseases and aphid infestation beside increased seed yield. Mixed application of these three chemicals proved most effective in achieving not only an excellent control of these two diseases and aphid infestation but also in getting the highest seed yield per plot, which was 123 per cent higher than that from control (Bhowmik and Singh, 1984).

Spraying of ridomil MZ (mancozeb + metalaxyl) alone resulted the lowest white rust severity with highest seed yield and benefit cost ratio (Bhargava *et al.,* 1997; Pandey *et al.,* 2000; Godika and Pathak, 2001; 2002; Khunti *et al.,* 2001; Singh and Singh, 2005 [a,b;] Singh and Bhajan, 2005). Efficacy of ridomil MZ was also evaluated with protectant fungicides to manage the white rust (*Albugo candida*) in different geographical location of the country and was noted most effective in reducing the foliar and floral infection and enhancing the seed yield (Khangura and Sokhi, 2000; Yadav, 2003; Singh and Singh, 2006). Application of apron (metalaxyl 35 ES 6 ml/

kg + ridomil gold (metalaxyl) at 0.25 g/liter resulted lowest incidence of white rust and highest seed yield (Meena *et al.,* 2003).

Kumar *et al.* (2009) treated the seed with metalaxyl (Apron) 35 SD @ 6 g/kg seed and sprayed the crop with different concentrations of Propiconazole, Iprobenphos and Azadiractin and reported minimum white rust intensity with three sprays of Iprobenphos 48 EC @ 0.10 per cent. Propiconazole 25 EC 0.05 per cent sprayed thrice, with highest benefit cost ratio proved most economical.

Powdery Mildew

The disease can be seen on oilseed *Brassicas* throughout the world. In India, Butler (1918) noticed first a severe outbreak of the disease in 1907 over large irrigated areas in Chenab canal colony of Punjab. However during last few years the disease could assume an epidemic form in *Brassica* growing areas of the country. In Rajasthan, Haryana and Punjab, its occasional outbreaks in epidemic form causes considerable losses in crop yield (Saharan and Kaushik, 1981; Singh *et al.,* 1984; Shankhla *et al.,* 1967; Sharma 1979). In Gujarat and in some non-traditional areas of the south, the disease is emerging as a serious threat to the crop in recent years, while in eastern states of the country; it is only sporadic in nature and economically unimportant (Bhowmik, 2003). In mid eastern India occurrence was noticed in late sown crops in late seasons (Bhowmik, 2003 and Singh and Singh, 2003).

Economic Importance

It is generally believed that the disease does not cause much damage in rapeseed and mustard crops except in occasional severe outbreaks, when all the leaves and siliquae get covered with the powdery growth of the fungus. In certain states of India such as Haryana and Rajasthan, the disease has been found to occur quite severe, resulting in considerable loss in yield. In mid eastern India observed that the crop is severely affected from powdery mildew disease, a minor disease earlier, , has become major one during recent years in late sown conditions and cause heavy yield losses up to 42.4 per cent (Singh and Singh, 2003; Singh and Singh, 2004, Singh *et al.,* 2016). Considering the differences in disease intensity from year to year, it appears the loss is proportional to the disease intensity, which varies conversably depending on the stage at which it occurs (Kolte, 1985).

Symptoms

All the above ground parts of the plant are attacked. In the beginning small, white circular powdery patches consisting mainly of conidiophores and conidia of the fungus appear on both surfaces of lower leaves and stem of the plant (Sharma, 1979). As the day temperature rises, the powdery patches increase in size, coalesce and rapidly spread to green stem, branches, leaves and pods giving the entire plant a dirty white appearance (Figures 8.7A and B). Leaves turn yellow and shed (Singh and Solanki, 1974), flowers abort and green siliquae, which in the initial stage of infection, show scattered white patches and later get completely covered white mass of mycelium and conidia. Such siliquae remain small in size and produce few seeds at the base only with twisted sterile tips (Figure 8.7C). Later, towards the end of March, cleistothecia in the form of small black bodies, scattered and in

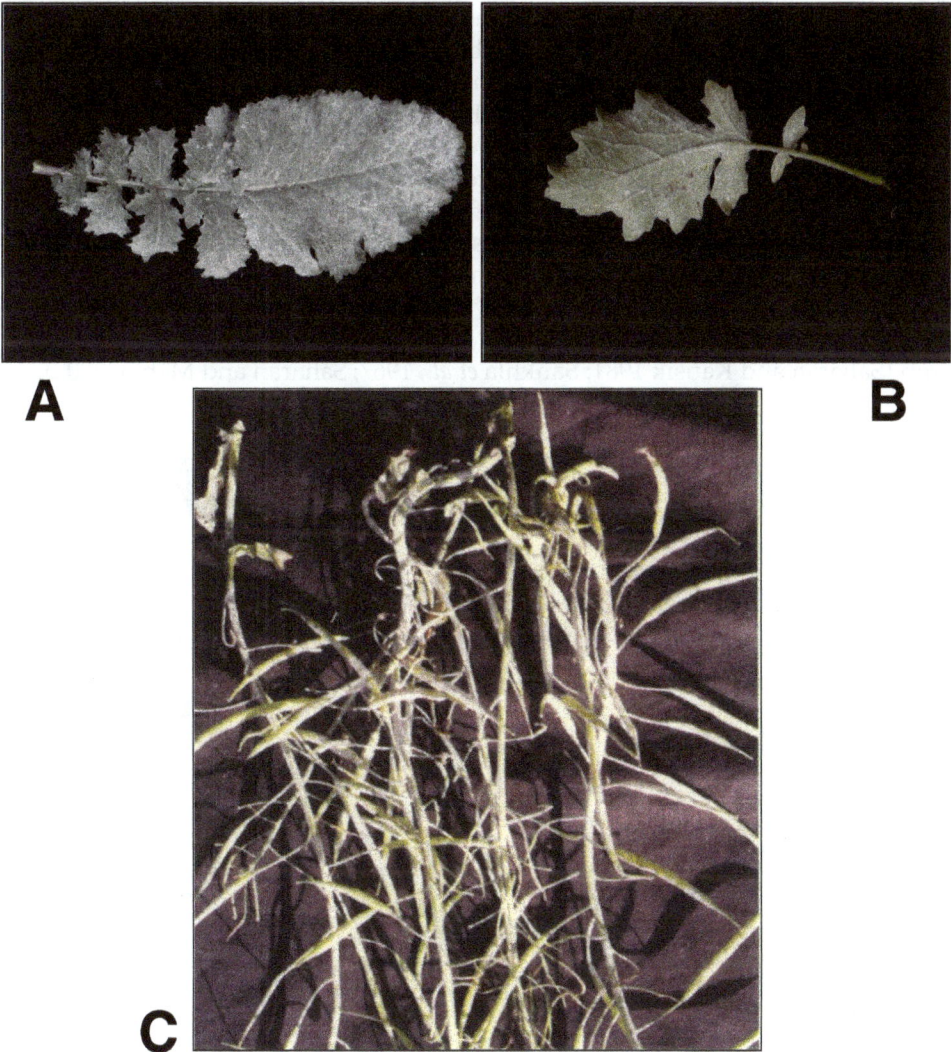

Figures 8.7A-C: A: Powdery Mildew Symptom on Upper Surface, B: Lower Leaf Surface and C: Powdery Infection on Siliquae.

small groups formed on the infected host surface (Saharan and Kausik, 1981; Singh and Singh, 2004).

Causal Organism

The disease is caused by *Erysiphe cruciferarum* (Opiz. Ex Junell). The pathogen belongs to division Eumycota, subdivision Ascomycotina, class Pyrenomycetes, order Erysiphales and family Erysiphaceae.

The pathogen is an obligate parasite. The mycelium of the fungus is superficial, amphigenous, white, dense, septate, branched, spreading and persistant. Conidia are hyaline, borne singly or in short chains, 25-45 x 12-16 μm in size and cylindrical in shape. Clestothecia are scattered, globose, at first yellowish orange becoming brown to dark brown and black with maturity, 90-130 μm in diameter. Cells of clestothecia are irregular, brown and 10-25 μm in diameter. Appendages are numerous, myceloid, present all over the surface of clestothecia, narrow hyaline to faintly coloured, seldom branched, often unequal in length up to three time the size of the clestothecia. Asci, oval to pyrifrom 3-12 in number, usually 6-8, with an indistinct stripe and 50-70 x 30-45 μm in size. Ascospores are ovoid, 2-7 in number and 16-22 x 11-14 μm in size. Conidia germinate only from one end (Figures 8.8A–F). The germ tube is normally branched having upto three branches. Perfect stage of the fungus develops rarely but it occurs quite abundantly on late sown mustard crop (Saharan and Kausik 1981; Sankhla *et al.*, 1967; Saharan and Mehta, 2002).

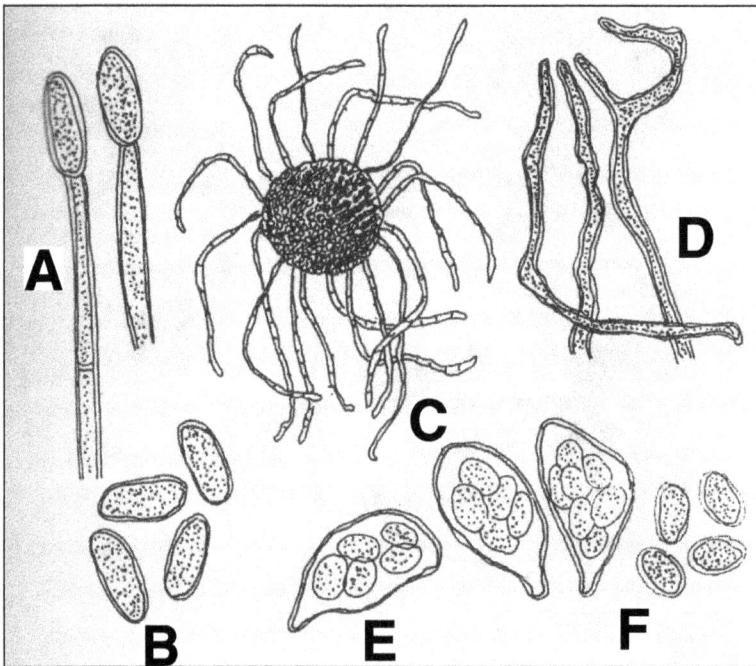

Figure 8.8A-F: *Erysiphe cruciferarum:* **A: Conidia on Conidiophores; B: Conidia; C: Perithecium; D: Appendages; E: Asci Containing Ascospores; F: Ascospores (Courtesy: Saharan and Mehta, 2002).**

Disease Cycle

The pathogen produces a large number of clestothecia on diseased plant tissues at the maturity stage of the crop. The possibility of clestothecia surviving on diseased plant debris and other weeds in the field to serve as source of primary inoculums. However it is yet to be confirmed. The secondary spread of the pathogen takes place through air borne conidia. Long distance dissemination of the pathogen takes place through wind currents under low humid conditions. Conidia fallen on the host tissues germinate. The germ tube enters the host and fungus grows and spreads in

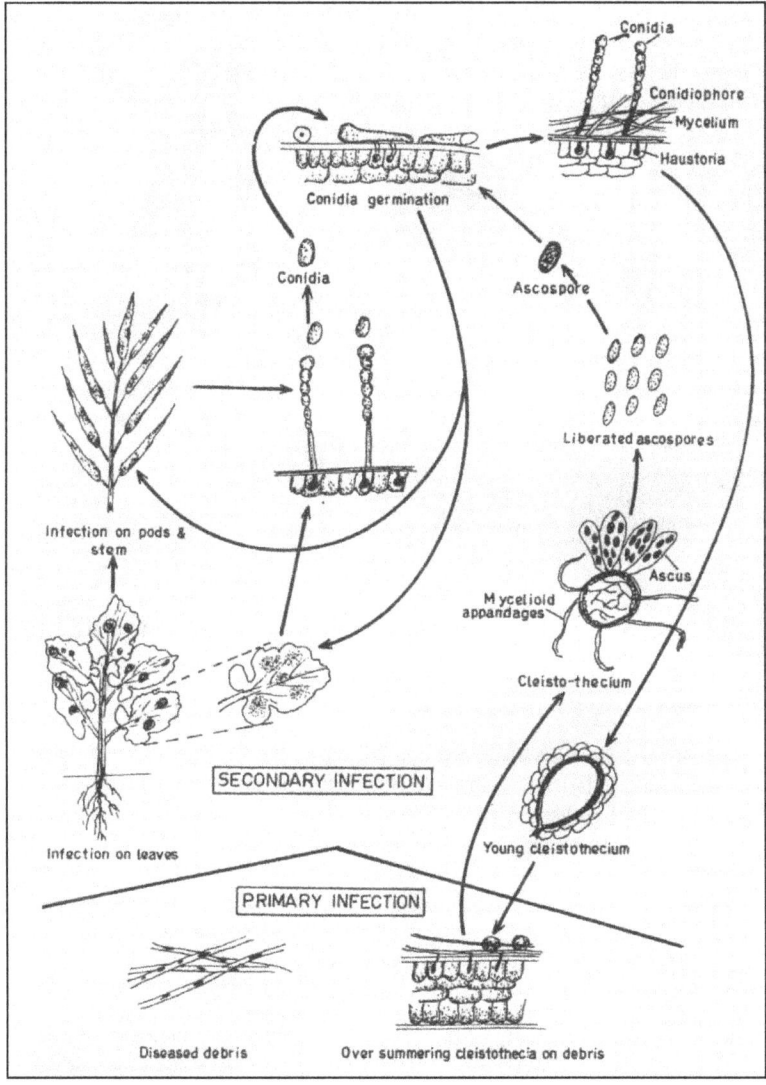

Figure 8.9: Disease Cycle of Powdery Mildew of Crucifers (Courtesy: Saharan and Mehta, 2002).

the form of mycelium on the host surface, later producing conidiophores and conidia in the form of white mildew growth (Figure 8.9). Clestothecial formation appears to be favoured by alternating low and moderate temperatures, low nutrition of host, low relative humidity, dry soil and aging of the host (Saharan and Mehta, 2002).

Epidemiology

The optimum temperature for the germination of conidia, germ-tube growth and appressorium formation is 20-25°C. Conidia could not germinate below 15°C and above 30°C. Maximum conidia germinate at 40 to 50 per cent RH. To initiate the spore germination in *Erysiphe cruciferarum* at least 30 per cent RH is essential and there is no conidial germination above 60 per cent RH. Conidial germination is not influenced by light and darkness. For onset and epidemic development of disease under field conditions, moderate temperature, low humidity, minimum rainfall or dry season during the months of February and March are more favourable in Haryana. Mean temp between 16-28°C, mean RH below 60 per cent and low or no rainfall are the most congenial weather factors for the development of the disease under field conditions. It has been observed that maximum cleistothecial formation is favoured by alternating low and moderate temperature. Heavy sporulation took place with low nutrition of the host, low relative humidity, dry soil and aging of the host (Singh and Singh, 2004 and Mehta, 2014). Dange *et al.* (2003) reported that early planted crop *i.e.* in the month of October, resulted in less severity as compared to late sown conditions. The average mean temp, average number of trapped conidia and crop age has been identified as most responsible factors for the disease development (Solanki *et al.,* 1999 and Bhowmik, 2003).

Disease Management

Cultural

Croping System

Rotate the crop with *rabi* cereals or pulses and sow it early after harvesting rice or other crops.

Field Sanitation

Collect and burn the infected mustard crop debris lying in the field post-harvest, again, to reduce soil inoculums. Remove self sown *Brassica* spp. and other volunteer plants from the field and surrounding areas.

Date of Sowing

Sowing of crop in the month of October to make it escape powdery mildew incidence. Late sown crops are attacked more severally. Timely sown crop escapes infection since it matures by the time pathogen assumes epidemic form.

Nutrition

Apply farmyard manure or compost to balance the nutrient supply to the crop. In their absence practice manuring, using sunhemp or *dhaincha (Sesbania.)* after harvesting mustard. These, in addition to nutrients, add organic matter to soil, which serve as substrate for the native bio-control microorganism. It increases the water holding and nutrient retention capacity of the soil also. Add NPK fertilizers, where necessary, to make up for a 120: 60: 60 kg dose per hectare.

Irrigation

Frequent irrigations prolong vegetative phase of the plant, thus delaying its maturity and increasing disease severity at later stage. Give only 2-3 irrigations, depending upon soil type and crop requirement.

Resistance Sources

There is a total lack of powdery resistant mustard varieties. Three *Brassica* species namely *B. alba*, *B. carinata* and *B. napus* are reported to be immune to the powdery pathogen. *B. juncea* genotypes, EC 129126-1, EC 129126-2, PR 8805, DIR 621, IJWHJ 001, PCR 10, PCR 9201, RK 8602, RAUD 101, RK 8615, YSPB 24 (Singh and Singh, 2003); *B. rapa* genotypes, Candle, Tobin, Torch; *B. carinata* genotypes, DHC 1, Wester, DHC 4 DHC 9601, CH1, PCC 1 and *B. napus* genotypes, GSL 1501 and N20-12-2 have been reported to be resistant to both powdery mildew and white rust under artificial field inoculation conditions (Saharan and Krishna, 1999). Single resistant gene present in HC 1 and PCC 2 controlled the resistance with complete dominance. Dang *et al.* (2000) found various genotypes, namely *B alba*, Gullivar, HC-1, DIR-1507, DIR-1522, TM-5-50, Tower, Gs-7027, HNS-4 and Midas resistant to powdery mildew. Mehta *et. al.* (2008) reported that four varieties *viz.*, HC-1, HC-2, HC-9605 and HC-9603 belonging to *B. carinata* and two genotypes *B. alba* and *S. alba* did not contracted disease whereas five varieties GSH-1, Midas, MNS-9605, YSPb-24 and TH-68 were considered as resistant. Singh *et al.* (2016) also screened 200 genotypes out of which, 20 genotypes, namely, EC-414309, PBC-9221, NPC-14, BAUSM-92-1-1, PBC- 2004-1, EC-339000, EC-338997, GSL-1, HNS-004, ONK-1, NRCDR-515, PBC-2002-2, NPC-15, OCN-3, EC-399299, NUDB-26-11, CAN-133, NPN-1, RGN-55, NRCR-837 showed consistently resistant reaction.

Chemical

The disease has been effectively controlled by using timely spray of foliar fungicides. Spraying of Karathane (0.1 per cent) or Calixin (0.1 per cent) or Sulfex (0.3 per cent) at an interval of 10 days immediately after appearance of the disease is very effective to minimize the disease intensity and increase the seed yield. Time and number of sprays are very important for effective and economical control of the disease. There is more than 95.0 per cent reduction in disease when Karathane (0.1 per cent) is sprayed twice at an interval of 10 days soon after the appearance of the disease. Singh and Singh (2003) and Singh *et al.* (2016) also reported three sprays of aqueous suspensions of Triademefon (0.1 per cent), Wettable sulphur (0.3 per cent), Dinocap (0.1 per cent) and Myclobutanil (0.1 per cent) at 15 days intervals, after disease appearance, controlled the disease effectively and increased the seed yield. Lomate *et al.* (2014) was also reported earlier in rapeseed-mustard.

Downy Mildew

The downy mildew disease of oil seed *Brassicas* causes significant damage to young plant up to 5-15 cm height, and then floral parts are also attacked which may become sterile. The disease causes 37-47 per cent losses in pods and 17-32 per cent in seed yield in *B. juncea* when occurs along with white rust as mixed infections (Bains and Joothy, 1979). It occurs in almost all *Brassica* growing areas of the country and is very common in mustard (Singh and Singh, 2005a).

Economic Importance

In order to assess the loss in yield due to downy mildew or white rust alone or due to both (Kolte *et al.*, 1981) in India considered the extent to which the raceme is affected. The length of the hypertrophied raceme (without silique), and seed yield from the healthy raceme in a corresponding length is considered as the loss in yield on that particular affected raceme. In this way, on an individual plant basis, loss in yield due to simultaneous occurrence of these two diseases in toria (*B. campestris* var. *toria*) is estimated to be about 34 per cent, if the average length of the individual hypertrophied raceme is 10cm (Sharma, 1980., Kolte *et al.*, 1981., Kolte, 1985).

Symptoms

The disease appears on all above-ground plant parts but usually on cotyledons, leaves and inflorescence. Generally a few days after sowing, small angular translucent light green lesions first appear on cotyledon or the first few true leaves in a seedling stage. Such lesions later enlarge and develop into grayish-white, irregular necrotic patches on the leaf, bearing downy growth of the fungus (conidia and conidiophores) on its under-surface. In a severe attack the affected leaf dries up and shrivels. The downy mildew symptoms are also found associated with white rust symptoms show the maximum alterations. Stems of the disease plant become deformed and distorted. The stalks are abruptly bent. Among floral parts sepals, petals and stamens become shrunken and thus more often floral buds are atrophied. This leads to sterility of the flowers. The downy mildew symptoms are also found associated with white rust symptoms on leaves (Figures 8.10A-C). Usually the downy growth of fungus appears in or around the white rust pustule but when the downy lesions appears first on the leaves, the appearance of white rust pustules in or around the downy mildew lesion is seldom seen (Kolte, 1985 and Saharan, 1992).

The most conspicuous and pronounced symptom is the thickening of the peduncle/inflorescence due to downy mildew, suggesting the hypertrophy of the affected cells. Vasudeva (1958) reported that the pith of the stems get comparatively more hypertrophied than the cortex. The affected inflorescence does not bear siliquae or produce abnormal siliquae, which are often curled without seeds. In the initial stage the affected inflorescence does not show the presence of oospore and remains without downy growth on its surface. But at the latter stage the affected plants show on its surface presence of the conidial fungal growth and formation of oospores takes place in the tissue as it dries. Under natural conditions, mixed symptoms of downy mildew and white rust are usually seen on the same inflorescence (Bians and Jhooty, 1979; Kolte and Tiwari, 1979). Usually mustard (*B. juncea*) plants show the mixed symptoms more frequently than the rapeseed (*B. campestris*) plants.

Downy mildew re-appears, occasionally in and around the white rust pustules on leaf and stems in the mid season (February) but almost invariably on hypertrophied floral organs and siliquae, in the end of season (March-April) when the weather gets little warmer and humid.

Mixed Infection and Association with White Rust

According to Bains and Jhooty (1985), conidia often appear first in combined infections. It is followed by infection with *P. prasitica*, which develops in and around *A. candida* colonies. *A. candida* predisposes the host tissues towards susceptibility to this pathogen. The development of hypertrophied tissues of the staghead phase is attributed to infection with *A. candida*. The intensity of mixed infection by

Figure 8.10A-C: A: Downy Mildew Infection Suppreses the Siliquae Formation; B: Mixed Infection of Downy Mildew and White Rust; C: Heavy Mixed Infection on Siliquae (Courtesy: Shukla *et al.*, 2003).

A. candida and *P. parsitica* on *B. juncea* inflorescence has been reported to be from 0.5 to 29.0 per cent under Punjab (India) conditions (Bains and Joothy, 1979 and 1985). The hypertrophropied malformed floral organs of mustard infected with *A. candida* are usually heavily covered with white sporulating fungal growth of *P. parasitica* consisting of conidida and conidiophores. Incidence and severity of mixed infections by *A. candida* and *P, parasitica* on *B. juncea* inflorescence is higher on detopped than on normal plants (Figure 8.10C). There is preferential parasitism of *P. prasitica* on galls of *B. campestris* caused by *A. candida*. Severity of mixed infections on leaves is not related to infections on inflorescence. It seems that greater susceptibility of new inflorescence and their availability over extended periods of time is associated with this phenomenon. The associations of downy mildew and white rust infection on oilseed *Brassicas*, vegetable *Brassicas*, wallflowers and stocks have long been observed.

In artificially inoculated leaves of mustard, the stimulatory effect of *A. candida* infection is more intense when *P. parasitica* is inoculated 7 days after *A. candida*. When *Peronospora* and/or *Albugo* are inoculated alone or in different combinations, the downy mildew infection takes 7 days, while white rust appears within 5-6 days inoculations. When both the pathogens are inoculated simultaneously in a 50: 50 spore concentration then there is delay in the expression of infections by *Peronospora* for 2-3 day (Mehta *et al.*, 1995[a]).

Histopathological studies carried out by Mehta *et al.* (1995a), indicated conidia of *peronospora* and sporangia of *Albugo* inoculated on mustard leaves germinate 24 hr after inoculation. Two days after inoculation, infection is normally confined to the host epidermis. The pathogens penetrate up to 1/3 of the mesophyll cells by the third day after inoculation. Six days after inoculation, the pathogens progress deeper into the tissues. When *P. parasitica* is inoculated prior or after *A. conidida*, mycelium can be seen in the intercellular spaces with globose to knob like haustoria in the mesophyll cells. When *A. candida* is inoculated alone or in combination with *P. parasitica*, the pathogen emerges from the lower epidermis and forms pustules. However, on its own *Peronospora* causes necrosis in the mesophyll cells. When both the pathogens are inoculated together, the infection is confined to the upper layer of the mesophyll with limited colonization of the cell and few haustoria or mycelium in the intercellular spaces. Nine days after inoculation, characteristic disease symptoms are visible. The white rust pustules show hyaline sporangiophore bearing globose to oval shaped soprangia in chains. The *Peronopora* mycelium is intercellular with lobe shaped haustoria in the distorted tissue of leaves. When both pathogens are inoculated together, infection is extended to mesophyll cells and there is development of pustule below the epidermis. Twelve days after inoculation, complete colonization of the host tissue by the pathogen is evident from the development of necrotic zone by *P. parasitica* and bursting of pustules releasing sporangia in case of *A. candida*. In the inflorescence, the mycelium passes through the epidermis, hypodermis and cortex and finally reaches the pith region. The mycelium is in abundance in the cortex and produces conidiophores bearing conidia above the epidermis layer. For *A. candida*, numerous sporangiophores bearing sporangia are observed below the epidermis layer in the form of pustules and knob like haustoria in the tissues. In the colonized tissues both pathogens cannot be distinguished based on somatic morphology (Saharan and Mehta 2002).

The Casual Organism

The fungus causing the disease, *Peronospora parasitica* (pers.) ex. Fr. (Chromista, Oomycota, Oomycetes, peronosporales, peronosporaceae) is considered species consisting of many smaller morphological species. The species occurring on *Brassica campestris* was often called *Peronospora brassiceae*. More recently, the fungus attacking crucifers has been named *Hyaloperonospora parasitica* on the basis of hyaline conidiophores, recurred conidiophores, branch tips and rDNA sequence comparisons (Constantinescu and Fatehi, 2002).

The hyphae are branched, coenocytic, hyaline, intra cellular with large clavate or finger shaped branched hostoria. Numerous more or less erect, branched sporangiophores emerged through stomata on the lower surface of the leaves. These are usually absent in floral parts. However, in some hosts including mustards conidial production on the floral parts is profuse covering the surface with a white downy growth which appears like powdery mildew. The conidiophores arise vertically and their base is flattened, somewhat twisted where it passes through the stomata opening. These structures are 100-300 µm long and unbranched major portion of their length. Dichotomous branching occurs, 6-8 times, at the tip. The strigmata are long, slender and pointed. They are at acute angles with each other. Single conidia are borne at the tip of each branch. The conidia are broadly oval, ellipsoidal, hyaline, and measure 24-27 x 15-20 µm in size (Singh, 2009).

Physiologic Specialization

As describe by Saharan and Mehta (2002), the isolates of *P. parasitica* differ in the range of cruciferous host that they can infect. Strains of *P. parasitica* are known to be both homothallic as well as heterothallic, therefore a good reason to suppose that the pathogen possesses substantial capability of genetic variation. Specialization of parasitism may be exhibited at generic and lower taxonomic levels of the host. Different pathogenic varieties or farmae specialis of the fungus have been recognized on species of *Raphanus* (var./f. sp. *raphani*), *Brassica* (*P. parasitica* var/f.sp. *brassicae*), and *Capsella* (var./f.sp. *capsellae*). Gaumamn (1926) also recognized three formae speciallis: one, *P. parasitica* f. sp. *brassicae* infecting and sporuling on various *Brassica* spp. (*e.g. B. oleracea, B. rapa, B. napus*) second, (f. sp. *sinapis*) infecting *Sinapis arvensis* and *S. alba* and a third (f.sp. *raphani*) infecting *Raphanus sativas* and *R. raphanistrum*. Physilogic specialization at generic level of the host is widely reported on *Brassica* and *Sinapis*. At the species level, Wang (1944) separated *P. parasitica* var. *brassicae* into six specialized forms, which differed in their ability to infect various *Brassica* spp. Chang *et al.* (1964), indicated the existence of separate races 1 and 2 of the fungus. According to Sherriff and Lucas (1987), isolates of *P. parasitica* from *Brassica* are in general, specifically adapted to their species of origin. Differential host resistance to homologous isolates of *P. parasitica* has been identified in *B. campestris, B. oleracea* and *B. napus*. Sherriff and Locas (1990) further obtained isolates from same host species but different geographic origins and showed that these differed in host range.

According to Natti *et al.* (1967), the predominant physiologic race of *P. parasitica* pathogenic to braccoli and other types of *B. oleracea* grown commercially in New York are race 1 and race 2. The later race is pathogenic to plant resistant to race-1. Dickinson and Greenhalgh (1977) observed a variation in the reaction of seedlings of different crucifer species to isolates of *Peronospora* derived from *Brassica* and *Raphanus* species.

In India, *P. prasitica* isolates from different host plants vary in host range. Isolates from *Brassica, Raphanus, Eruca* and *Sisymbrium* are not cross infective (Bains and Joothy, 1983). Mehta and Saharan (1994) tested the host range of 9 isolates of *P. parasitica* collected from the leaves and stagheads of 6 plant speceis on 17 host differentials. Isolates from oilseed *Brassica* infected all species except *B. alba*, whereas isolates from cauliflower leaves did not infect *B. carinata, B. nigra, B. chinensis. B. perkinensis* and *B. napus*. There was no significance difference among the conidial size of the isolates collected from leaves and stagheads but significant differences were observed among these groups. The isolates were classified into two distinct pathotypes, one from cauliflower and other from oilseed *Brassicas*. There was no significant difference between the isolates in percentage of spore germination.

In the United Kingdom, differential host resistance in relation to pathogenic variation of isolates derived from the same host species has been identified in *B. rapa* (Moss, 1991, Silue *et al.*, 1996) *B. napus* (Nashaat and Rawlinson, 1994), *B. junea* (Nashaat and Awasthi, 1995[a]) and *B. oleracea* (Silue *et al.*, 1996). Isolates from different *Brassica* species found to be most virulent on their species of origin were nevertheless able to grow on other *Brassica* species to a lesser extent (Sherriff and Lucas, 1990). Nashaat and Awasthi (1995[b]) identified five groups of *B. juncea* accessions with differential resistance to U.K. isolates R1 and P003 derived from oilseed rape (*B. napus sp. oleifera*) and Indian isolates IP01 and IP02 derived from mustard (*B. juncea*). All *B. juncea* accessions were resistant to isolates from *B. napus* but at the same time *B. napus* cv. Arian was resistant to isolates from *B. juncea*. Twenty-one differential responses to *P. parasitica* isolates from *B. olreacea* and two from *B. rapa* were identified. Of the seven isolates tested, four were from crops of cauliflower in France, two from oilseed rape in U.K. and one was from mustard in India. All *Raphanus sativas* accessions were resistant to all seven isolates (Silue *et al.*, 1996).

Disease Cycle

The pathogen perennates in the soil through oospores that are formed in abundance in the malformed tissues of the infected plants. Seeds may be contaminated with plant trash containing oospores during threshing operation. Infection originated when such seeds are sown after getting suitable temperature and relative humidity. Oospores formed in malformed and senesced host tissues constitute an important means of survival of *P. parasitica* over periods of unfavourable conditions. It is also known to survive through mycelium and conidia (Figure 8.11). The presence of *P. parasitica* mycelium in the seed coat in case of Chinese cabbage has been recorded by *Chang et al.* (1963). Oospores formation is abundant in the infected tissues of all crucifers, which form primary source of the pathogen. In radish and rapeseed-mustard, there is abundant production of oospores in infected leaf tissues, on the seed surface and pericarp and embryo of seeds. However, in rapeseed and mustard, seed transmission is low and may be non-systemic, ranging from 0.4 to 0.9 per cent in the seedlings grown from infected seeds (Saharan *et al.*, 2005 and Naresh, 2014).

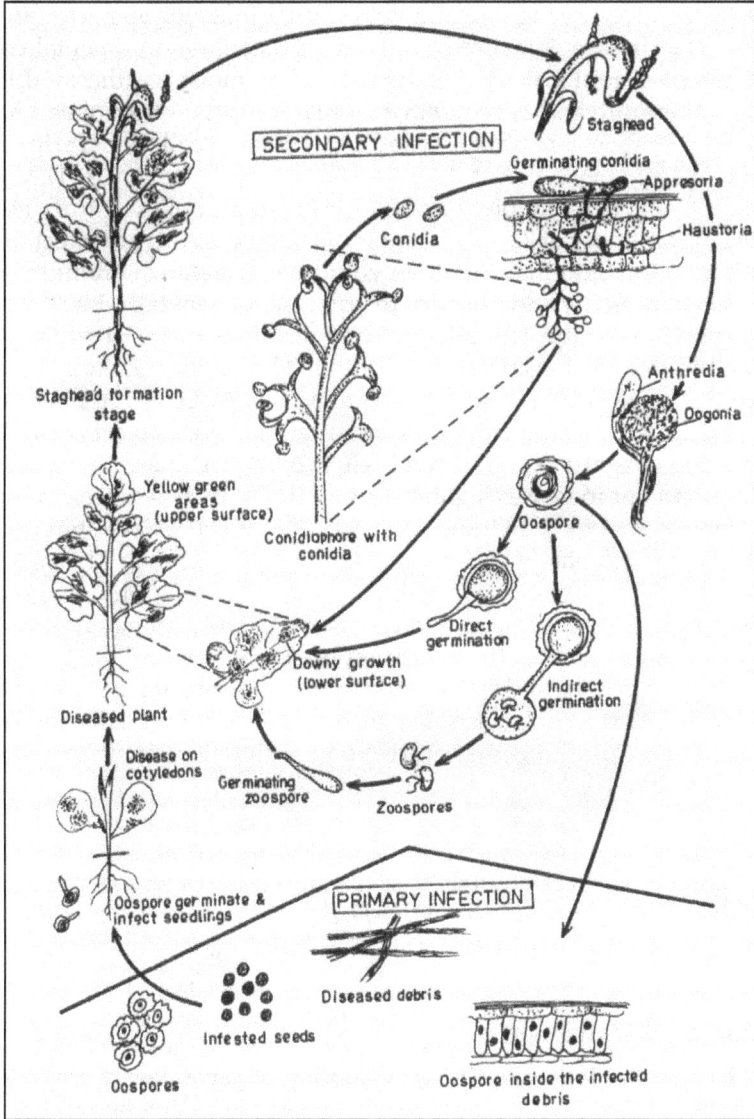

Figure 8.11: Disease Cycle of Downy Mildew of Crucifers (Courtesy: Saharan and Mehta, 2002).

Epidemiology

Mixed infection with white rust pathogen; *Albugo candida* and downy mildew pathogen, *Peronospora parasitica* on *Brassica juncea* inflorescence, because of incubation period of *A. candida* is 6-7 days and that of *P. parasitica* 3-4 days is common to find the same plant infection. *Peronospora* produces both sexual (oosperes) and asexual (conidia) spores, which are helpful in survival and dissemination of the pathogen. The nature of the disease is infectious, infected plants are seen randomly distributed throughout the field. Floral infection increases in the late sown crops. The disease is favored by damp and cool environmental condition. More infections occur at low temperature (8-16°C), moist weather and low light intensity. Temperature and rainfall have a great impact on the appearance of the disease. At low temperature of 14.3°C and 151.9 mm rainfall the severity increases and may go upto 26 per cent but it declines to only 0.5 per cent at 17°C and low rainfall of 50-80 mm. penetration of the host occurs most rapidly at 16°C. Haustorial development is optimum at 20-25°C. Sporangia exposed to air are infectious for six weeks but direct sun light may kill them within 5-6 h., relative humidity above 70 per cent helps in rapid development of the disease. In general, 15-20°C have been found to be best temperature for infection and development of downy mildew. At this temperature regime infection occurs within 24 hr of inoculation. The infection frequency is reduced at 25°C with no infection at 30°C. Leaf wetness duration of 4-6 hr at 20°C and for 6-8 hr at 15°C is essential for severe infection and disease development on mustard (Saharan and Mehta, 2002). The infection frequency and disease development increases significantly with the increase in duration of leaf wetness (Mehta *et al.*, 1995[b]). In India, infection of mustard foliage starts by the end of October (Cotyledon stage) and progresses up to November. The crop planted after mid November may not be infected. However, downy Mildew growth as a mixed infection with white rust on floral parts can be seen up to March (Mehta and Saharan, 1998). Bains and Joothy (1978) reported that downy mildew pathogen produces heavy infections and sporulation in plants of *Brassica juncea* systemically infected with mustard mosaic virus. Sangeetha and Siddaramaiah (2007) reported 26-29°C range of maximum temperature along with relative humidity more than 65 per cent favoured the downy mildew development.

Host Range

All the cruciferous vegetables show severe infection at seedling stage. However, *B. rapa* var. yellow *sarson*, brown *sarson*, *toria*, *B. nigra*, *B. rapa*, *B. juncea*, *Eruca sativa*, *Raphanus sativus*, *Cardamine impatiens* and *Sisymbrium iris* show infection at various stages.

Evaluation of Germplasm

As described by Bhowmik, (2003) evaluation of germplasm/lines carried out both at the seedling and adult stages for their reaction to downy mildew disease. Seedlings are inoculated at the cotyledonary stage with zoospore suspension of *P. parasitica* and then rated for their disease reaction according to either 0-5or 0-9 rating scale as described for white rust disease. Grontoff (1993) used a rapid screening method for testing large number of breeding lines of *Brassica*. Several hundred breeding lines *Brassica* species were sown on green house tables filled with soil. After one week, the cotyledons were artificially inoculated and covered with a plastic tunnel. About two weeks later, the average percentage of wilted leaf area was estimated and the disease response of the lines evaluated. In India, the adult plants are rated for their reaction to mixed infection of white rust and downy mildew as they occur together under field conditions according to the method described for white rust disease.

Genetics of Host-Parasite Interaction

As describe by Saharan and Mehta (2002) resistance derived from the broccoli Introduction No. PI 189028 to *P. parasitica* race 1 is found to be governed by one dominant gene. The distribution of resistant plants in populations segregating for both downy mildew resistance and waxless foliage indicates that resistance is independent of foliage wax. Resistance to race 1 and race 2 obtained from the cabbage introduction PI 245015 is found to be inherited independently and is governed by one dominant gene for each race (Natti *et al.*, 1967). However, in later study, Hoser-Krauze *et al.* (1991) found that in broccoli (*B. oleracea* var. *botrytis*) resistance to a Polish isolate of *P. parasitica* at the 4-5 leaf stage is determined by a single recessive gene different from genes determining resistance at the colyledon stage. Subsequently they found that in broccoli resistance to downy mildew is governed by 3 or 4 dominant complimentary genes. Both seedling and mature plant resistance has been reported in *B. oleracea* with the later quantitative.

Differential host resistance to homologus isolates of *P. parasistica* has been identified in *B. rapa (B. campestris) B. nopus, B. juncea* and *B. oleracea*. In *B. napus*, resistance in the oilseed rape cultivar Cresar is controlled by a single dominant allele (Lucas *et al.*, 1988). In *B. oleracea*, differential resistance has been located in a land race cauliflower "Pelermo Green".

According to Nashaat (1995[a,b] and 1996) resistance of the Res-01-1-4 and Res-26 lines of *B. napus* to isolate P003 at *P. parasitica* is conditioned by a single dominant resistant gene, whereas resistance of Res-02 is conditioned by two independent dominent resistant genes. Later, Nashaat and Awsathi (1995) selected differential putative homozygous resistant lines from seedling populations of accessions that exhibited a hetrogeneous reaction to the isolates from *B. juncea* (Saharan and Mehta, 2002).

Disease Management

Cultural and Chemical

Clean and healthy seeds should be sown. Crop rotation and deep ploughing in summer help in reducing the primary inoculums load in soil. Weed host should be eradicated. Spraying of fungicides like Diflolatan (0.2 per cent), Indofil M-45 (0.2 per cent), Dithane Z-78 (0.2 per cent), Thiovit (0.2 per cent) and Ridomil MZ (0.25 per cent) at an interval of 15 days after appearance of the disease have been reported to reduce disease intensity and increase in yield (Saharan and Mehta, 2002). Three to four sprays of the chemical can effectively manage the disease. First spray should be given soon after the appearance of the disease on cotyledonary leaves, since it may initiate the systemic infection, which becomes unmanageable. Chemicals reported to be effective against white rust are also good for downy mildew disease management. Seed treatment with Apron 70SD (35 per cent Metalaxyl and 35 per cent Captan) (1g kg^{-1}), Apron SD-35 (2g a.i. kg^{-1}) along with foliare sprays of Ridomil MZ-72 at 30 days intervals give best disease control (Mehta *et al.*, 1996 and Singh and Singh, 2005[b]).

Sources of Resistance

Brassica alba, *B. carinata* (HC-1), *B. juncea* (DIR-1507, DIR-1522) and *B. napus* (GS-7027, Midas and Tower) posses multiple disease resistance to downy mildew, while genotypes PBC-9221, MC-9605, RL-1359, DIR-447, GS-52, RGNC-1, HNS-9601, NUDB-38, NUDB-42, PBN-2001, RC-17, RC-41, RC-89, B-115, B-229 were reported resistant only to downy mildew (Singh and Singh, 2005ª, and Kumar *et al.*, 2009).

Club Root of Crucifers

The club-root or finger and toe disease is mo st destructive disease of crucifers, especially cabbage and closely related plants. The disease is widespread in those areas having temperate climate. However, it is not much prevalent in India. It was observed in Kalimpong hills and West Bengal and is wide spread in neighboring Sri Lanka. It is an important disease of oilseed brassicas in France, Germany, Canada, Sweden, Newzeland, U.K. and other countries. In mustard, the loss may be as high as 50 per cent or more. Once the disease appear in the area, it becomes difficult to raise economically important crops of crucifer in that locality. Affected plants completely damage and give total loss (Singh, 2009). In Nepal, the disease is severe and widespread.

Symptoms

The disease symptom generally noted after about 60 days of sowing. Affected plants show normal healthy growth at the initial stages, but as the disease progresses, the plants become stunted showing pale-green or yellowish leaves. The plants died within a short time. When such plants are pulled, the most characteristic symptoms of the disease, appearing the main and lateral roots become visible in the form of small and large spindle and spherical knobs, called clubs. Depending on the type of root of a crop species, the shape of the club varies. The swelling may be few and isolated or they may be coalese together to cover the entire root system (Figure 8.12). The infected roots contain large numbers of resting spores. The older, and particularly larger, clubbed roots disintegrate before the end of the season due to invasion of bacteria and other saprophytic fungi.

Casual Organism

The disease is caused by *Plasmodiophora brassicae* Worn. It is an obligatle biotrophic fungus. They are hyaline and spherical and each germinates to produce a single biflagellate zoospore and thallus gives rise to zoosporangia. Swollen root contain large number of spherical, small resting spores and when the roots decay, they are released into the soil after decay of the roots.

The life cycle of *Plasmodiophora brassicae* completes into two phases *i.e.* primary and secondary phase, which occurs in the root hairs and cortical cells of root, respectively. Pathogen survives as resting spores on decayed roots. Resting spores are hyaline, spherical measures up to 4µm diameter. It germinates and produces zoospores with two flagella of unequal length, both of whiplash type and the usual 9+2 arrangement of fibrils. The primary zoospores attach to root hairs and infect them. The flagella coil around the body which becomes flattened against the

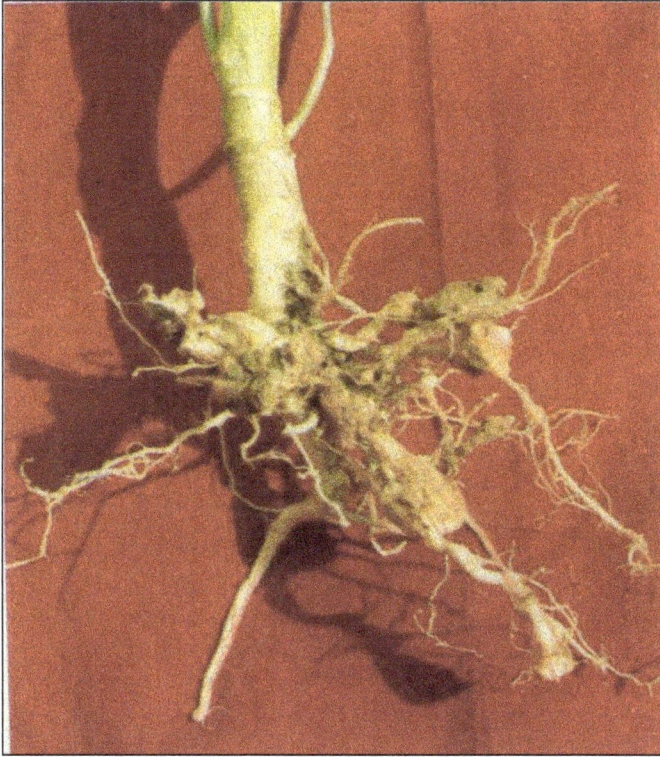

Figure 8.12: Symptom of Club Root Affected Root of the Plant.

host wall, and pseudopodium-like extensions- adhesoria of zoospores develop. The flagella are withdrawn and zoospore encysts, attached to the root hair. The myzxamoeba is now injected into the root hair cell and is caught by cytoplasmic streaming. Within the infected root hair, the amoeba divides to give rise to several uninucleate amoebae. Later the nuclei divide to form small, multinucleate plasmodia. These are sporangial or primary plasmodia. They divide to form variable number of roughly spherical, thin-walled zoosporangia, lying packed in the host cell. Each zoosporangium finally contains 4-8 nucleate zoospores. The mature zoosporangium becomes attached to the host cells wall and a pore develops at this point through wide the zoospores escape in the soil. Occasionally the zoopores are released into the lumen of the host cell. Further behaviour of the released zoospores is not completely known, but it is believed that they function like gametes and fuse in pairs to form quadriflagellate, binucleate zoospores. Karyogamy does not follow plasmogamy (Sharma, 2006 and Alexopolous, *et al.*, 2007).

The binucleate quadric flagellate zoospores re-infect the root to form binucleate plasmodia which penetrate the root cortex. Plasmodium enlarges, repeated nuclear divisions take place and the host cells become hypertrophied. Plasmodia are then transformed into masses of resting spores. The multinucleate mature plasmodia contain many haploid nuclei associated in pairs. Prior to resting spore formation, karyogamy occurs to give diploid zygote and after that meiosis takes place. Thus, resting spores, obtained from plasmodial cytoplasm are haploid which later surrounded by a thin cell wall and are closely packed together inside the host cell which released into the soil as the roots decay (Sharma, 2006 and Alexopolous, *et al.*, 2007).

Disease Cycle

The pathogen survives as resting spores lying freely in soil or in crop debris. In the field, the disease spreads by drainage water, farm implements, wind blown soil, moving animals and most important, seedlings raised in infested soil. In areas where pond water is used for irrigation, this pond water serves as source of primary inoculum, as the ponds receive runoff water from infested fields. Resting spores may remain viable for as long as 10 years. They germinate poorly or not at all in alkaline soils. The pathogen is a potent soil invader. The pathogen can also survive on some crucifers weeds, like wild mustard, which may serve as primary source of inoculums (Sharma, 2006).

Disease Management

Cultural practices are an important method, which help in reducing the disease incidence. Eradication of cruciferous weeds like wild mustard, which may serve as collateral hosts and follow very long term crop rotation with non-cruciferous crops (more than 5 years). Field sanitation by burning the root debris of crop after harvest of disease has occurred in crop season. By deep ploughing in summer season, use of recommended varieties for specific region, balance dose of fertilizer, proper drainage in the field, need base irrigation, reduced the incidence of disease

Some volatile chemicals, such as vapam, methyl dibromide, benomyl *etc.* reduce the resting spores in soil. Use of smaller particle size lime in soil to maintain soil pH 6.7 to 7.9 or above by addition lime gives good control. Lime is added about six weeks before planting. Application of lime @ 10-20 tone/ha increased soil pH from 6.7 to 7.9 and reduced disease incidence. By using Gypsum (calcium sulphate) and sodium carbonate, similar results can be obtained. Workers in Taiwan reported good control of the disease by amending the soil with S-H mixture @ 0.5 to 1 per cent by weight of soil consisting of sugarcane bagasse, rice husk, oyster shell powder, urea, potassium nitrate, calcium superphosphate and mineral ash (Sharma, 2006).

Soil application of PCNB (Pentachloronitrobenzene) formulations, like Terraclor, Quintozone and Brassicol has some suppressive effect against the pathogen. PCNB (75 per cent) should be applied @ 30 kg/ha. The soil application should be done a week before planting. Bhattacharya and Pramanik (1998) reported *Pseudomonas flourescens* H_{237} and *Streptomyces graminofaciens* GIH reduced the size and growth of clubs as well as rate of root hair infection process. Neem oil suspension applied through soil was phytotoxic but neem cake and nemin reduced the disease severity.

Sclerotinia Rot

Sclerotinia diseases occur worldwide and effect plants at all stage of growth, including seedlings, mature plants and harvested products. In north India also many vegetable and oilseed crops suffer from rot and wilt diseases caused by *S. sclerotiorum* during cool weather with heavy dew in winter. The pathogen has a very wide host range attacking more than 350 plant species belonging to more than 60 families. The losses vary with the percentage of the plants infected and the stage of the growth of the crop at the time of infection and host susceptibility. Plant infected at the early flowering stage produce little or no seeds and those infected at the late flowering stage will set seed and may suffer little yield reduction (Bhowmik, 2003, Sharma, 2006 and Singh, 2009).

Symptoms

The disease attacks all the above ground parts of the plants. Even roots may be infected (Saxena and Rai, 1986, 1988). Infection at the plant base leads to collar rot (Figure 8.13A). Usually, the first symptoms are seen on leaves or stem as light brown water soaked spots which rot in moist weather. In case of stem, infection starts as elongated water soaked lesions, which latter on covered by a cottony mycelial growth of the fungus (Figure 8.13B). When the stem is completely girdled by such lesions, the plants wilts and dies. The leaves arresting from the infected patches droop down and shed. Numerous sclerotia appear embedded in the external mycelium and within the rotted stem. Infected pods become similarly covered with superficial mycelium, resulting no seed formation but often contains mycelium and sclerotia. The severely affected parts rot and shred into fibres, the stem may break at the point of infection or the entire plant collapses and becomes covered with mycelium and sclerotium (Figures 8.13C and D). The sclerotia fall on the ground and persist in the soil or get harvested with seeds. In dry weather infections remain restricted as necrotic spots (Davies, 1986 and Bhowmik, 2003).

Pathogen

The pathogen is *Sclerotinia sclerotiorum* (Lib.) De Bary (*Syn. S. libertiana* Fuckel; Whetz- elinia *sclerotiorum* (Lib.) Korf and Dumont). Mycelium is thin, 9 to 18 µm in diameter with lateral branches of smaller diameter than the main hyphae. The vegetative hyphae are multinucleate with the haploid having chromosome number eight. The sclerotia are black, round or semi-spherical in shape measure 30 to 10 mm (Figure 8.14). which, are formed terminally and produced one or two or more concentric rings on agar media. The fungus produces some micro-conidia in culture media.

On germination the sclerotia form stalked apothecia. One to several apothecia may grow from a single sclerotium. Apothecium produces asci and paraphyses. The asci measure 119 to 162.4 X 6.4 to 10.9 µm in size. Each ascus formed eight ascospores of uniform size, with the haploid chromosome number being eight. They measure 10.2 to 14.0 X 6.4 to 7.7 µm in size. Each ascospore is hyaline, ellipsoid, and has smooth walls. The paraphyses are about 100 µm in diameter, slightly swollen at their tips; multinucleate, sparsely septate, and occasionally branched at the bases also affect the disease incidence and development (Kolte, 1985 and Sharma, 2006).

Figures 8.13A-D: A: Sclerotinia Stem Rot Affected Plants Sowing Fungal Growth on Girth; B: Sclerotinia Stem Rot Affected Fungal Growth on Stem; C: Sclerotinia Stem Rot Affected Plant having Sclerotia in the Pith; D: Sclerotia in the Pith in Large View (Courtesy: Shukla *et al.*, 2003).

Disease Cycle

The pathogen survives as sclerotia, on soil surface in unploughable fields or in crop debris or as a mixture with the seed. The fungus sclerotia may remain viable in the soil for at least 8 years (Davis, 1986 and Mehta, 2014). In the next season the crop infected by ascopores produced after germination of the sclerotia. The pathogen spreads from field to field or from one geographic area to another by wind-borne ascopores and soil adhering to seedlings, farm equipment, animals or man. The pathogen may also spread through manure, if the sheep or cattle are fed with

Figure 8.14: *Sclerotinia sclerotiorum.*

contaminated feed. The pathogen has a vast host range such as vegetables, forage, ornamentals crops, trees and shrubs and numerous herbaceous weeds.

Sclerotia are the main source of primary inoculum. Survival of sclerotia is influenced by source of and location on the substrate, soil type, soil moisture and temperature, and soil treatments like flooding, fumigation *etc*. In the plains of north India, sclerotia remain viable till next season. However, in cool and wet *Tarai* areas of U.P. sclerotia present on or in soil decompose rapidly, and only those present in stem of infected plant debris, germinate in next season. Soil temperature, moisture and depth influence the germination of sclerotia. Sclerotia lying upper surface of the soil and in soil with much sand, and moisture content of 30 per cent, germinate very quickly. In north India outbreaks of the disease may occur by ascospores from germinating sclerotia during mid-winter (Sharma, 2006). Secondary infection can also be induced by the natural contact of disease and healthy plants in close planting of the crop (Figure 8.15).

Epidemiology

Species of *Sclerotinia* can function as either soil borne or air borne pathogen. Infection of above ground plant parts results from ascosporic inoculum whereas soil borne infection may result from either ascospores or sclerotia. Below ground infection however, results from mycelial germination of soil borne sclerotia. Continuous moisture for about 10 days are required for apothecial development and even a slight moisture tension prevent apothecial formation. Apothecia of *S. sclerotiorum* are produced at an optimum temperature of 15°C and ascospores survive at a wide range of conditions but high temperature and humidity reduce the viability (Clarkson *et al.*, 2001). No apothecial initials are produced at either 30°C

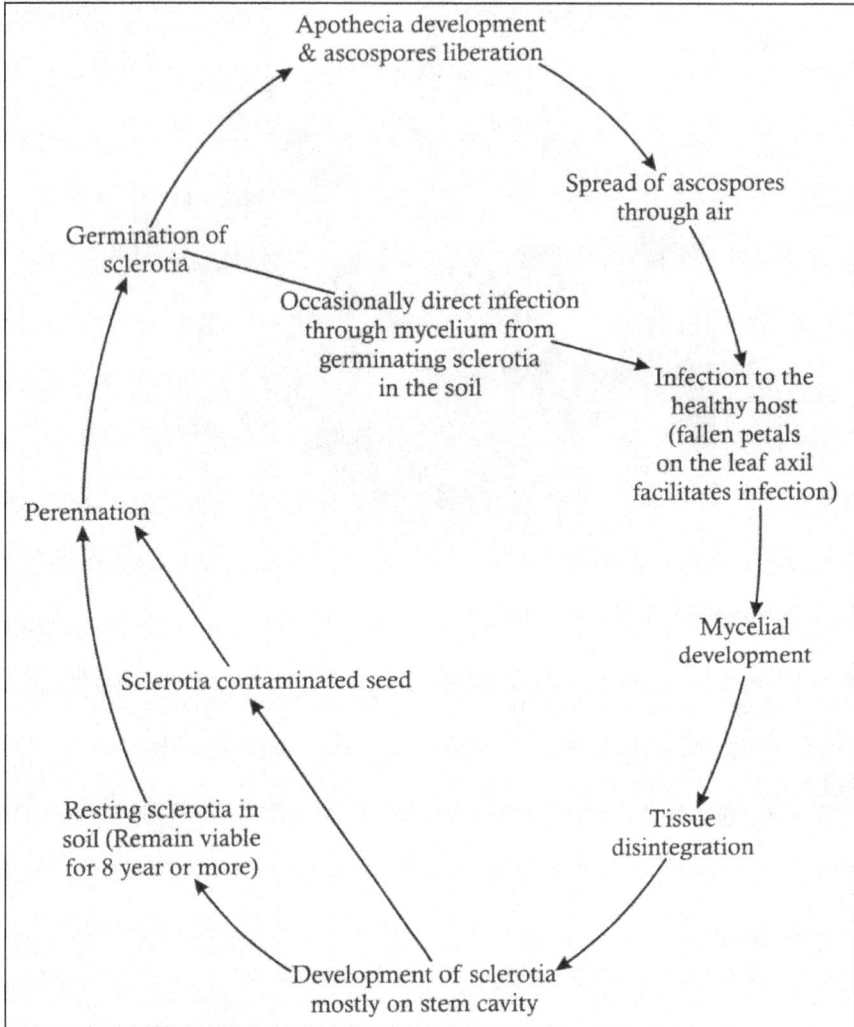

Figure 8.15: Disease Cycle of Sclerotinia Stem Rot (Courtesy: Shukla *et al.*, 2003).

or 5°C. Approximately 48-72 h of continuous leaf wetness is required for infection by ascospores. The infection by *S. sclerotiorum* on yellow *sarson* and in *B. campestris* var. toria got aggravated by low temperature, heavy rainfall and close spacing in Uttar Pradesh and Bihar (Saxena and Rai, 1988). The study conducted in Canada revealed that increase in seeding rate increased the disease intensity. The lodging of plant also increased when seeding rate exceeded 6.7 kg/ha. This is due to fact that higher seeding rates modify the micro environment and increase the potential of lodging and may be responsible for plant-to-plant spread of the disease (Jurke and Fernando, 2006). Temperature of 6 to 10°C during March and April and high soil moisture until the apothecia have developed, with subsequent changing weather favours infection. Ascospores released and petal fall should occur at the same time (Kruger, 1980). Morrall and Dueck (1982, 1983) have reported severe infections in the fields with few or no apothecia.

The pattern of petal fall and petal deposit on leaves suggests that the crop is most vulnerable to infection towards the end of flowering about 25days after the beginning of flowering in the UK (Mc Cartney *et al.,* 2001[a]; 2001[b]). The role of extrinsically produced ascospores in causing disease in rapeseed fields may therefore be of considerable importance (Morrall and Dueck, 1982; 1983; Williams and Stelfox, 1979; 1980). Accordingly, ascospore concentrations above the crop canopy and on plant surfaces might reflect the disease potential in a crop better than the density of apothecia in the field. Gugel and Morrall (1986) demonstrated a significant positive relationship between petal infestation at early bloom and disease incidence. Infested petals and disease incidence regularly found when apothecia were absent, thereby demonstrating the rate of extrinsically produced ascospores in the infestation of crops. Flowers of rapeseed from the time expand and retained on their petals on an average for 6days. During this period, the petals "in situ" get contaminated by ascospores of *Sclerotinia*. Infection takes place preferentially on senescent petals because young petals are resistant to a certain extent. The senescent petals have been reported to be most easily colonized (McLean, 1958) and do provide the ascospores with a source of carbon, which permits their germination (Purdy, 1958). The hyphae, which, develop subsequently, play a very important role in the initiation of infection (Kapoor, 1983). Dead petals often stick to leaves and this allows the disease to become established (Brun *et al.,* 1983).

While studying clonal dispersal and spatial mixing of *S. sclerotiorum* isolates from rape fields in Canada, it was observed that spatial mixing of ascospore inoculums from resident or immigrant sources took place (Kohli *et al.,* 1995).

Management

Cultural Practices

The crop debris and refuse of the crops should collect and burn. Cleaned seeds without the presence of sclerotia should be used. Deep ploughing, crop rotation and weed control in non-host crops are essential for preventing the soil population of sclerotia and flooding of the soil also minimize the disease incidence. Reducing the crop density and planting crops with open canopy structures are effective ways to minimize disease development. The type of the soil as well as frequency and amount of irrigation has important role in the germination of sclerotia as well as development of apothecium. It has been observed that least number of apothecia was recorded in the sandy soil whereas sandy loam soil resulted in maximum number of apothecium production (Mehta *et al.,* 2009[b]). Further, it was observed that flooding of the field (once in week) prior to sowing resulted in least disease incidence and minimum lesion length. The optimum irrigation applied once in 3 or 7 days intervals also had low disease intensity as compared to control (Mehta *et al.,* 2009[b]). Seed treatment, crop rotation between vegetable and cereals, optimum fertilizers, rational close planting, pruning of old and infected leaves reduced the disease up to 50 per cent (Mehta, 2014).

Resistant Varieties

Brassica napus and *B. juncea* cv. Rugosa have been reported to possess resistance in the field as well as in greenhouse conditions (Singh *et al.*, 1994). Nine genotypes *viz.*, Cutton, ZYR-6, PSM-169, PDM-169, Wester, PYM-7, Parkland, Tobin and Candle showed resistance to stem rot in India out of 34 *Brassica* genotypes belonging to four species (Shivpuri *et al.*, 1997). Four genotypes *viz.*, PCR-10, RW-8410, RW- 9401 and RGH-8006 had resistance against *S. sclerotiorum* (Pathak *et al.*, 2002). Sharma *et al.* (2009) evaluated indigenous and exotic genotypes of *B. napus* and *B. juncea* in infected field and observed that genotypes EC-597328 of *B. juncea* were tolerant to this disease. Sharma *et al.* (2013) observed genotypes/lines BLN 3630 (EC597274) of *B. napus* and Berry (EC597329) of *B. juncea* as resistant to disease.

Biological Control

Trichoderma harzianum in the presence of vermi-compost combined with EPTC treatment has been reported to be most promising control strategy for *S. sclerotiorum* *T. viride T. harzianum and Gliocladium virens* plus carbendazim (Bavistin) improved the plant growth and reduced the disease (Pathak *et al.*, 2001). Sen (2000) reported that application of Kalisena SL formulation in the sick plot of *S. sclerotiorum* could save the cauliflower seedlings whereas *Aspergillus niger* AN27 application parasitized and killed the existing population of pathogen and checked the sclerotial formation in the soil. The *Streptomyces* sp. isolate 422 significantly reduced the disease incidence. Antagonists applied @ 15g/kg soil as wheat bran inoculum was superior in reducing Sclerotinia stem rot (Mehta *et al.*, 2012 and Mehta, 2014).

Chemical Fungicides

Seed treatment with captan and thiophanate-methyl completely eradicate the fungus from the infected seed. Seed treatment with carbendazim was also found effective in minimizing the disease incidence (Singh, 2009).

Various chemicals applied as foliar application *viz.*, azoxystrobin, benomyl, boscalid, iprodione, prothioconazole, tebuconazole, thiophanatemethyl, trifloxystrobin and vinclozolin significantly reduced the disease incidence and improved seed yield as compared to control (Bradley *et al.*, 2006). Pereira *et al.* (1996) have reported that integrated effect of vermi-compost, soil solarization, herbicide (EPTC) fungicide (procyniodene), soil solarisation through coverage of transparent polythene (0.1 mm) for 45d has been observed to be a good control strategy. EPTC treatment significantly increased the degree of control irrespective of the depth of the sclerotia in the soil.

Mehta *et al.* (2010[a]) has demonstrated that application of SAR chemicals (isonicotinic acid or salicyclic acid @ 100 ppm) as seed treatment resulted in significant less seedling mortality, lesion length, disease intensity and disease incidence. Soil amendment with Bougainvillea leaves was effective in reducing seedling mortality, number of apothecia production, lesion length and disease intensity in mustard (Mehta *et al.*, 2011). Soil amendment with mustard cake, sesamum cake and jamun seed powder (*Syzygium cuminii*) and poultry manure has also been found effective in reducing seedling mortality, number of apothecia production, lesion length and disease intensity in mustard (Mehta *et al.*, 2010[b] and Mehta, 2014).

Black Leg or Phoma Stem Canker

Black leg is one of the most serious disease of cabbage and oilseed *Brassicas* (canola, *B. napus, B. Juncea, B. rapa*) and globally it is reported from 49 countries. On rapeseed (*Brassica napus* var. *oleifera)*, the disease is common in many temperate climate countries such as Canada, United States, Europe and Australia where the damage to the oilseed crop may be as high as 60 per cent through basal canker and logging. However, epidemics of the disease in Asia including India are rare (Singh, 2009).

Symptoms

The plant may be infected by the pathogen at any stage of their growth. In the seedling stage, the disease appears on cotyledons as small, round to irregular, grayish white lesions bearing numerous black tiny fruiting bodies or pycnidia. On leaves, lesions appear as pale to light yellow coloured, circular spots which gradually turn gray-white and loaded with pcyindia. Pink masses of pycnidiospores oozing out of pcynidia dispersed and incite further leaf infections which increasing with age of leaves (Hammond and Lewis, 1986). On susceptible cultivars of oilseed rape (*B. napus* f. sp. *oleifera)*, infection becomes systemic on leaf lamina and petiole resulting in a stem cankers near the petiole base (Hammond *et al.*, 1985; Hammond and Lewis, 1986). At later stage of plant growth, stem cankers occur at the base of the stem near to the attachment of the lower leaves (Gladders and Musa, 1980). Dry rot or dry cankers as favoured by 15-24°C and wet rot usually develop at 7-12°C temperature (Bhowmik, 2003; Rimmer and van der Berg, 2007).

Cankers on the stem are similar in appearance to leaf lesions except that they are sunken in the plant tissue and have a black or purple border. In case of severe infection, the canker expands girdle the stem base and restricted the flow of nutrients and water, which result in pre-maturity of plants. The basal lesions sometimes damage the plant completely from the root system and cause it to lodge. On the stalk and pods the lesions are elliptical; seeds are shriveled, light and discolored (Bhowmik, 2003; Rimmer and van der Berg, 2007).

Casual Organism

The disease is caused by *Leprophetia maculans*, a heterothallic loculoascomycetes, producing its asci in pseudothecia. *Phoma lingam* (syn *Plenodomus lingam*), a coeflomycete with pyccnidia the anamorphic state of the fungus. The mycelium is septet, branched hyaline when young and becoming dark walled in later stage. Pseudothecia are emerged, becoming erumpent, globose and black, with protruding ostioles and 300-500μm in diameter. Asci are bitunicate, cylindrical to clavate, sessile or with stipes, 80-125 x 15-22 μm in size. Each ascus contains eight biseriate, multi-cellular, cylindrical to ellipsoidal ascopores with rounded ends, and are yellow brown in colour, measuring 35-70 x 5-8 μm in size. Pycnidia produced are of two types- globular and sclerotioid. Globular develops in live tissues and in agar cultures, which sclerotioid develops on dead woods tissues and serve as source of survival for primary infection (Bhowmik, 2003; Sharma, 2006 and Rimmer and van der Berg, 2007).

Disease Cycle

The disease is monocyclic and the pathogen is facultative saprophyte. Primary infections starts from infected seeds or infected crop debris. Infected seeds produce plants that develop lesions with pycnidia. In some cases, pcynidiospores may cause secondary infection of the disease through splashing rain or irrigation water. The conidia can spread only up to 1 meter from their original foci. Pods infected with pycnidiospores and ascospores, can lead to seed infection. Dead infected plant debris is the substrate for saprophytic survival of the pathogen. Ascopres are produced in plant debris, and are forcibly discharged in moist conditions and can be carried by wind a long distances. Pycnidiopores as well as ascospores may cause infections, but latter are more efficient. Infected seed is the primary source of inoculums for disease epidemics in transplanted brassica vegetable crops like cabbage. Ascospores are the primary source of inoculum for blackleg epidemics in soil for seed brassicas, as rapeseed (*Brassica napus* sub sp. *oleifera*) (Kolte, 1985, Bhowmik, 2003, Sharma, 2006 and Rimmer and van der Berg, 2007).

Disease Development

The optimum temperature require for maturation of pseudothecia is 14-15°C and symptoms quickly develop at 15-28°C. Their development is much delayed at temperatures below 12°C. The production of number of leaf spot lesions develop upon leaf wetness duration and temperature. The conditions that produced the greatest numbers of leaf spot lesions area leaf wetness duration of 48 hours at 20°C. Numbers of lesions decreased with decreasing leaf wetness duration and increasing or decreasing temperature. At 20°C with 48 hours of leaf wetness, one out of 4 spores infected leaves to cause a lesion whereas with 8 hours of wetness only one out of 300 spores caused a lesion. As temperature increased from 8°C-20°C the incubation period decreased from 15-5 days. Sosnowski *et al.* (2005) have emphasized that in infection of resistant and susceptible cultivars at *Brassica napus* by pycnidiospores of *L. maculans* day/night temperatures, wetness duration and spore concentration in the inoculum have significant effects. The greatest number of leaf lesions developed on plants exposed to a day/night temperature of 18/15°C with a 96 hours wetness period. Development of stem infection is maximum at 23/20°C with a wetness period at 48-72 hours (Kolte, 1985, Bhowmik, 2003, Sharma, 2006; Rimmer and van der Berg, 2007 and Singh, 2009).

Disease Management

The disease can be managed in on integrated manner by using resistant varieties, cultural practices and chemical applications. Cultural practices are used to minimize or eliminate the ascospore production in infected crop debris. A three to four year crop rotation between susceptible crucifer and non host crops proved most effective method for disease management. Ascospore production can also be minimized by burning or burying infected plant debris. Susceptible crucifer weeds harbouring the strains of the pathogen should eradicated by herbicides. Foliar applications of fungicides are often ineffective. Fungicide can be applied during early stages of plant development. Application of zinc is found to alter the severity or delay the development of stem cankers by reducing the activity of sirodesmin toxin. Seed borne transmission can be controlled through exclusion and eradication. Inspection of seed fields and testing of commercial seeds will help excluding the pathogen and its entry to new areas. A lot of seed should be treated with aqueous or slurries of benomyl or iprodione dust @ 450-2000 g per quintal of seeds (Kolte, 1985, Bhowmik, 2003, Sharma, 2006 and Singh, 2009).

Black Rot

The disease began to alarm researchers in Europe by 1899, and today this is the most important disease of crucifers worldwide. The incidence of the disease was observed under natural condition in *Brassica juncea* in India in the year 1949. The disease has spread to all cultivated brassicas, radishes and numerous cruciferous weeds throughout the world. Black rot is present on all the crucifers in Sikkim, India and was detected on seeds of radish, rayosag (*B. juncea*) and cauliflowers (Gupta and Chaudhary, 1995). In eastern part of Uttar Pradesh the disease was first observed by Singh *et al.*, 2008, 2016 at Faizabad and its adjoining region.

Symptoms

Symptoms appear when the plants are about two months old. At initial stages, dark-coloured streaks of varying length are observed near the base of the stems above the ground level, which gradually enlarge and girdle the stem making very soft and hallow due to severe rotting and results in total collapse of the plant. Lower leaves show midrib cracking 'V' shape yellowing on the leaf margin, browning of veins and withering of the leaves (Figure 8.16). Exudation of yellowish fluid from the affected stems and leaves may also occure. The disease does not cause much disagreeable odor as in bacterial stock rot (Shukla *et al.*, 2003 and Singh *et al.*, 2016).

Causal Organism

Black rot is caused by *Xanthomonas compestris* pv. *compestris* (Pammel) Dowson. The bacterium isolate of *B. juncea* makes best growth on Nutrient Dextrose Agar. The colonies are dark yellow, circular, non-fluicid, convex, entire and opaque. The bacterium is rod-shaped, motile with a single polar flagellum, gram-negative, non-acid fast, encapsulate, aerobic, hydrolyses starch and gelatin (Figure 8.17). The optimum temperature for growth is 20 to 25°C and pH 6.6. In culture on PDA varying in age from 1-3 weeks, it measures 1.17 to 2.02µ × 0.54 to 0.99 µ. It is positive for H_2S production, catalase test. Reaction on milk is alkaline accompanied by peptonization. Produces acid but no gas from glucose, fructose, lactose, sucrose, ribose, raffinose, xylose, cellobiose, monitor and starch. Thermal death point is 58°C. (Gandhi and Parashar, 1977., Bhowmik, 2003 and Shukla *et al.*, 2003).

Disease Cycle

The bacteria survive in diseased plant debris only for some time. The survival in plant debris in soil depends on temperature and decomposition of the plant debris.

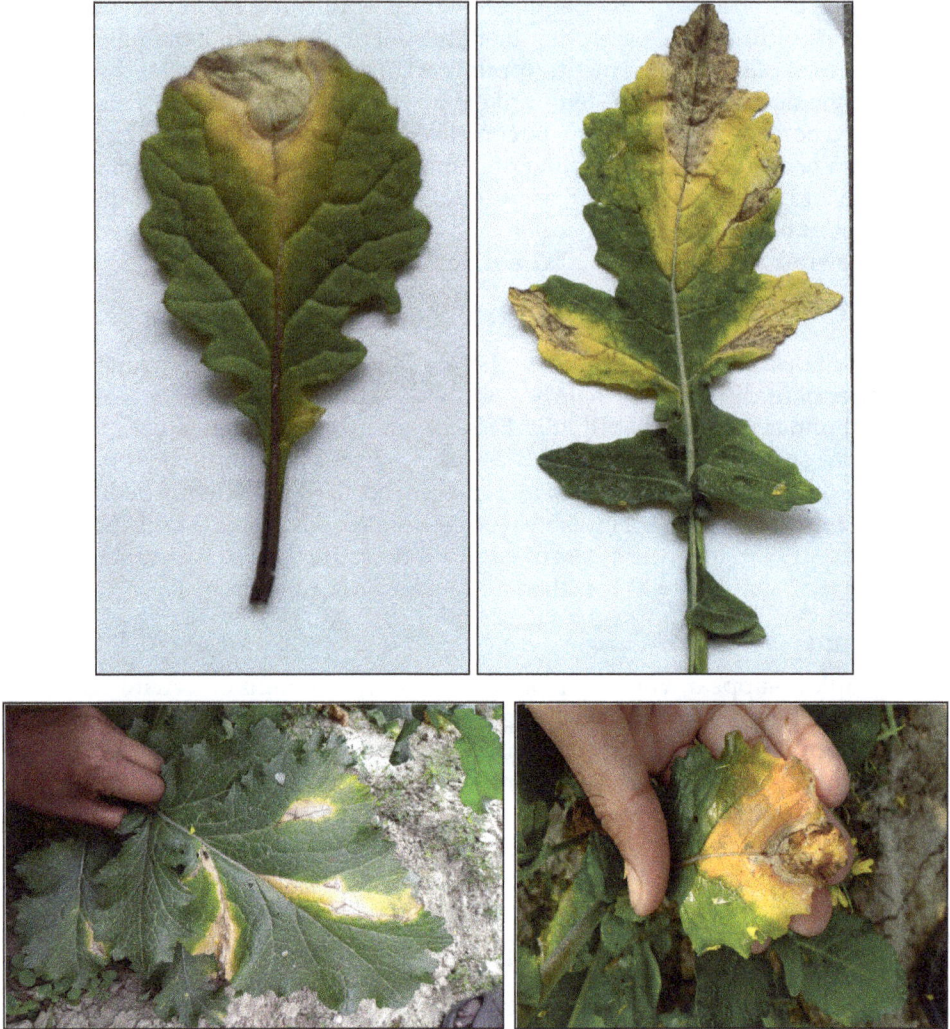

Figure 8.16: Typical Symptoms of Bacterial Rot Affected Leaf.

The bacteria can also survive as epiphytes on symptomless infected leaves of *Brassica compestris*. It survives as epiphytotic form 9 to 48 days depending on the plant species. The main source of survival of the bacterium is seed where it can survive for 3 years. When germination of infected seed occurs, the seed coat that comes out along with cotyledons serves as the source of primary inoculum. The bacteria enter the cotyledons through stomata on the margins. Then from cotyledons the bacteria pass on to true leaves. The true leaves are invaded through the hydathodas in the morning hours. On entry into the host, the bacteria move to the xylem vessels where they multiply rapidly and move up and down the plant. The bacterial cells and the (*Xanthomonas*) gum block the vessels. The secondary inoculum is spread by raindrop splashes, blowing of detached leaves, and occasionally by insects.

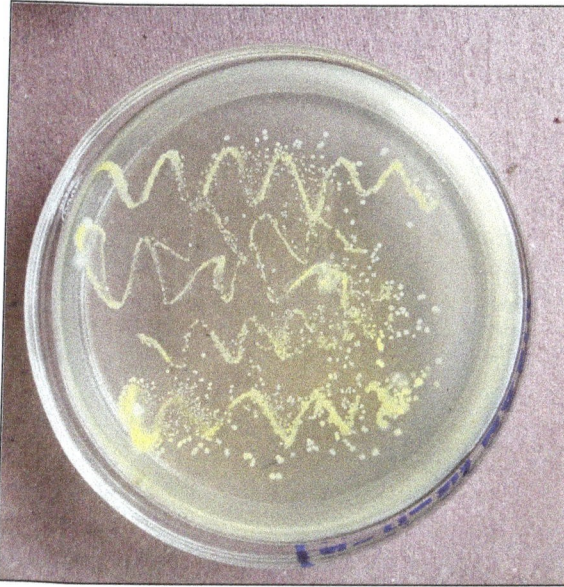

Figure 8.17: Pure Culture of *Xanthomonas compestris* pv. *compestris*.

Rains and sprinkler irrigation spread the bacteria among the plants. The optimum temperature for growth of the bacterium is about 26.5°C-30°C, minimum being 5°C and maximum 36°C. At optimum temperature incubation period is 7-14 days. At temperature below 18-20°C, symptoms do not develop. Boron status of the soil plays a role in susceptibility of plant to black rot. Susceptibility is higher in boron deficient soil (below 0.4 mg/kg) and boron-excess soil (above 1.6 mg/kg) significant reduction in black rot incidence is obtained by exogenous application of boron alone (up to 6.4 kg/ha) or with nitrogen (Bhowmik, 2003, Sharma, 2006 and Singh, 2009)

Disease Management

Black rot of mustard can be managed by eradication of the pathogen in the seed, and reduction of eradication of the inoculum in the soil as well as through breeding or developing disease-resistant cultivars. Seed borne inoculum can be reduced or eradicated by hot water treatment at 50°C for 20-30 min. and by use of antibiotics such as agrimycin (0.01 per cent), streptomycin and aureomycin (0.01 per cent) have been used. Soaking of seeds in 0.5 per cent, sodium hypo chloride solution for 30 min provide good control. Since bacteria can survive in soil for over 8 to 9 months, its eradication from the plant debris could provide only little success. Use of Copper oxychloride @ 0.2 per cent at initiation of the disease can be sprayed ~t the crop. Repeat it after 15-20 days interval if required. Singh *et al.*, 2016 ~d treatment with *Pseudomonas fluorescens* @ 5g/kg seed + spraying ~ (100ppm) thrice at 15 days interval followed by Seed treatment ~ppm) + two sprays with Streptocycline (100ppm) + Copper 15 days interval), respectively.

As free living, the black rot pathogen is very poor survivor in the soil. Rotation of crucifers with non-cruciferous crops over long period of time may help reducing the inoculum potential in the soil. Use of resistant cultivars is the most effective for the management of black rot. Early Fuji, U.S.A. and South America indicate that early Fuji is a good resistance source against the pathogen. In India, some resistant cultivars have been developed. These are 294bc, Puakea, MGS-2, -3, SN 445, Avans and Iglory. In Himanchal Pradesh the cultivars Symphony (From UK) was found resistant to black rot (Shukla *et al.*, 2003; Sharma, 2006). Singh *et al.* (2016) also reported ten genotype namely T-27, GSL-1, PHR-2, DRMR-243, PBC-9221, JMTA-9, EC-399299, EC-38899, HNS-9605 and HNS-004 as resistant from Faizabad (U.P.). Weed may be also carrier of disease. Clean crop will help in reducing the disease buildup.

Bacterial Stalk Rot

Stalk rot of rapeseed-mustard caused by *Erwina carotovora* (Jones) Holland (*E. carotovora* pv. *carotovora*) was first noticed in the Pali district of Rajasthan in 1979 up to 60 per cent (Bhowmik and Trivedi, 1980, Shukla *et al.*, 2003). The disease appeared in epiphytotic form on the popular *Brassica juncea cv.* Varuna during mid-October. On an average, about 40 to 60 per cent plants may be affected with the disease. The incidence was relatively more in crops sown during the first week of September than those planted later. Vigorously growing succulent plants, due to an extra dose of nitrogen as well as those grown on poorly drained moist soil suffered most (Kolte, 1985).

Symptoms

The symptoms of disease are characterized by the appearance of water-soaked lesions at the collar region of plants which is usually accompanied by a white frothing. The lesions advance rapidly upwards affecting the tender branches. The foliage shows signs of water stress and withering. The pith tissues of the affected stem and branches become soft, pulpy and often produce dirty white ooze with foul smell. The infected collar region becomes sunken, turns buff white to pale brown in colour. Badly affected plants topple over at the basal region within a few days (Kolte, 1985, Bhowmik, 2003 and Shukla *et al.*, 2003).

Causal organism

Erwinia carotovora (Jones) Holland

The bacterium is Gram negative, rod shaped with blunt ends, capsulated motile with peritrichous flagella. The bacterium does not produce florescence or green pigment on King's medium B. The bacterium grows on Nutrient Agar (NA) producing colonies which are grayish, circular, translucent, shining, and smooth, with raised centre and wavy margin. Nutrient broth becomes uniformly turbid within 24 hours without formation of pellicle even on prolonged incubation.

Disease Cycle

The pathogen over winters the infected plant debris and penetrating host plant either through stomata or hydathodes and establishing the infection ulti-

plant showing the bacterial stock rot symptoms (this is known as primary infection). Later on the pathogen spreads through various cultural operations, irrigation, wind, splash rain and dew to healthy crop plant and spreading the disease in larger areas. This is known as secondary infection called repeating cycle. The bacterium also infects *Lycopersion esculentum, Daucus carota, Brassica oleracea* var. *botrytis* and *Nicrotiana* sp. (Kolte, 1985 and Shukla *et al.*, 2003).

Intregated Disease Managenent

1. **Burn debris:** Burn the debris of crop after if the disease has occurred during the crop season.

2. **Deep ploughing:** Deep plough the land after harvest during summer season, which help in elimination of pathogen.

3. **Crop rotation:** The crop rotation help in minimizing inoculums buildup, as for as possible take mustard after 3-4 years in the same field.

4. **Use appropriate improved varieties:** Recommended varieties for specific region be used and choose at least two or more varieties and change them every two years, so that pathogen get diverse genotypes and ultimately disease spread would be checked.

5. **Use healthy seed (pathogen free seed):** Sowing with healthy, clean and certified seeds may be used avoid initial inoculums of the disease.

6. **Sow the crop right time:** Timely sown crop is likely to be less affected with the disease. Sowing in mid of October is found the most appropriate time of sowing.

7. **Remove weeds:** Weeds could be a carrier of disease. Clean crop will help in reducing the disease buildup.

8. **Excess water drain out:** Provide only need based light irrigation. Excess of water may be drain out for proper growth of the plants, which in turn reducing relative humidity. High humidity favours bacterial development.

9. **Be watchful:** As soon as the weather condition becomes favorable, disease may occur and therefore, it is important to be watchful and take immediate step to minimize the effect of disease.

10. **Roguing and destruction of affected plants:** As soon as infection noticed in the field, rouging of the affected plant should be taken up immediately and destroyed by burning or burrowing in soil, so that further spread of pathogen would be checked.

Phytoplasmal Disease

The disease has been reported to occur in India on toria (*B. compestris* var. *toria*), the sarson (*B. compestris* var. yellow *sarson*) in the states of Punjab, Haryana and Uttar Pradesh. Reduction in plant height, pod length, seed yield and seed oil content in *toria* due to phyllody.

Economic Importance

Yield loss due to the disease appears to be quite high by considering the symptoms on individual plants, Bindra and Bakhetia studied the yield data on toria (*B. compestris* var. *toria*), and the average yield reported by them is only 0.63g per diseased plant as compared to 5.62g per healthy plant. Thus, yield loss comes to about 86.38 per cent. The loss in yield in ITSA, Synthetic-65 and Shyamgarh varieties of *toria* appears to be 78.8, 90.8 and 69.3 per cent, respectively. Losses in yield over a large area would be tremendous if the average percentage of diseased plants were high (Kolte, 1985, Bhowmik, 2003 and Shukla *et al.*, 2003).

Symptoms

Transformation of floral parts into leafy structures, green and sepaloid petals and indehiscent stamens is the characteristic symptom of phytllody. The gynoecium is borne on distinct gynophore and produces no ovules in the ovary. In addition, there are some leafy structures attached to the false septum. The affected plants may show varying degrees of severity of the disease and the affected part of the raceme do not form siliquae (Figure 8.18). Some plants may show only terminal portion of the branches affected with the disease, whereas in others the whole branches show the symptoms (Bhowmik, 2003 and Shukla *et al.*, 2003).

Figure 8.18: Phyllody Afftected Plants in the Field (Courtesy: Shukla *et al.*, 2003).

Causal Organism

Phytoplasmas are Mollicutes, which are bound by a triple-layered membrane, rather than a cell wall. The phytoplasma cell membranes studied to date usually contain a single immunodominant protein of unknown function that constitutes most of the protein in the membrane. The disease is caused by the jassid transmissible Mycoplasma like organism.

Survival of Pathogen

The causal agent from the phyllody-affected plants of *toria* and *sarson* could be transmitted to *sesamum* plants, successfully producing phyllody symptoms in *sesamum*. The pathogen is transmitted through insect vector like the leaf hopper [*Orosius albicinctus (Deltocephalus sp.)*]

Disease Cycle

The pathogen survives on alternate host like sesam and other host plants and serve as a source of primary inoculums. This disease is transmitted/spread principally to healthy plant through insect vector like the leaf hopper and develop disease symptoms to the plant. Later on the disease spread by repeated cycles of secondary infection through the process of transmission (insect vector) of the pathogen and disease development under favourable conditions (Shukla *et al.*, 2003).

Favourable Weather Condition

Warm and dry climate favours disease development. Early planting of *toria* in late August or at its normal planting time in September has been shown to favour the development of the disease in *toria* under Indian conditions. The reason for low incidence of the disease in late planted crops is attributed to low temperature, which reduces the population and activity of the incest vector (Kolte, 1985).

Disease Management

Crop rotation helps in minimizing inoculums buildup, as far as possible take mustard in alternate years in the same field. If as soon as infection notice in the field, rouging of the affected plant should be taken up immediately and destroy by burning or burrowing in soil, so that further spread of pathogen would be checked. Spray with insecticide *i.e.* Methyldematon (Metasystox) 25EC or Dimethioate (Rogor) 30EC @ 0.1 per cent (1ml/liter of water) can be sprayed to protect the crop at initial appearance of the disease and insect vectors. Repeat it after 15-20 days interval if required.

References

AICRP-RM (All India Coordinated Research Project on Rapeseed- Mustard) (1990-2016). Annual Reports. National Research Centre on Rapeseed and Mustard (ICAR), Bharatpur-321303 (Rajasthan), India.

Alexopolous, C.J., Mims C.J. and Blackwell, M. (2007). *Introductory Mycology*. 4th Edition. Wiley Eastern Limited, New Delhi.

Anonymous. (2010). Proceedings of All India Coordinated Research Project on Rapeseed-mustard. Directorate of Rapeseed-mustard Research (ICAR), Sewar, Bharatpur, Rajasthan, India.

Anonymous (2013). Agriculture Outlook and Situation Analysis Reports. Under the Project Commission by the National Food Security Mission. Ministry of Agriculture (NCAER), April –June, p. 24.

Ansari, N.A., Khan, M.W. and Muheet, A. (1989). Effect of some factors and growth and sporulation of *Alternaria brassicae* causing Alternaria blight of rapeseed and mustard. *Acta Bot. Indica*. **17** (1): 49-93.

Awasthi, R.P., Nashaat, N.I. and Kolte, S.J. (1995). Interaction between *Peronospora parasitica* and *Albugo candida* in relation to development of satgheads in *Brassica juncea*: 36. In: National symposium on Detection of Plant pathogens and their Management, Faizabad, India.

Awasthi, R.P., Nashaat, N.I., Heran, A., Kolte, S.J. and Singh, U. (1997). The *Albugo candida* on the resistance to *Peronospora parasitica* and *vice versa* in rapeseed-mustard. In: ISHS symposium on *Brassicas*, 10th Crucifer Genetics Workshop, Rennes, France.

Bains, S.S. and Joothy, J.S. (1978). Increased susceptibility of virus infected *Brassica Juncea* to *Peronospora parasitica*. *Plant Dis. Rep.* 62: 1043-1046.

Bains, S.S. and Jhooty, J.S. (1979). Mixed infection of *Albugo candida* and *Peronospora parasitica* on *Brassica juncea* inflorescence and their control. *Indian Phytopath.,* **32**: 268-271.

Bains, S.S. and Joothy, J.S. (1983). Host range and morphology of *Peronospora parasitica* from different sources. *Indian J. Mycol. Pl. Pathol*. 13- 372-375.

Bains, S.S. and Joothy, J.S. (1985). Association of *Peronospora parasitica* with *Albugo conidida* on Brassica juncea leaves. *J Phytopathol.,* 112: 28–31

Barbetti, M.J. (1981). Effects of sowing date and oospore seed contamination upon subsequent crop incidence of white rust (*Albugo candida*) in rapeseed. *Australian Plant Path,* **10:** 44-46.

Begum, H.A., Meah, M.B., Howlider, M.A.R. and Kasthen, M.A. (1993). Effect of Alternaria infection on husk constituents in mustard. *Bangladesh J. Plant Pathol.* **9** (1-2): 31-34.

Berkenkamp, B. (1972). Diseases of rapeseed in central and northern Alberta in 1971. *Can. Plant Dis. Surv.,* **52**: 62-63.

Bernier, C.C. (1972). Diseases of rapeseed in Manitoba in 1971. *Can. Plant Dis. Surv.,* **52**: 108.

Bhardwaj, C.L. and Sud. A.K. (1993). Reaction of *Brassica* cultivars against Albugo *candida* isolates from Kangra Valley. *Indian Phytopath.,* **46:** 258-260.

Bhargava, A.K.; Sobti, A.K.; Bhargava, L.P. and Nag, A. (1997). Chemical control of white rust of mustard through seed treatment-cum spray and their economics. *Indian Phytopath,* **50** (2): 242-245.

Bhatia, J.N. (1994). Field evaluation of mustard germplasm resistant to white rust. *Cruciferae Newsletter*, **16**: 119.

Bhattacharya, I. and Pramanic, M. (1998). Effect of different antagonistic rhizobacteria and *neem* products on club root of crucifers. *Indian phytopathol.* 51-87.

Bhowmik, T.P. (2003). *Oilseed Brassicas Constraints and Their Management.* CBS Publication and Distributors, 4596/1-A, 11 Darya Ganj, New Delhi.

Bhowmik, T.P. (1985). Disease management for stabilizing productivity of oilseed crops, pp. 433-435. In: H.C. Srivastava, S. Bhaskaran, B. Vatsya and K.K.G. Menon. Eds., Oilseed Production: Constrains and Opportunities. Oxford and IBM Publishing Co., New Delhi.

Bhowmik, T.P. and Munde, P.N. (1987). Identification of resistance in rapeseed and mustard against *Alternaria brassicae* (Berk.) Sacc. and some resistance sources. *Beitrage trop. Landwirtsch. Veterinarmed.* **25**: 49-53.

Bhowmik, T.P. and Singh, A. (1984). Efficacy of metalaxyl against white rust (*Albugo candida*) of rapeseed and mustard and its compatibility with captafol and metasystox. Symp. Chemical Control of Plant Diseases, Feb. 16-18, 1984. Indian Phytopath. Society, New Delhi (Abstr.).

Bhowmik, T.P. and Trivedi, B.M. (1980). A new bacterial stalk rot of Brassica. *Curr. Sci.,* **49** (17): 674-675.

Bonnet, A. (1981). Resistance to white rust in radish (*Raphanus sastivus* L.). *Cruciferae Newsletter*, **6**: 60.

Bradley, C.A., Lamey, H.A., Endres, G.J., Hensen, R.A., Mc Kay, K.R., Halvorson, M., Le Gare, D.G. and Porter, P.M. (2006). Efficacy of fungicides for the control of Sclerotinia stem rot of canola. *Pl. Dis.* 90: 1129-1134.

Bremer, H., Ismen, H., Karel, G., Ozkan, H. and Ozkan, M. (1947). Contribution to the knowledge of parasitic fungi. *Abstr. Rev. Appl. Mycol.,* **26**: 532.

Brun, H., Jouan, B., Plessis, J. and Tribodet, M. (1983). Properties preventives et curatives de la procymadone et du benomyl dams la lutte centre *Sclerotinia sclerotiorum* sur colza. In: *6th Rapeseed Conf,* Paris. May, 17-19, 1983. p 187.

Butler, E.J. (1918). *Fungi and diseases in plants.* Thaker Spink and Co., Calcutta, pp. 547.

Butler, E.J. (1918). White rust (*Cystopus candidus* Pers. Lev.): 291-297. In: *Fungi and diseases in plants.* Thacker Spink and Co., Calcutta, p. 547.

Chahal, A.S. (1982). Diseases of rapeseed and mustard. *Indian Fmg.* **32** (6): 27-36.

Chahal, A.S. (1986). Losses and chemical control of Alternaria blight in rapeseed-mustard in Punjab. *Plant Dis. Res.* **1** (1-2): 46-50.

Chahal, A.S. (1991-92). Sources of multiple diseases resistance in *Brassicae.* Abstr. *Indian Phytopath. Suppl.,* **45**: LXXVIII. (Abstr.)

Chahal, A.S. and N.S. Jaura (1994). Sources for Alternaria resistance in *Brassica. Indian Phytopath.* **47** (3): 306 (Abstr.).

Chahal, A.S. and Sekhon, J.S. (1980). How to control Alternaria blight of rapeseed and mustard? *Prog. Fmg.* **17** (3): 15.

Chang, I.H., Xu, R.F. and Chiu, W.F. (1963). On the primary sources of infection of downy mildew of Chinese cabbage caused by *Peronospora parasitica* (Pers.) Fr. and the limited systemic infection of seedlings. *Acta Phytopathol Sinica.* **6**: 153-162.

Channon, A.G. (1981). Downy mildew of *Brassicas*: 636. In: The Downy Mildew (ed.) D.M. Spencer, Academic Press, London.

Chatterjee, S.C., Gopal, M. and Mukherjee, I. (1997). Bio-efficacy of Iprodione against Alternaria blight and its residue in mustard. P. 354 Golden Jubilee Int. Conf., Integrated Plant Disease Management for Sustainable Agriculture, Nov. 10-15, 1997. *Indian Phytopath. Soc.*, New Delhi.

Chattopadhyay, A.K. (1989). Relationship of phenols and sugars in Alternaria blight resistance of rapeseed-mustard *Indian J. Mycol. Res.* **27** (2): 195-199.

Chattopadhyay, A.K. and Bagchi, B.N. (1994). Relationship of disease severity and yield due to leaf blight of mustard and spray schedule of mancozeb for higher yield. *J. Mycopath. Res.* **32** (2): 83-87.

Chattopadhyay, A.K. and Bhunia, C.K. (2003). Management of Alternaria blight of rapeseed-mustard by chemicals. *J. Mycopathol. Res.*, **41** (2): 181-183.

Chowdhary, S. (1944). Some fungi from Assam I. *Indian J. Agric. Sci.*, **14**: 230-233.

Chup, C. (1925). Manual of Vegetable Garden Diseases. MacMillan, New York, pp. 530.

Clarkson, J.P., Whipps, J.M. and Young, C.S. (2001). Epidemiology of *Sclerotinia sclerotiorum* on lettuce. In: *Proc XI Int Sclerotinia Workshop* Central Science Lab. York, UK, pp. 79-80.

Conn, K.L., Tewari, J.P. and Awasthi R.P. (1990). A disease assessment key for Alternaria black spot in rapeseed and mustard. *Can. Plant. Dis. Surv.* **70**: 19-22.

Conners, I.L. (1967). All allotted index of plant diseases in Canada and fungi recorded on plants in Alaska, Canada and Greenland. Queen's Printer, Ottawa, Canada. Research Branch, Can. Dep. Agric. Publ. No. 1251: II pp. 39, 43, 61, 69, 70, 99, 100, 111, 149, 154, 178, 240, 245, 266, 304.

Constantinescu, O. and Fatehi, J. (2002). Peronospora-like fungi (Chrmista, Peronosporales) parasitic on Brassiceae and related hosts, Nova Hedwigla 74: 291.

Damle, V.P. (1944). A new species found on *Cardamine subumbelata. J. Univ. Bombay,* **12**: 42-45.

Dang, J.K., Kaushik, C.D. and Sangwan, M.S. (1995). Quantitative relationship between Alternaria leaf blight of rapeseed-mustard and weather variables. *Indian J. Mycol. Plant Pathol.* **25** (3): 184-185.

Dang, J.K., Sangwan, M.S. Mehta, N. and Kaushik, C.D (2000). Multipul disease resistance four fungal foliar diseases of rapeseed mustard. *Indian Phytopathol.* **53**: 455-458.

Dange, S.R.S., Patel, R.L, Patel, S.I. and Patel, K.K. (2003). Effect of planting time on the appearance and severity of white rust and powdery mildew disease of mustard. *Ind J Agric Res.* **37**: 154-156.

Darpoux, H. (1945). A contribution to the study of the diseases of oleaginous plants in Frallce. *Ann. Epihyt., N.S.,* **11**: 71-703.

Dasgupta, B. (1991). Effect of fungicidal sprays on the incidence of Alternaria blight and yield of mustard. *Natn. Bot. Soc.* **45**: 35-37.

Davies, J.M.L. (1986). Diseases of oilseed rape. pp. 195-236. In: D.H. Scarisbrick and R.W. Danials eds., *Oilseed rape* Collins London.

deBary, A. 1866. Morphologie und physiologia der Pilze, Flechten und Myxomyceten. Wilhelm Engelmann, Leipzig. deBruyn, H.L.G. (1937). Heterothallism in *Peronospora parasitica. Genetica,* **19**: 553-558.

Dickinson, C.H. and Greenhalgh, J.R. (1977). Host range and taxonomy of *Peronospora* on crucifers. *Trans. Br. Mycol. Soc.* 69: 111-116.

Dikshit, R.K. and A.N. Srivastava (1991). "Vardan" mustard is suited to late planting on irrigated lands. *Indian Farming.* **40** (12): 19.

Dubey, S.C. (1996). Chemical control of white rust of mustard. *Plant Disease Res.* 11 (2): 155-158.

Dubey, S.C. and Mishra, B. (1994). Evaluation of fungicides against white rust of mustard. *Indian J. Mycol. Pl. Path.,* **24** (2): 150.

Dueck, J. and Stone, J.R. (1979). Evaluation of fungicides for control of *Albugo candida* in turnip rape. *Can. J. Plant Sci.,* **59**: 423.

Ellis, M.B. (1971). *Dematiaceous Hyphomycetes.* Comonwealth Mycological Insitute, Kew, Surrey, England, UK.

Fox, D.T. and Williams, P.H. (1984). Correlation of spore production by *Albugo candida* on *Brassica campestris* and a visual white rust rating scale. *Can. J. Plant Pathol.,* 6: 175-178.

Gadewadikar, P.N., Bhadouria, S.S. and Bartaria, A.M. (1993). Inheritance of resistance to white rust (*Albugo candida*) disease in Indian mustard (*Brassica juncea* (L.) Czern and Coss.). In: National Seminar on oilseed Research and Development in India: Status and Strategies, Aug. 2-5, 1993.

Gadre, U.A., Joshi, M.S. and Manokhot, A.M. (2002). Effect of weather factors on the incidence of Alternaria blight, white rust and powdery mildew of mustard. *Ann. Pl. Protec. Sci.,* **10** (2): 337-339.

Gandhi, S.K. and Parashar, R.D. (1977). Bacterial rot of Raya (*Brassica juncea*). *Indian phytopath.* 30(1): 24-27.

Gangasaran and Giri, G. (1984). Agronomic manipulations for increasing yields of edible oilseeds, pp. 285-292. In: H.C. Srivastava, S.Bhaskaran, B. Vatsya and K.K.G. Menon. Eds., Oilseed Production: Constrains and Opportunities. Oxford and IBM Publishing Co., New Delhi.

Gaumamn, E. (1926). On the specialization of downy mildew (*Peronospora brassicae* Gaum.) on cabbage and related species. *Landw. Jahrb. Deschewiz.* 40: 463-468.

Gladders, P. and Musa, T.M. (1980). Observation on epidemiology of *Leptosphaeria maculans* stem canker in winter oilseed rape. *Plant Pathol.* **29** (1): 28-37.

Godika, S. and Pathak, A.K. (2001). Evaluation of fungitoxicants against Alternaria blight and white rust diseases of Indian mustard (*Brassica juncea*). *Indian J. Agric. Sci.*, **71** (7): 497-498.

Godika, S. and Pathak, A.K. (2002). Efficacy of some fungicides against white rust and alternaria blight diseases of mustard. *Plant Disease Res.*, **17** (1): 138-140.

Godika, S. and Pathak, A.K. (2009). Efficacy of botanicals against major diseases of rapeseed-mustard. *Indian J. Agric. Sci.*, **79** (5): 404-405.

Godika, S., Jain, J.P. and Pathak, A.K. (2001). Evaluation of fungitoxicant against Alternaria blight and white rust of Indian mustard. *J. Agricultural Sci.* **71**: 497-498.

Greelman, D.W. (1963). New and noteworthy diseases. *Can. Plant Dis. Surv.*, **43**: 61-63.

Grontoff (1993). A rapid screening method for testing the resistance of cotyledons to downy mildew in *Brassica napus* and *B. compestris*. *Plant Breed.* 110 (3): 207-211.

Gugel, R. K. and Morrall, R.A.A. (1986). Inoculum disease relationship in Sclerotinia stem rot of rapeseed in Saskatchewan. *Can. J. Plant Pathol.*, 8: 89-96.

Gupta D.K. and Chaudhary K.C.B. (1995). Infection of radish and rayosag seeds by *Xanthomonas compestris* pv. *compestris*. *Indian J. Mycol. Plant Pathol.* **25**(3): 332.

Gupta, I.J. and Sharma, B.S. 91978). Chemical control of white rust of mustard. *Pesticides*, **12** (12): 45-46.

Gupta, J.F., Sharma, B.S. and Dalela, D.G. (1977). Control of white rust and Alternaria leaf spot of mustard. *Indian J. Mycol. Plant. Pathol.*, **7** (2): 163-164.

Gupta, R., Awasthi, R.P. and Kolte, S.J. (2003). Influence of sowing dates and weather factors on development of Alternaria blight of rapeseed-mustard. *Indian Phytopath.* **56**: 398-402.

Gupta, R.B.L. and Singh, M. (1994). Source of resistance to white rust and powdery mildew of mustard. *Int. J. Trop. Plant. Dis.*, **12**: 225-227.

Gupta, R; Awasthi, R.P. and Kolte, S.J. (2004). Identification of pathotypes of *Albugo candida* with stable characteristic symptoms on Indian mustard. *J. Mycol. Pl. Pathol.*, **32** (1): 46.

Gupta, S.K., Gupta, P.P. and Kaushik, C.D. (1995). Changes in leaf peroxidase, polyphenol oxidase, catalase and total phenol due to Alternaria leaf blight in *Brassica* spp. *Indian J. Mycol. Plant Pathol.* **25** (3): 175-180.

Gupta, S.K., Kumar, P., Yadav, T.P. and Sharan, G.S. (1984). Changes in phenolic compounds, sugars and total nitrogen in relation to Alternaria leaf blight in Indian mustard. *Haryana Agric. Univ. J. Res.* **14**: 535-537.

Gupta, S.P., Singh, B.R. and Tripathi, D.P. (1997). Management of blight disease of mustard. P. 346. *Golden Jubilee Int. Conf., Integrated Plant Disease Management for Sustainable Agriculture,* Nov. 10-15, 1997, Indian Phytopath. Soc. New Delhi.

Hammett, K.R.W. (1969). White rust disease. *N. Z. Gardner,* **26**: 43.

Hammond, K.E. and Lewis, B.G. (1986). The timing and sequence of events leading to stem canker disease in population of *Brassica napus* var. *olerifera* in the field. *Plant Pathol.* 35: 551-564.

Hammond, K.E., Lewis, B.G. and Musta (1985). A systemic pathway in the infection of oilseed rape plants by *Leptosphaeria maculans. Plant Pathol.* **34** (4): 557-565.

Hegde, V.M. and Anahosur, K.H. (1993). Chemical control of white rust of mustard. *Karnakata J. Agri. Sci.,* **6** (3): 263-267.

Hegde, V.M. and Anahosur, K.H. (1994). Influence of sowing date of mustard on the epidemiology of white rust. *Indian Phytopath.,* **47** (4): 391-394.

Hirata, S. (1954). Studies on the phytohormone in the malformed portion of the diseased plants. I. The relation between the growth rate and the amount of free auxin in the fungus galls and virus- infected plants. *Ann. Phytopath. Soc. Japan,* **19**: 33-38.

Hiura, M. (1930). A simple method for germination of oospores of *Sclerospora graminicola. Science* (Wash.), **72**: 95.

Hong, C.X. and Fitt, B.D.L. (1995). Effects of inoculums concentration, leaf age and wetness period on the development of dark leaf spot and pod spot (*Alternaria brassicae*) on oilseed rape (*Brassica napus*). *Ann. Appl. Biol.* **127** (2): 283-295.

Hoser-Krauze *et al.* (1991). Resistance to some *Brassica oleracea* L. plant introduction to downy mildew (*Peronospora parasitica*). *Cruciferae Newsletter.* N. 14/45: 144-145.

Humaydan, H.S. and Williams, P.H. (1976). Inheritance of seven characters in *Raphanus sativus* L. *Hort. Sci.,* **11**: 146-147.

Humpherson-Jones, F.M. and Phelps, K. (1989). Climatic factors influencing spore production in *Alternaria brassicae* and *Alternaria brassicicola. Ann. Appl. Biol.* **114** (3): 449-458.

Hussain, A. and Thakur, R.N. (1963). Some sources of resistance to Alternaria blight of rapeseed and mustard. *Indian Oilseeds J.* **7**: 259-264.

Jain, J.P. (1995). Management of white rust and Alternaria blight diseases of mustard. Proceedings of the Global Conference on Advances in Research of Plant Diseases and their Management, Udaipur, Rajasthan, India.

Jain, K.L., Gupta, A.K. and Trivedi, A. (1998). Reaction of rapeseed mustard lines against white rust pathogen, *Albugo Candida. J. Mycol. Pl. Path.,* **28**: 72-73.

Jat, R.R. (1992). Screening of mustard/toria germplasm against white rust high altitude intermediate zone of J&K. *Indian J. Mycol. Pl. Pathol.*, **22**: 95.

Jurke, C.J. and Fernando, G. D. (2006). Effects of seeding rate and plant density on Sclerotinia stem rot incidence in canola. *Arch Phytopath Pl Prot.* 1-14.

Kapoor, A.S. and Sugha, S.K. (1995). Efficacy of some fungicides in controlling white rust of mustard. *Indian J. Mycol. Plant. Pathol.*, **25** (3): 285-286.

Kapoor, K.S. (1983). Some aspects of the host parasite relations between *Sclerotinia sclerotiorum* (Lib.) de Bary and rapeseed. *6th Int Rapeseed Conference* Paris, Fr. May 17-19: 188.

Kaushik, C.D. and Saharan, G.S. (1980). Screening of raya germplasm against foliar diseases. Ann. Prog. Report of Rapeseed-mustard, HAU, Hisar, India.

Kaushik, C.D., Kaushik, J.C. and Saharan, G.S. (1983). Field evaluation of fungicides for control of Alternaria blight of *Brassica juncea*. *Indian J. Mycol. Plant Pathol.* **13** (3): 262-264.

Kaushik, C.D., Saharan, G.S. and Kaushik, J.C. (1984). Magnitude of losses in yield and management of Alternaria blight in rapeseed-mustard. *Indian Phytopath.* **37** (2): 398.

Khan, M.M., Khan, R.U. and Mohiddin F.A. (2007[a]). Variation in occurrence and morphology of *Alternaria brassicae* (Berk.) Sacc. causing blight in rapeseed and mustard. *Ann. Pl. Protec. Sci.* **15**(2): 414-417.

Khan, M.M., Khan, R.U. and Mohiddin F.A. (2007[b]). Studies on the cost effective management of Alternaria blight of rapeseed and mustard (*Brassica* spp.). *Phytopathol. Mediterr.* **46**: 201-206.

Khangura, R.K. and Sokhi, S.S. (2000). Efficacy of ridomil MZ and protectant fungicides to control white rust (*Albugo candida*) and their effect on seed yield, oil content and fatty acid composition in mustard. *Austr. J. Experil. Agric.*, **40** (5): 699-706.

Khunti, J.P.; Khandar, R.R.; Bharaniya, M.F. (2001). Fungicide control of white rust (*Albugo cruciferarum* S.T. Gray) of mustard (*Brassica juncea* L. Czern and Coss.). *Agric. Sci. Dig.*, **21** (2): 103-105.

Klemm, M. (1938). The most important diseases and pests of colza and rape. *Abstr. Rev. Appl. Mycol.*, **17**: 717.

Kohli, Y., Brunner, L. J., Yoell, H., Milgroom, M. G, Anderson, J. B., Morrall, R. A.A. and Kohn, L. M. (1995). Clonal dispersal and spatial mixing in population of plants and pathogenic fungus *Sclerotinia sclerotiorum*. *Mol Ecology.*, **4**: 69-77.

Kolte, S.J. (1985). *Diseases of Annual Oilseed Crops*. Volume-II: Rapeseed-Mustard and Sesame Diseases. CRC Press, inc. Boca Raton Florida. Pp. 1-127.

Kolte, S.J. (1987). Important disease of rapeseed-mustard in India: present research progress and future research needs. Oil crops: niger, rapeseed/mustard: 91-106. Proceedings of the 3rd Oil crops Network Workshop (ed.). A. Omran, Addis Ababa, Ethiopia.

Kolte, S.J. and Tewari, A.N. (1980). Note on the susceptibility of certain oleiferous *Braccicae* to downy mildew and white blister diseases. *Indian J. Mycol. Pl. Path.*, 10: 191-192.

Kolte, S.J., Awasthi, R.P. and Vishwanath. (1987). Assessment of yield losses due to Alternaria blight in rapeseed and mustard. *Indian Phytopath.* 40 (2): 209-211.

Kolte, S.J., Awasthi, R.P. and Vishwanth. (1985). Field performance of improved toria (*Brassica campestris* variety *toria*) varieties against Alternaria blight, downy mildew and white rust diseases. *Indian J. Mycol. Pl. Path.*, 15: 211-213.

Kolte, S.J., Awasthi, R.P. and Vishwanth. (1986). Effect of planting dates and associated weather factors on staghead phase of white rust and downy mildew of rapeseed and mustard. *Indian J. Mycol. Pl. Path.*, 16 (2): 94-102.

Kolte, S.J., Awasthi, R.P., Vishwanth, Sawant, S.D. and Thakur, R. (1985). Plant Pathology. In: Oilseed Research at Pantnagar. Direct. Expt. Stn., G.B. Pant University. Gric. Tech., Pantnagar, *Tech. Bull.*, 111: 36-62.

Kolte, S.J., Sharma, K.D. and Awasthi, R.P. (1981). Yield losses and control downy mildew and white rust of rapeseed and mustard, *Abstr., 3rd Int. symp. Plant pathol.* New Delhi, December 14 to 18, 70.

Kruger, W. (1980). On the effect of calcium cyanamide on the development of apothecia of *Whetzelinia Sclerotiorum* (Lib.) Korf and Dumont, the agent of stalk rot of rape. *Rev. Plant Pathol.* 59: 5438.

Kumar, A. (1996). Efficacy of different fungicides against Alternaria blight, white rust and stag head infection of mustard. *Plant Dis. Res.* 11 (2): 174-177.

Kumar, A. and Chauhan, J.S. (2005). Status and future thrust areas of rapeseeds–mustard in India. *Indian J. Agric. Sci.* 75 (10): 621-635.

Kumar, K. and Singh, D.P. (1986). Control f *Alternaria brassicae* infection in mustard and rapeseeds. *Pesticides* 20 (6): 22-23.

Kumar, N. and Kumar, A. (2006). Effect of cultural practices on Alternaria blight of *Brassica juncea* and *B. napus*. *Indian J. Agric. Sci.* 76 (6): 389-390

Kumar, S. and Kalha, C.S. (2005). Evaluation of rapeseed-mustard germplasm against white rust and Alternaria blight. *Annl of Biol*, 21 (1): 73-77.

Kumar, S. and Singh, R.B. (2008). Multiple disease resistance against three major foliar diseases of rapeseed-mustard. *Indian Phytopath.* 61 (3): 405.

Kumar, S., Singh, R.B. and Shakyawar, R.C. (2014). Epidemiological studies on Alternaria blight of Brassica campestris L. var. Yellow *sarson* Prain caused by *Alternaria* spp., *Ann. Pl. Protec. Sci.,* 22 (2): 398-402.

Kumar, S; Singh, R.B. and Singh R.N. (2009). Fungicides and genotypes for the management of foliar diseases of rapeseed-mustard. *Proc. Nat. Acad. Sci. India, sect. B.* 79 (II): 189-193.

Kumar, V., Kaushik, C.D. and Gupta, P.P. (1995). Role of various factors in the development white rust disease of rapeseed-mustard. *Indian J. Mycol. Pl. Path.*, 25 (3): 145-148.

Lakra, B.S. and Saharan, G.S. (1988[b]). Efficacy of fungicides in controlling white rust of mustard through foliar sprays. *Indian J. Mycol. Pl. Path.*, **18** (2): 157-163.

Lakra, B.S. and Saharan, G.S. (1989). Sources of resistance and effective screening techniques *Brassica Albugo* system. *Indian Phytopathology*, **42**: 293.

Laura, S and Juha, H. (2004). Do inspection practices in organic agriculture serve organic values? A case study in Finland. *Agriculture Human Values*, 1-13

Li, C.X.; Sithamparam, K.; Waltor, P. and Salisbury. (2007). Expression and relationship of resistant to white rust (*Albugo candida*) at cotyledonary, seedling and flowering stage in *Brassica juncea* germplasm from Australia, China and India. *Aust. J. Agric. Res.*, **58** (3): 259.

Liu, J.Q. and Rimmer, S.R. (1993). Production and germination of oospores of *Albugo candida*. *Can. J. Plant Pathol.*, **15**: 265-271.

Lomate, C.B., Mate, G.D. and Kalaskar, R.R. (2014). Management of powdery mildew of mustard with chemicals and bio-agents. *Internat. J. Plant Protec.* **7**: 122-124.

Louvet, J. (1958). The black spot disease of colza *Alternaria brassicae*. *C.R. Acad. Agric. Fr.* **44**: 694.

Lucas, J.A., Crute, I.R., Sherriff, C. and Gordon, P.L. (1988). Identification of gene for race-specific resistance to *Peronospora parasitica* (Downy mildew) in *Brassica napus* Var. *oleifera*. (oilseed rape) *Plant Pathol.* **37**: 538-545.

Mathur, S., Bhatnagar, M.K. and Mathur, S. (1991). Comparative tolerance of *Albugo candida* and *Peronospora parasitica* to metalaxyl. *Int. J. Tropical Plant Diseases*, **9** (2): 195-199.

McCartney, A, Heran, A. and Li, Q. (2001[a]). Infection of oilseed rape (*Brassica napus*) by petals containing ascospores of *Sclerotinia sclerotiorum*. In: *Proc Sclerotinia 2001, the XI Intl Sclerotinia Workshop* (Young, C.S., Hughes, K.J.D., Eds.), York 8 -12, July 2001, Central Sci. Lab, York, England. pp. 183-184.

McCartney, A., Heran, A., Li, Q. and Freeman, J. (2001[b]). Petal fall, petal retention and petal duration in oilseed rape crops. In: *Proc Sclerotinia 2001, the XI Intl Sclerotinia Workshop* (Young, C.S., Hughes, K.J.D., Eds.), York 8 -12, July 2001, Central Sci. Lab, York, England. pp. 185-186.

McDonald, W.C. (1959). Gray leaf spot of rape in Manitoba, *Can. J. Plant Sci.* **39**: 409.

McLean D.M. (1958). Role of dead flower parts in infection of certain crucifers by *Sclerotinia sclerotiorum* (Lib.) de Bary. *Pl. Dis. Reptr.* **42**: 663-666.

Meena, P.D., Chattopadhyay, C. and Meena, R.L. (2008). Ecofriendly management of Alternaria blight of *Brassica juncea*. *Indian Phytopath.* **61**(1): 65-69.

Meena, P.D., Chattopadhyay, C., Singh, F., Singh, B. and Gupta, A. (2002). Yield loss in Indian mustard due to white rust and some cultural practices on Alternaria blight and white rust severity. *Brassica*. **4**: 18-24.

Meena, P.D., Chattopadhyay, C., Vijay, R. K. Meena, R.L., and Rana, U.S. (2005). Spore behaviour in atmosphere and trends in variability of *Alternaria brassicae* population in India. (In: abstract, Global Confrence II on Plant Health – Global Wealth Organised by the Indian Soc. Mycol. Plant. Pathol., MPUAT, Udaipur, 25-29 Nov. 2005, pp. 91). *J. Mycol. Pl. Pathol.* **35**: 511.

Meena, P.D., Meena, R.L., Chattopadhyay, C., and Kumar, A. (2004). Identification of critical stage for disease development and biocontrol of Alternaria blight of Indian mustard (*Brassica Juncea*). *J. Phytopathology*, **152**: 204-209.

Meena, P.D., Awasthi, R.P., Chattopadhyay, C., Kolte, S.J. and Kumar, A. (2010). Alternaria blight: a chronic disease in rapeseed and mustard. *J. Oilseed Brassica*. **1**(1): 1-11.

Meena, R.L. and Jain, K.L. (2002). Fungicides and plant products in managing white rust of Indian mustard caused by *Albugo candida* (Pers. Ex. Lev.) Kuntze. *Indian. J. Pl. Prot.*, **30**(2): 210-212.

Meena, R.L.; Meena, P.D. and Chattopadhyay, C. (2003). Potential for biocontrol of white rust of Indian mustard. *Indian J. Pl. Protec.*, **31** (2): 120.123.

Mehta N., Saharan, G.S. and Sharma, O.P. (1995[a]). Sequence of events in the pathogenesis of *Peronospora* and *Albugo* on mustard. *J. Indian Bot. Soc.* 74: 299-303.

Mehta, N. (2014). Epidemiology and forecasting for the management of rapeseed-mustard diseases. *J. Mycol. Plant Pathol.* **44** (2): 131-147.

Mehta, N. and Saharan, G.S. (1994). Morphological and pathological variations in *Peronospora parasitica* infecting *Brassica* species. *Indian Phytopathol.* 47: 153-158.

Mehta, N. and Saharan, G.S. (1998). Effect of planting dates on infection and development of white rust and downy mildew disease complex in mustard. *J. Mycol. Pl. Pathol.* 28: 259-265.

Mehta, N., Hieu, N.T. and Sangwan, M.S. (2009[b]). Influence of soil types, frequency and quantity of irrigation on development of Sclerotinia stem rot of mustard. *J. Mycol. Pl. Pathol.*, **39**: 506-510.

Mehta, N., Hieu, N.T. and Sangwan, M.S. (2010[a]). Influence of systemic acquired resistance chemicals on the management of white stem rot of mustard caused by *Sclerotinia sclerotiorum* (Lib.) de Bary. *Pl. Dis. Res.*, **25**: 171-173.

Mehta, N., Hieu, N.T. and Sangwan, M.S. (2010[b]). Management of white stem rot (*Sclerotinia sclerotiorum*) of mustard with organic soil amendments. *J. Mycol. Pl. Pathol.* 40: 238-243.

Mehta, N., Hieu, N.T. and Sangwan, M.S. (2011). Efficacy of some botanicals against *Sclerotinia sclerotiorum* inciting white stem rot of rapeseedmustard. *Pl. Dis. Res.* 26: 82-86.

Mehta, N., Hieu, N.T. and Sangwan, M.S. (2012). Efficacy of various antagonistic isolates and species of *Trichoderma* against *Sclerotinia sclerotiorum* causing white stem rot of mustard. *J. Mycol. Pl. Pathol.* 42: 244-250.

Mehta, N., Saharan, G.S. and Kaushik, C.D. (1996). Efficacy and economics of fungicidal management of white rust and downy mildew complex in mustard. *J. Mycol. Plant Pathol.* **26**: 243-346

Mehta, N., Saharan, G.S. and Sharma, O.P. (1995[b]). Influence of temperature and free moisture on the infection and development of downy mildew on mustard. *Pl. Dis. Res.* 10: 114-121.

Mehta, N., Singh K. and Sangwan, M. S. (2008). Assessment of yield losses and evaluation of different varieties/genotypes of mustard against powdery mildew in Haryana. *Pl. Dis. Res.* 23: 55-59.

Morrall, R.A.A. and Dueck, J. (1982). Epidemiology of *Sclerotinia* stem rot of rapeseed in Saskatchewan. *Can. J. Pl. Pathol.,* **4**: 161-168.

Morrall, R.A.A. and Dueck, J. (1983). *Sclerotinia* stem rot of spring rapeseed in Western Canada. *Proc 6th Int Rapeseed Conf Paris,* France: 957-962.

Moss (1991). Evidence for differential response to isolates of *Perospora parasitica* (Downy Mildew) in *Brassica* rapa. Test Agro. Chem. Cvs. 12. *Annals Appl. Boil.* 118: 96-97.

Mridha, M.A.U. (1983). Virulence of different isolate of *Alternaria brassicae* on winter oilseed rape cultivars. *6th Intl. rapeseed Congr.,* Paris, France, 17-19 may 1983, pp.1025-1029.

Mridula, K., Mohanty, A.K., Acharya, N.N. and Sethi, P.N. (1994). Efficacy of some selected fungicides against *Alternaria brassicae* causing leaf blight of mustard. *Orissa J. Agric. Res.* 7 Suppl.: 90-91.

Mukherjee, I., Gopal, M. and Chatterjee, S.C. (2003). Persistence and effectiveness of iprodione against Alternaria blight in mustard. *Bull. Environ. Contami. and Toxicol.,* **70** (3): 586-591.

Munde, P.N. and T.P. Bhowmik (1985). A source of morphological resistance to leaf blight disease of rapeseed and mustard caused by *Alternaria brassicae* (Berk.) Sacc. *Curr. Sci.* **54** (11): 514-515.

Nashaat (1995[a]). Genetic analysis of a noval factor for resistance to *Peronospora parasitica* in *Brassica napus* sp. oleifera. *Indian J. Mycol. Pl. Pathol.* 25: 35.

Nashaat (1995[b]). Inheritance of two new factors for resistance to downy mildew (*Peronospora parasitica*) in oilseed rape. (*Brassica napus* sp. oleifera). Proc. 9th Inter. Congress. Cambridge, July 4-7, 1995, 4: 1283-1285.

Nashaat (1996). Resistance to downy mildew and its interaction with in rapeseed-mustard. 2nd International crop science congress, New Delhi, India. Nov. 17-24. pp. 199.

Nashaat and Awasthi (1995). Evidence for differential resistance to *Peronospora parasitica* (Downy mildew) in accessions of *Brassica jucea* (mustard) at the cotyledon stage. *J. Phytopathology.* 143: 157-159.

Nashaat and Rawlinson (1994[a]). The response of oilseed rape (*Brassica napus sp. Oleifera*) accessions with different gluconsinolate and erucic acid contents to four isolates of *Peronospora parasitica* (Downy mildew) and the identification of new sources of resistance. *Pl. Pathology*, **43**: 278-285.

Natti, J.J. Dickson, M.H. and Atkin, J.D (1967). Resistance of *Brassica oleracia* varieties to downy mildew. *Phytopathol.* **57**: 144.

Nigam, R., Kuswaha, A. Srivastava, A. and Srivastava, N. (2011). Bio-efficacy of botanicals against *Alternaria* blight of Indian mustard caused by *Alternaria brasscae. Ann. Pl. Protec. Sci.* **19** (1): 245-246.

Pace, M.A. and Campbell, R. (1974). The effect of saprophytes on infection of leaves of *Brassica* spp. by *Alternaria brassicicola. Trans. Brit. Mycol.Soc.* **63**: 193-196.

Pandey, M.K., Kumar, N., Singh, H.K. and Kumar, S. (2018). Effect of mancozeb on disease severity, infection rate and seed weight of mustard [*Brassica juncea* (L.) Czen and Coss.] caused by *Alternaria* spp. *Int. J. Curr. Microbiol. App. Sci.* 7(02): 3689-3699.

Pandya, R.K.; Tripathi, M.L. and Singh, Reeti (2000). Efficacy of fungicides in the management of white rust and Alternaria blight of mustard. *Crop Research*, **20** (1): 137-139.

Parham, B.E. (1942). White rust of cruciferae (*Albugo candida*). *Agric. J. Fiji.*, **13**: 27-28.

Pathak, A.K., Godika, S., Jain, J.P. and Muralia, S. (2001). Effect of antagonistic fungi and seed dressing fungicides on the incidence of stem rot of mustard. *J Mycol Pl Pathol.* **31**: 327-329.

Pathak, A.K., Godika, S., Jain, J.P. and Muralia, S. (2002). Screening of *Brassica* genotypes against stem rot disease of mustard caused by *Sclerotinia sclerotiorum* (Lib.) de Bary. *J. Mycol. Pl. Pathol.* 32: 111-112.

Patni, C.S. and Kolte, S.J. (2006). Effect of some botanicals in management of Alternaria blight of rapeseed -mustard. *Ann. Pl. Prot. Sci.*, **14**: 151-156.

Patni, C.S. Kolte, S.J. and Awasthi, R.P. (2006). Cultural variability of *Alternaria brassicae,* causing *Alternaria* blight of mustard. *Ann. Pl. Physiol.* **19** (2): 231-242.

Pereira, J.C.R., Chaves, G.M., Matsouok, K., Silva, A.R. and Vale, F.X.R.D. (1996). Integrated control of *Sclerotinia sclerotiorum. Fitopathol Brasil.* **21**: 254-260.

Perwaiz, M.S., Moghal, S.M. and Kamal, M. (1969). Studies on the chemical control of white rust and downy mildew of rape (*Sarson*). *W. Pak. J. Agric. Res.*, **7**: 71-75.

Petrie, G.A. (1973). Diseases of Brassica species in Saskatchewan, 1970-72. I. Staghead and Aster yellows. *Can. Plant Dis. Survey*, **53**: 19-25.

Petrie, G.A. (1975). Diseases of rapeseed and mustard: 399-413. In: Oilseed and Pulses Crops in Western Canada – A symposium (ed.) J.T. Harapiak, Western Co-operative Fertilizer Ltd., Calgary, Alberta, Canada.

Petrie, G.A. and Vanterpool, T.C. (1974). Fungi associated with hypertrophies caused by infection of cruciferae by *Albugo cruciferarum. Can. J. Plant Dis Surv.*54: 37

Petrie, G.A. and Verma, P.R. (1974). A simple method for germinating oospores of *Albugo candida. Can. J. Plant Sci.*, **54**: 595-596.

Prasad, R., Saxena, D. and Dixit, R.K. (2003). Effect of sowing dates, varieties and fungicides on the incidence of Alternaria blight of mustard. *J. Oilseed Res.* **20**(2): 310-311.

Purdy, L.H. (1958). Some factors affecting penetration and infection by *Sclerotinia sclerotiorum*. *Phytopathology.*, 48: 605-609.

Rai, B., Kolte, S.J. and Tewari, A.N. (1976). Evaluation of oleiferous Brassica Germplasm for resistance to Alternaria leaf blight. . *Indian J. Mycol. Plant Pathol.* **6**: 76-77.

Rayss, T. (1938). A new contribution to the study of the Palestinian mycoflora. *Palest. J. Bot. Jerusalem Series*, 1: 143-160.

Rimmer, S.R. and van der Berg, C.G.J. (2007). Black leg (Phoma stem canker). Pp. 19-22 in: Compendium of Brassica Diseases. S.R. Rimmer, V.I. Shattuck, and L. Buchwaldt. (eds.), APS Press, St. Paul, MN. 117 pp.

Saharan, G.S. and Kaushik, C.D (1981). Occurrence and epidemiology of powdery mildew of Brassica. *Indian Phytopath.* **34**(1): 54-57.

Saharan G.S., Mehta, N. and Sangwan, M.S. (2005). *Diseases of oilseed crops.* Indus Publication Co., New Delhi, pp.-643

Saharan, (1992). Management of rapeseed and mustard diseases. In: *Advances in Oilseed Research.* Vol I Rapeseed and Mustard. Editors: Kumar, D. and Rai, M., Scientific Publisher 5A New Pali Road, P.O. Box 91, Jodhpur-342001, India.

Saharan, G.S. (1984). A review of research on rapeseed and mustard pathology in India. Annu. Workshop. All India Coordinated Research Project Oilseeds, I.C.A.R. Jaipur, Aug. 6-10, 1984.

Saharan, G.S. (1996). Studies on physiologic specialization, host resistance and epidemiology of white rust and downy mildew disease complex in rapeseed and mustard: 1-83. Final Report of an Adhock Research Project of ICAR, Dept. of Plant Pathology, CCS HAU, Hisar, India.

Saharan, G.S. (1997). Disease resistance. In: Recent Advances in Oilseed *Brassicas* (eds.) H.R. Kalia and S.C. Gupta, Kalyani Pub., Ludhiana, India.

Saharan, G.S. (2000). Multiple diseases resistance in rapeseed-mustard. *Indian Phytopath.*, **53**: 342.

Saharan, G.S. and Chand, J.N. (1988). Diseases of Oilseed Crops. Harayana Agric. Univ. Hisar, pp. 268.

Saharan, G.S. and Mehta, N. (2002). Fungal diseases of rapeseed-mustard. Diseases of field crops- Editors: V.K. Gupta and Y.S. Paul. Indus Publishing Company, New Delhi. pp. 193-228.

Saharan, G.S., Kaushik, J.C. and Kaushik, C.D. (1981). Progress of Alternaria blight on Raya cultivars in relation to environmental conditions. P. 136. *Third Int. Symp. Plant Pathol.*, New Delhi, Dec. 14-18.

Saharan, G.S., Kaushik, C.D. and Kaushik, J.C. (1984). Sources of resistance and epidemiology of white rust of rapeseed-mustard. *Indian J. Mycol. Pl. Path.*, **14**: 11.

Saharan, G.S., Kaushik, C.D. and Kaushik, J.C. (1988). Sources of resistance and epidemiology of white rust of mustard. *Indian Phytopath.*, **41**: 96-99.

Saharan, G.S., Kaushik, C.D., Gupta, P.P. and Tripathi, N.N. (1984). Assessment of losses and control of white rust of mustard. *Indian Phytopath.*, **36**: 503-507.

Saini, J.S. (1982). Production technology for mustard. *Indian Fmg.* 32 (5): 7-10.

Sangeetha, C.G. and Siddaramaiah, A.L. (2007). Epidemiological studies of white rust, downy mildew and Alternaria blight of Indian mustard (*Brassica juncea* L. Czern. and Coss.). *African J. Agricultural Res.* 2(7): 305-308.

Sangwan, M.S. and Mehta, N.K. (2001). Rapeseed-mustard diseases and their control. *Indian Farming*, **51** (4): 17-21.

Sankhla H.C., Singh, H.G., Dalela, G.C. and Mathur, R.L. (1967). Occurrence of perithecial stage of *Erysiphe polygoni* on *Brassica compestris* var. Sarson and *Brassica juncea. Plant Dis. Reptr.* **51**: 800.

Savulescu, O. (1946). A study on the European species of the genus *Cystopus* Lev. with special reference to the species found in Rumania. Thesis-213. University of Bucarest, Romania (Abstract in Review of Applied Mycology 27: 542, 1948).

Saxena, V.C. and Rai, J.N. (1988). Survey and occurrence of white rot of crucifers caused by *Sclerotinia sclerotiorum* in UP and Bihar. *Ind. J. Mycol. Pl. Pathol.* 17: 89-91.

Sempio (1940). Contributo alla Conosceaza dele azion esercitata de vari fattori anbientali su al cune malattie parasitarie di pinate colbivate. *Riv. Patol. Veg.* 30: 29-64.

Sen, B. (2000). Biological control: A success story. *Ind Phytopath.*, **53**: 243-249.

Sharma, K.D. (1979). Powdery mildew of some crucifers from J&K state. *Indian J. Mycol. Plant Path.* 9: 29-32.

Sharma, K.D. (1980) Symptomatology, yield losses and control of Downy mildew and White rust of rapeseed and mustard, M.Sc. (Agriculture) Thesis, G.B. Pant Univ. Agric. Tech., Pantnager (India).

Sharma, K.D. and Kolte, S.J. (1985). Metalaxyl in the control of downy mildew and white rust of rapeseed and mustard. *Pestology*, **9** (1): 31-35.

Sharma, P. D. (2006). Plant Pathology. Narosa Publishing house Pvt. Ltd., New Delhi

Sharma, P., Chauhan, J.S. and Kumar A. (2013). Evaluation of Indian and exotic germplasm for tolerance to stem rot caused by *Sclerotinia sclerotiorum. J. Mycol. Plant Pathol.* **42**: 297-302.

Sharma, P., Kumar, A., Meena, PD., Goyal, P., Salisbury, P., Gurung, A., Fu, T.D., Wang, YF., Barbette, M.J. and Chattopadhyay, C. (2009). Search for resistance to *Sclerotinia sclerotiorum* in exotic and indigenous *Brassica* germplasm. Proc of 16 Australian Res Assembly on Brassicas, Ballarat Victoria. pp. 1-5.

Sharma, T.R. and Singh, B.M. (1992[a]). Parameters of resistance to *Alternaria brassicae* in some *Brassica* species. *Indian Phytopath.* **45** (suppl.) P. LIX (Abstr.).

Sherriff, C and Lucas, J.A. (1987). Variation in host specificity in the *Brassica* population of *Peronospora parasitica*. In: Genetics and plant pathogenesis. (Eds. P.R. Day and G.J. Jellis). Blackwell Scientific Pub. Oxford, pp. 333-335.

Sherriff, C. and Lucas, J.A. (1990). The host range of isolates of mildew, *Peronospora parasitica* at the cotyledon stage. *Plant Pathol.* 39: 77-91.

Shivpuri A., Chhipa, H.P., Gupta, R.B.L. and Sharma, K.N. (1997). Field evaluation of mustard genotypes against white rust, powdery mildew and stem rot. *Annals Arid Zone.*, 36: 387-389.

Shivpuri, A., Siradhana, B.S. and Bansal, R.K. (1988). Management of Alternaria blight of mustard with fungicides. *Indian Phytopath.* **41** (4): 644-646.

Shukla, A.K., Kumar, A., Singh, N. B. and Kolte, S.J. (2003). Manual on Management of Rapeseed-Mustard Diseases. NRCR-M (ICAR), Sewar Bharatpur, Rajasthan. pp. 29-31.

Silue, D., Nashaat, N.I., and Tirilly, Y. (1996). Differential differences of *Brassica oleracea* and *B. rapa accessions* to seven isolates of *Peronospora parasitica* at the cotyledon stage. *Plant Dis.* **80**: 142-144.

Singh, A. and Bhowmik, T.P. (1985). Persistance and efficacy of some common fungicides against *Alternaria brassicae*, the causal agent of leaf blight of rapeseed and mustard. *Indian Phytopath.* 38: 35-38.

Singh, B.D. and Kolte, S.J. (1981). Report on the observational study tour to rapeseed research, centres in Canada, United Kingdom and Sweden, July 10 to August 8, 1981. Submitted to International Development Research Centre Cmlada, Singapore (mimeographed).

Singh, D. and Maheshwari, V.K. (2003). Effect of Alternaria leaf spot on seed yield of mustard and its management. *Seed Res.* **31** (1): 80-83.

Singh, H.K. and Singh, R.B. (2007). Integrated management of Alternaria leaf blight of rapeseed and mustard. *Indian Phytopath.* **60** (3): 396.

Singh, H.K. and Singh, R.B. and Bhajan, R. (2006). Identification of resistant sources against Alternaria blight of rapeseed-mustard. *Indian Phytopath.* **59**(3): 385.

Singh, H.K., Chauhan, M.P., Srivastava, T. and Singh, M.K. (2016). Rapeseed-mustard germplasm evaluation and management against black rot (*Xanthomonas campestris* pv. *campestris*)-A Potential Threat. *J AgriSearch*, **3** (4): 220-222.

Singh, H.K., Kumar, K., Kumar, P., Singh, S., Singh, R.B., Maurya, K.N. and Chauhan M.P. (2016). Management of powdery mildew of rapeseed-mustard. *Indian Phytopath.* **69** (4s) 394-396.

Singh, H.K., Singh, R.B. and Kumar, K. (2007). Evaluation of rapeseed-mustard genotypes for resistance to Alternaria blight. *Indian Phytopath.* **60** (3): 395-396.

Singh, H.K., Singh, R.B. and Verma, O.P. (2008). Occurrence of bacterial blight of rapeseed-mustard in Eastern Uttar Pradesh. *Indian Phytopathology.* 61(3): 408.

Singh, H.K., Singh, R.B., Kumar, K. and Verma O.P. (2008). Screening of brassica genotypes against Alternaria blight of rapeseed-mustard caused by *Alternaria brassicae* (Berk.) Sacc. *Cruciferae Newsl.* **27**: 35-36.

Singh, H.K., Singh, R.B., Kumar, P., Singh, M., Yadav, J. K., Singh, P.K., Chauhan, M. P., Shakywar, R.C., Maurya, K.N., Priyanka, B. S., Srivastava, T., Yadav, S.K. and Maurya M.K. (2017). Alternaria blight of rapeseed-mustard–A Review. *Journal of Environmental Biology*. **38**: 1405-1420.

Singh, H.K., Yadav, J.K., Maurya M.K. and Singh, S.K. (2018). Management of Alternaria blight through genotypes, fungicides, bio-agents and botanical in rapeseed-mustard. *Int. J. Curr. Microbiol. App. Sci.* 7(02): 2463-2469.

Singh, J., Srivastava, S.K., Singh, R. Singh, J., Singh, R. (1998). Effect of planting dates on occurrence of disease-pest in mustard varieties. *J. Oilseed Res.* **15**: 329-333.

Singh, J.P. Singh, H.K. and Singh, R.B. (2008). Integrated management of foliar disease of mustard. *Indian Phytopath.* **61** (3): 408-409.

Singh, R. and Singh, V.K. (2007). Evaluation of fungicides against Alternaria blight of *Brassica campestris* var. yellow *sarson*. *Ann. Pl. Protec. Sci.*, **15** (1): 266-267.

Singh, R., Tripathi, N.N., Kaushik, C.D. and Singh, R. (1994). Management of Sclerotinia rot of Indian Mustard [*Brassica juncea* (L.) Czern and Coss] by fungicides. *Crop Res.* **7**: 276-281.

Singh, R.B. and Bhajan, R. (2004). Effect of chemicals on the severity of Alternaria leaf spot, seed yield and vigour index in Indian mustard. *Farm Sci. J.* **13** (1): 6-7.

Singh, R.B. and Bhajan, R. (2005). Occurrence, avoidable yield loss and management of white rust, *Albugo candida*, in late sown mustard *Brassica juncea* (L.) Czern and Coss. *J. Oilseeds Res.*, **22** (1): 111-113.

Singh, R.B. and Chauhan, Y.S. (1997). Integrated management of Alternaria blight of mustard. P. 358. *Golden Jubilee Int. Conf., Integrated Plant Disease Management for Sustainable Agriculture*, Nov. 10-15, 1997, *Indian Phytopath.* Soc. New Delhi.

Singh, R.B. and Singh, H.K. (2011). Management of Alternaria blight of Indian mustard. Paper presented in National Conference on "Climate Change and Biodiversity", Department of Botany, Saraswati Vidya Mandir P G College, Arya Nagar (North), Gorakhpur from January 21-22, pp. 80-81.

Singh, R.B. and Singh, H.K. and Verma, O.P. (2011). Epidemiology based management of foliar diseases of rapeseed-mustard in mid-eastern India. Paper presented in National Conference on "Climate Change and Biodiversity", Department of Botany, Saraswati Vidya Mandir P G College, Arya Nagar (North), Gorakhpur from January 21-22, pp. 68.

Singh, R.B. and Singh, R. N. (2003). Management of powdery mildew of mustard. *Indian Phytopath.* **56** (2): 147-150.

Singh, R.B. and Singh, R. N. (2004). Managing powdery mildew- a threat to late sown mustard in eastern India. *Indian Farming*, pp. 11-12.

Singh, R.B. and Singh, R. N. (2005ᵃ). Status and management of foliar diseases of timely sown mustard in mid-eastern India. *Pl. Dis. Res.* **20** (1): 18-24.

Singh, R.B. and Singh, R.N. (2005ᵇ). Fungicidal management of foliar diseases of mustard in mid-eastern India. *Indian Phytopath.* **58** (1): 51-56.

Singh, R.B. and Singh, R.N. (2006). Spray schedule for the management of Alternaria blight and white rust of Indian mustard (*Brassica juncea*) under different dates of sowing. *Indian J. Agricul. Sci.* **76** (9): 575-579.

Singh, R.B. and Singh, R.N. (2007). Integrated management of foliar diseases of rapeseed- mustard. *Indian Farmers Digest.*, Semtember, 2007. pp. 29-32.

Singh, R.M.; Agarwal, D.K. and Sarbhoy, A.K. (1993). HCIO descriptions of plant pathogenic fungi, Set11, Nos. 61-66, IARI, New Delhi.

Singh, R.S. (2009). *Plant Diseases.* Oxford and Publishing Company Pvt. Ltd., New Delhi.

Singh. H.K. and Singh, R.B. (2013). White rust of Rapesee-Mustard and its Managemen (Plant Disease Management for Food Security. Edt. By C.R. Prajapati P.K. Singh, and A.K. Gupta). LAP LAMBERT Academic Publishing GmbH and Co. KG Heinrich-Bocking-Str. 6-8, 66121,Saarbrucken, Germany. pp. 359-409.

Singh. H.K., Singh, R.B. and Shakywar, R.C. (2013). Alternaria Blight of Rapesee-Mustard and its Management. Edt. By A.K. Singh and B.P. Bhatt.). LAP LAMBERT Academic Publishing GmbH and Co. KG Heinrich-Bocking-Str. 6-8, 66121, Saarbrucken, Germany. PP-410-466.

Sinha, R.K.P., Rai, B. and Sinha, B.B.P. (1992). Epidemiology of leaf spot of rapeseed-mustard caused by *Alternaria brassicae*. *J. Appl. Biol.* **2** (1/2): 70-75.

Solanki, V.A., Patel, B.K. and Sheikh, A.M. (1999). Meteorological variables in relation to an epiphytotic of powdery mildew disease of mustard. *Indian Phytopath.* **52**: 138-141.

Sosnowski, M.R., Scott, E.S. and Ramsey, M.D. (2005). Temperature, wetness period and inoculum concentration influence the infection of canola (*Brassica napus*) by pycnidiophores of *Leptosphaeria maculans Aust. Plant Pathol.* **34** (3): 359

Srivastava, L.S. and Verma, R.N. (1987). Field screening of mustard cultivars against white rust in Sikkim. *Indian J. Mycol. Pl. Path.*, **17**: 46.

Srivastava, L.S. and Verma, R.N. (1989). Fungicidal control of white rust of mustard in Sikkim. *Indian Phytopath.*, **42** (1): 84-86.

Stankova, J. (1972). Varietal variability of winter rape in its susceptibility to dark leaf spot and the factors influencing the development of the disease. *Rostlinna Vyroba* 18 (6): 625-630.

Stovold, G.E., Mailer, R.J. and Francis, A. (1987). Seed borne levels, chemical seed-treaement and effects on seed quality following a severe outbreak of *Alternaria brassicae* on rapeseed in New South Wales. *Plant Prot. Quart.* **2** (3): 128-131.

Takeshita, R.M. and Linn, M.B. (1953). Possible means of over wintering of the horse radish white rust fungus *Albugo candida*. *Dis. Abstr.*, **13**: 483-487.

Tripathi, N.N., Kaushik, C.D., Yadav, T.P. and Yadav, A.K. (1980). Alternaria leaf spot resistance in raya. *Haryana Agri. Univ. J. Res.* **10** (2): 166-168.

Tripathi, N.N., Saharan, G.S., Kaushik, C.D., Kaushik, J.C. and Gupta, P.P. (1987). Magnitude of losses in yield and management of Alternaria blight of rapeseed and mustard. *Haryana Agri. Univ. J. Res.* **17** (1): 14-18.

Vanterpool, T.C. (1959). Oospores germination in *Albugo candida*. *Can. J. Bot.*, **37**: 169-172.

Vasudeva, R.S. (1958). Diseases of rapeseed and mustard, pp. 77-86. In: D.P. Singh ed. *Rape and Mustard*. Indian Central Oilseeds Committee, Hyderabad.

Verma, P.R. and Petrie, G.A. (1980). Effect of seed infestation and flower bud inoculation on systemic infection of turnip rape by *Albugo candida*. *Can. J. Plant Sci.*, **60**: 267-271.

Verma, P.R. and Sharan, G.S. (1993). *Alternaria brassicae* (Berk.) Sacc., *A. brassicicola* (Schwein.) Wiltsh., and *A. raphani* Groves and Skolko: Introduction, bibliography and subject index. *Agric. Canada Res. Stn. Saskatoon Tech.* Bull. No. 93-002: 81p

Verma, P.R., Spurr, D.T. and Petrie, G.A. (1983). Influence of age and time of detachment on development of white rust on detached *Brassica campestris* leaves at different temperatures. *Can. J. Plant Pathol.*, 5: 200-205.

Verma, U. (1987). Studies on the white rust of *Brassica* caused by *Albugo candida* (Pers. ex. Lev.) Kunze. Ph. D. Thesis, IARI, New Delhi, p. 115.

Verma, U. and Bkowmik, T.P. 1986. A simple method of inoculating white rust on rapeseed and mustard. *Int. J. Tropical Plant Dis.*, **4**: 41-43.

Verma, U. and Bkowmik, T.P. (1988). Oospores of *Albugo candida* (Pers. ex. Lev.) Kunze.-its germination and role as the primary source of inoculum for white rust diseases of rapeseed and mustard. *Int. J. Tropical Plant Dis.*, **6**: 265-269.

Verma, U. and Bkowmik, T.P. (1989[b]). Epidemiology of white rust disease of mustard, *Brassica juncea*. *Indian Phytpath.*, **42**: 274-275 (Abstr.).

Vishwanath and Kolte, S.J. (1997). Varibility in *Alternaria brassicae*: Response to host genotypes, toxin production and fungicides. *Indian Phytopathol.*, **50**: 373-381

Vishwanath., Kolte, S.J., Singh, M.P. and Awasthi, R.P. (1999). Induction of resistance in mustard (*Brassica juncea*) against Alternaria black spot with an avirulent *Alternaria brassicae* isolate-D. *Europian J. Pl. Pathol.*, **105**: 217-220.

Walker, J.C. (1957). White rust of crucifers: 707. In: Plant Pathology, McGraw-Hill Book Co. Inc., New York, U.S.A.

Wang, C.M. (1944). Physiological specialization in *Peronospora parasitica* and reaction of host *Chinese J. Sci. Agric.* **1**: 249-257.

Williams, J.R. and Stelfox, D. (1979). Dispersal of ascospores of *Sclerotinia sclerotiorum* in relation to Sclerotinia stem rot of rapeseed. *Pl Dis Reptr.*, **63**: 395-399.

Williams, J.R. and Stelfox, D. (1980). Influence of farming practices in Alberta on germination and apothecium production of sclerotia of *Sclerotinia sclerotiorum*. *Can. J. Pl. Pathol.* 2: 169-172.

Williams, P.H. and Pound, G.S. (1963). Nature and inheritance of resistance to *Albugo candida* in radish. *Phytopath.*, **53**: 1150-1154.

Woods, D.L. and Petrie, G.A. (1989). Scimitar brown mustard. *Can. J. Plant Sci.*, **69**: 247-248.

Yadav, M.S. (2003). Efficacy of fungitoxicant in the management of Alternaria blight and white rust of mustard. *J. Mycol. Pl. Pathol.*, **33** (2): 307-309.

Yadav, M.S. (2009). Biopesticidal effect of botanicals on the management of mustard diseases. *Indian Phytopath.*, **62** (4): 488-492.

Yadav, M.S. and Barar, K.S. (2003). Relationship between meteorological factors and influence of Alternaria blight and white rust of Indian mustard in south-western Punjab. *Pl. Dis. Res.* **18** (1): 80-82.

Yadav, Y.P. and Singh, H. (1992). The potential exotic sources for white rust resistance in Indian mustard [*Brassica juncea* (L.) Czern and Coss.]. *Oil Crops Newsletter*, **9**: 18.

Zhang, Z.Y., Wang, Y.X. and Lill, Y.L. (1984). Taxonomic studies on the family Albuginaceae of China. II. A new species of *Albugo* on Cruciferae. *Acta Amycologica Sinica*, **3**: 65-71.

9 Breeding for Disease Resistance in Indian Mustard (*Brassica juncea* L.)

H.S. Meena, V.V. Singh, P.D. Meena, B.L. Meena, P.K. Rai and Dhiraj Singh

The family Brassicaceae includes 338 genera and 3709 species (Warwick *et al.*, 2006) comprising crops, weeds and ornamental plants (Love *et al.*, 2005). In this family, there exists a vast diversity of economically important crop forms that are the major source of edible and non-edible oil, vegetable, forages, fodder and condiments. In the genus *Brassica* of the family Brassicaceae, rapeseed-mustard play an important role in human and animal diet, and generally crushed to yield oil (vegetable oil for cooking) and residual protein-rich meal, which is used mainly for animal feed.

In India, rapeseed-mustard (*Brassica* species) are placed at second position in total acreage (23.7 per cent) and production (26 per cent) after soybean among oilseeds crops. In India, three ecotypes of *Brassica* (2n = 20) *i.e. Brassica campestris* ssp. Oleifera *viz.*, yellow sarson, brown sarson and toria, collectively called rape (syn. *B. rapa* L.), *B. juncea* or brown mustard (2n = 36) and *Eruca sativa* or Taramira (2n = 22) are commercially cultivated. Among these, Indian-mustard (*Brassica juncea*) accounts for nearly 75-80 per cent of the area under rapeseed-mustard crops in the country.

Although, the breeding system for a crop species mainly depends upon it's mode of pollination. Therefore, in plant breeding, breeders need to be aware of the centre of origin, existing genetic variability in terms of nature and magnitude, and wild relatives of a crop species, its reproductive behavior, adaptation to environments and cropping systems, and usage for the objective and methods chosen for its genetic improvement. Oilseed Brassicas include a number of crop species ranging from completely cross-pollinated (diploid cultivated species) to self-pollinated (tetraploid species). However, self-pollinated species also behave as often cross-pollinated species with cross pollination to an extent of 14-30 per cent due to the stray pollen contamination and visit by various types of bees particularly honey bees (Rai and Singh, 1976; Rakow and Woods, 1987; Chauhan and Kumar, 1991). Here, we will emphasize primarily on the breeding of Indian mustard (*Brassica juncea*) for disease resistance.

Centre of Origin of *Brassica juncea*

Although the centre of origin of *B. juncea* is not entirely clear, but it is believed to have been derived from natural interspecific hybridization between *B. nigra* (n=8) and *B. rapa* (n=10), since, *B. juncea* is an amphidiploid with a chromosome number of 2n = 36. Several views have been expressed for the centre of origin of *B. juncea*. According to Prakash and Hinata (1980) the species originated in the middle-east Asia and neighboring regions where distributions of *B. nigra* and *B. rapa* overlap. Contrarily, Hemingway (1976) argued that *B. juncea* has probably arisen by hybridization between different *B. campestris* and *B. nigra* genotypes at several different times and localities resulting in secondary centres of origin in China, northeastern India and the Caucasus. Recently, Spect and Diederichsen (2001) support the view that *B. juncea* probably evolved somewhere between Eastern Europe and China, where the progenitor species are sympatric. Although, *B. juncea* species occurs in the temperate as well as tropical regions of the world such as a weed in the south of the European part of Russia, the Caucasus, central Asia and southern Siberia and as a casual or feral plant in south and southeast Asia, Africa and America.

Cytological analysis of chromosome pairing in the progeny of interspecific crosses has clearly established that the three species with higher chromosome numbers are amphidiploids, derived from the mongenomic or diploid species (U, 1935) (Figure 9.1).

Breeding Objectives

Increased seed yield with acceptable seed quality and stability in yield performance across environments is the major breeding objective not only in Indian mustard but also in all crop improvement programmes. Stability in production is sought by incorporating resistance or tolerance to major biotic stresses, such as diseases like Alternaria blight [*Alternaria brassicae* (Berk.) Sacc.], white rust [*Albugo candida* (Pers. Ex Lev.) Kuntze], powdery mildew [*Erysiphe cruciferarum*] and Sclerotinia rot [*Sclerotinia sclerotiorum* (Lib.) de Bary] and the insect pests like mustard aphid [*Lipaphis erysimi*], painted bug (*Bagrada cruciferarum* [*Bagrada hilaris*]), sawfly [*Athalia lugens proximo*], pea leaf miner [*Chromatomyia horticola*] and flea beetle [*Phyllotreta cruciferae*]), parasitic weeds (broomrape [*Orobanche*]) and such abiotic stresses as frost, early or terminal heat, drought, acid, and saline toxicity.

At national level, Indian-mustard is grown in a diverse array of cropping systems. Therefore, it requires the nomination of clear objectives with respect to general and/or specific adaptation of the crop to the existing or new cropping

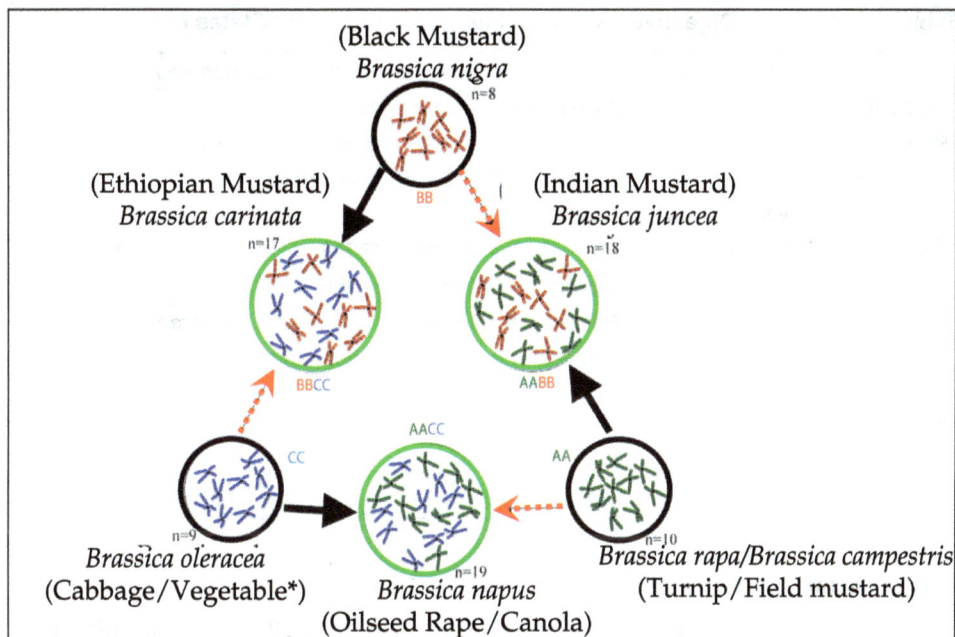

Figure 9.1: The "Triangle of U" Diagram, Showing the Genetic Relationships among Species of the Genus *Brassica*. Chromosomes from each of the genomes A, B and C are represented by different colors. Black and green circles represent diploids and allopolyploids species, respectively. Black (Solid) and Red (broken) arrows in the allopolyploid represent female and male parents, respectively.

systems. The adaptation of improved genotypes of different maturity groups to such systems as; early, medium or full maturity is an important objective in Indian mustard improvement. In addition, comprehensive understanding of different plant attributes such as appropriate duration, rapid early growth, total biomass production and/or high harvest index, and resistance to biotic and abiotic stresses peculiar to the region are specifically needed to realize yield gains and stability. Some objectives will be common across systems but others will be region specific. Major objectives relevant to Indian mustard growing regions are given in Table 1. The effectiveness and efficiency of selection is greatly advanced when the magnitude and nature of genetic variation is understood, and rapid and reliable screening techniques are available.

Resistance to Major Diseases

Among the different types of diseases in Indian mustard, mainly fungal diseases such as Alternaria blight (*Alternaria brassicae*), white rust (*A. candida*) and stem rot (*Sclerminia sclemtionan*) are the major problem that causes considerable yield losses. However, powdery mildew (*Erysiphe cruciferarum*) is also a constraint in some mustard growing areas particularly Gujarat.

Table 9.1: Breeding Objectives in Indian Mustard for different States of India

Objective	Relevance to Potential Production Regions
Yield enhancement	All mustard growing regions
Resistance to Alternaria blight	Punjab, Uttarakhand, UP, Bihar and Jharkhand
Resistance to white rust	Haryana, Punjab, MP, Delhi and Rajasthan
Resistance to powdery mildew	Gujarat
Resistance to Sclerotinia rot	Haryana, Punjab, Delhi, UP, Rajasthan, MP, Bihar and WB
Resistance to mustard aphid	Eastern UP, MP, Assam and WB
Resistance painted bug	Rainfed area of southern Haryana and Rajasthan
Resistance to broomrape	Southern Haryana and Northern Rajasthan
Resistance to frost	Southern Hayana and Northern Rajasthan
Resistance to salinity	Haryana, UP and Bihar
Resistance to drought	Haryana, Rajasthan and Gjurat
Short duration	WB, Bihar, Jharkhand, Orissa and North Eastern States

White Rust

White rust, caused by *Albugo candida*, is an economically important disease of Indian mustard (*B. juncea*), particularly in India. The disease is characterized by two types of symptoms. Initially isolated white pustules or blisters are formed on the leaves and often appear in a circular arrangement around a big central pustule. In later stages, the young stem, inflorescence, floral parts and young siliquae will be infected. The pathogen becomes systemic in infected tissues and causes various types of malformations with prominent hypertrophy and hyperplasia that appears swellings or distortions and finally form "stagheads". These malformations and deformities are the result of hormonal imbalance induced by the pathogen.

Specialized races of *A. candida* are recognized that attack on specific host species. Biological races 1 and 2 are reported from Kangra Valley in Himachal Pradesh. Eleven pathotypes have been reported on different species in North America, which are AC1 (*Raphanus sativus*), AC2 (*Brassica juncea*), AC 2V (*B.napus*), AC3 (*Armoracia rusticana*), AC4 (*Capsella bursa-pastoris*), AC5 (*Sysmbrium officinale*), AC6 (*Rorippa islandica*), AC7 (*Brassica rapa*), AC7V (*Brassica rapa* cv. Reward), AC8 (*Brassica nigra*), AC9 (*B.oleracea*), AC10 (*Sinapis alba*) and AC11 (*Brassica carinata*). Two additional pathotypes AC12 (*Brassica juncea*) and AC13 (*B rapa* var Toria) were reported from India. Nine of 11 reported from North America also exist in India (Verma *et al.*, 1999). Four new distinct pathotypes AC-14, AC-15, AC-16 and AC-17 have been reported by Gupta and Saharan (2002) based on differential interaction of 11 host differentials.

Kaur *et al.* (2011) collected *A. candida* isolates from *B. juncea, B. rapa, B. oleracea, B. tournefortii, Raphanus raphanistrum, R. sativa, Eruca vesicaria* subsp. *sativa, Capsella bursa-pastoris* and *Sisymbrium irio* from different locations in Western Australia (W.A.) to characterize their pathogenicity on different cruciferous host differentials. They reported a new distinct pathogenic strain within race 7 infecting *B. rapa* that had characteristics different from races 7A or 7V. The strains from *B. tournefortii* and *S. irio* being highly host specific, failing to be pathogenic on any other differentials. *B. tournefortii* was host to a strain attacking *B. juncea* and *E. vesicaria* subsp. *sativa*. The strain from *R. raphanistrum* showed a relatively wide host range among the differentials tested. The *B. oleracea* isolate (race 9) was pathogenic to *B. juncea* 'Vulcan' while isolate from *B. juncea* (race 2V) was not pathogenic on *B. oleracea*. Further, the strain from *C. bursa pastoris* (race 4) was pathogenic on *B. juncea* Vulcan but *B. juncea* strain was not pathogenic on *C. bursa pastoris*. In contrast, the strain from *R. sativus* (race 1) was pathogenic on *B. juncea* and the *B. juncea* strain was also pathogenic on *R. sativus*. Field isolates from *B. rapa*, *B. tournefortii*, *E. vesicaria* subsp. *sativa*, and *S. irio* were all nonpathogenic on *B. juncea*. Isolates from *B. juncea* and *R. raphanistrum* were pathogenic on *B. napus* (FAN 189). The study highlighted the pathogen (*A. candida*) diversity and the particular crucifer weeds involved in pathogen inoculum carryover between successive Brassica crops.

White Rust Resistance

The major challenge in breeding for white rust resistance in Brassicas is the prevalence of large number of pathotypes of *Albugo* parasitizing different cruciferous species (Pound and Williams, 1963 and Liu *et al.*, 1996). Race 2 of *A. candida* infects *B. juncea* (Lakra and Saharan, 1988 and Petrie, 1988). Genetic analysis of available white rust resistance through biometrical techniques has revealed a digenic mode of inheritance with duplicate gene action in *B. napus* (Verma and Bhowmik, 1989 and Fan *et. al.*, 1983) and monogenic dominant resistance in *B. juncea* (Tiwari *et al.*, 1988; Bansal *et al.*, 1999 and Sachan *et al.*, 2000) as well as in *B. rapa*, *B. carinata* and *B. nigra* (Delwiche and Williams, 1974 and 1981).

Inheritance of white rust (*Albugo candida*, race 2) resistance in Indian mustard (*Brassica juncea*) was studied by Tiwari *et al.* (1988) using crosses between one resistant and two susceptible cultivars. The reaction of F_1 (all resistant) and segregation of resistant and susceptible plants in F_2 (3: 1) and backcrosses (1: 1) indicated the resistance as dominant monogenic and controlled by nuclear genes and can be easily transferred to adapted susceptible genotypes *via* backcrossing.

Vignesh *et al.* (2009) investigated the mode of inheritance using indigenously developed resistance source and allelic relationship of genes for white rust resistance in two different sources *viz.*, Bio-YSR (*Brassica juncea*) and NPC-12 (*B. carinata*). The inheritance of resistance in donors when crossed with two highly susceptible cultivars, Varuna and Bio-902 indicated the presence of a single dominant gene for white rust resistance. The cross between two resistant sources from *B. juncea* and *B. carinata* segregated in 15: 1 (resistant: susceptible) ratio in F_2 indicated the involvement of two different genes governing white rust resistance in these sources.

Adhikari *et al.* (2003) reported a avirulence gene AvrAc1 in *A. candida* and suggested that a single dominant gene controls avirulence in race Ac2 to *B. rapa* cv. Torch.

Changes in race composition of pathogen have often resulted in short-lived efficiency of host resistance in improved cultivars. Therefore, it is necessary to identify new sources of white rust resistance for durable resistance. White rust resistant donor Bio-YSR has been developed in *B. juncea* which is better under Indian conditions can be utilized in breeding programmes. Some new white rust resistant genotypes of *B. juncea viz.*, DRMR 2019 and DRMR 2035 has been developed at ICAR-DRMR, Bharatpur and registered by NBPGR, New Delhi.

Breeding Methods

The effort have also been made through inter-specific hybridization between *B. juncea* and *B. carinata* to transfer white rust resistance into well adapted high yielding back-ground of *B. juncea* from *B. carinata* through pedigree selection. The monogenic resistance with complete dominance in *B. carinata* could be partially introgressed into *B. juncea* cultivars through selection of disease-free plants in segregating gene-rations grown under heavy disease pressure and repeated back crossing. With the use of resistant gene (L 6) from Canada, many white rust resistance lines have been developed in the genetic background of high yielding varieties. The varieties Basanti, JM-1, JM-2 and JM-3 have been released for general cultivation in white rust prone areas of India.

Molecular Studies

In the genus Brassica, molecular mapping of genes has been reported for various traits like seed coat colour, growth habit, oil content, fatty acid content and resistance to diseases, including white rust (Quiros *et al.*, 2001; Lakshmikumaran *et al.*, 2003). Sources of resistance against white rust are available in east European *B. juncea* gene pool. Genetic analysis of white rust resistance in *B. juncea* has been undertaken at the molecular level to locate gene/s and to identify markers for marker-assisted introgression of the traits using RFLP (Cheung *et al.*, 1998), RAPD (Prabhu *et al.*, 1998; Mukherjee *et al.*, 2001), AFLP (Somers *et al.*, 2002), CAPS (Varshney *et al.*, 2004) and IP markers (Panjabi-Massand *et al.*, 2010). A locus (ACA1) controlling resistance to white rust has been mapped in *B. napus* (Ferreira *et al.*, 1995) and *B. rapa* (Kole *et al.*, 1996), using restriction fragment length polymorphism (RFLP) markers.

Tanhuanpaa (2004) tagged a locus for white rust resistance in a F_2 population of *Brassica rapa* ssp. *oleifera* employing bulked segregant analysis with random amplified polymorphic DNA (RAPD) markers, linkage mapping and a candidate gene approach based on resistance gene analogs (RGAs) using Finnish line Bor4109 as resistance source. The reaction against white rust races 7a and 7v was scored in 20 seedlings in F_2. The proportion of resistant plants among F_3 families varied from 0 to 67 per cent. Bulked segregant analysis did not reveal any markers linked with resistance, thus a linkage map with 81 markers was created. A locus accounting for 18.4 per cent of the variation in resistance for white rust was mapped to linkage group (LG) 2 near the RAPD marker Z19a. During the study, a bacterial resistance gene homologous to *Arabidopsis* RPS2 and six different RGAs were sequenced. RPS2 and five of the RGAs were mapped to linkage groups LG1, LG4 and LG9. Unfortunately, none of the RGAs could be shown to be associated with white rust resistance.

Singh *et al.* (2015) effectively applied the already identified *Arabidopsis*-derived intron polymorphic (IP) markers At5g41560 and At2g36360, which were highly linked with AcB1-A4.1 and AcB1-A5.1, respectively and validated in a set of 25 genotypes of Indian Mustard and three F_2 populations. The relationships between the variation of PCR products of the two markers with the percent disease index (PDI) of the tested genotypes, and the co-segregation analysis of the markers with disease phenotype in F_2 populations clearly indicated that At5g41560 and At2g36360 are genotype-nonspecific markers and are closely linked to white rust resistance loci AcB1-A4.1 and AcB1-A5.1, respectively. It also became evident from the study that AcB1-A4.1 and an another white rust resistance locus Ac(2)t are likely the same gene locus. These markers can further be used in marker assisted breeding for gene pyramiding.

Alternaria Blight

Alternaria Blight (*Alternaria brassicae*) disease has been reported from all the continents of the world and is one of the important diseases of Indian mustard causing up to 47 per cent yield losses (Kolte, 1985). In India, three distinct isolates of *A. brassicae* namely A (highly virulent), C (moderately virulent) and D (avirulent) has been reported (Vishwanath and Kolte, 1997). In Indian mustard, no proven source of resistance against *Alternaria* blight reported yet, but some sources of tolerance against this disease have been identified (Gupta *et al.*, 2001). Moreover, some resistance sources have been identified in related and wild species of *Brassica* like *Sinapis alba* L. (Hansen and Earle, 1997), *Camelina sativa*, *Capsella bursa-pastoris*, *Neslia paniculata* and taramira (Tewari and Conn, 1993), *Doplotaxis berthautii*, *D. catholica, D. cretacea, D. erucoides* and *Erucastrum gallicum* (Sharma *et al.*, 2002). Inheritance studies suggested that tolerance to *Alternaria* blight is governed by multigenes or cluster of genes with additive or partial dominance effects (Zhang *et al.*, 1997; Krishnia *et al.*, 2000).

Inter-specific Variation

As mentioned above adequate source of resistance is not yet available in *B. juncea*. However, large variation among various genera and species has been reported for disease reaction. Weedy and wild taxa may become rich reservoir of genes for disease resistance. Lesion type, number and size has been used as selection criterion in breeding for Alternaria blight resistance. Among cultivated species, *B. carinata* and *B. napus* reported to be more resistant followed by *B. nigra, B. juncea* and *B. campestris* in that order (Bhowmik and Munde, 1987). Sharma and Singh (1992) reported *B. hirta* as most resistant among *B. juncea, B. carinata, B. napus, B. rapa* and *B. hirta*. Contrary to these observations, Tewari (1991) reported that there is hardly any variability between resistant and susceptible host at the early stage of infection. Among the related species, *Camelina sativa*, Capsella *bursa-pastoris* and *Neslia paniculata* showed no disease symptoms mainly because of their inhibition capacity of fungal growth on leaf surfaces by producing phytoalexin shortly after infection. *Eruca sativa* (Miller) Thell showed localized flacks and *Armoracia rusticana* Gaertn., show slow necrosis and chlorosis. Pre and post fertilization barriers restrict the transfer of resistant genes from these wild species into *B. juncea* and *B. campestris*. Embryo rescue technique can help in transfer of resistance from these species.

Resistance against *Alternaria*

Host resistance against *Alternaria* species has various components and it is multilayered. Inheritance of resistance in inter and intra-specific crosses of *B. juncea* and *B. carinata* to *A. brassicae* is governed by additive genes, dominant genes, additive x additive type epistatic genes, additive x dominance and dominance x dominance type of non-allelic interaction genes. Inter-mating between tolerant plants helps in increasing the level of resistance against *A. brassicae* by pyramiding of resistant genes. High level of horizontal resistance in genotypes of oilseed *Brassicas* has been recorded (Saharan and Krishnia, 2001). Genotypes PR-8988, PR-9024, PAB-9511, PAB-9534, EC-399296, EC-399299, and PHR-2 show higher degree of tolerance/ partial resistance or slow blighting. Epicuticular wax (Candle, Tobin, and Tower), low number and narrow stomatal aperture (Tower, RC-781) provide resistance to *Alternaria* infection in *Brassica* species (Saharan and Kadian, 1983; Tewari, 1991). The concentration of phenolic compounds, activation of polyphenol oxidase and catalase is higher in tolerant genotypes. Chitinase modifying proteins (cmps) are secreted by fungal pathogens of crucifers, which interfere with fungalysin cmp activity to improve plant resistance against multiple fungal diseases (Naumann and Wicklow, 2013). GLIP1 in association with ethylene signaling may be a critical component in plant resistance (*A. thaliana*) to *A. brassicicola* (Oh *et al.*, 2005). *B. juncea* plants transformed with chitinase gene tagged with an over expressing promoters, 35S CaMV give defense response by degrading the cell walls of invading fungi (Mondal *et al.*, 2003). β–aminobutyric acid treatment leads to proper balance of oxidant and antioxidants suitable for expression of resistance in *B. carinata* against *A. brassicae* by curtailing pathogens entrance during early stages of colonization (Chavan *et al.*, 2013). Zeatin a cytokinin up regulates plant immunity via an elevation of MAPK-4 and antagonizes the effects of *A. brassicae* (Marmath *et al.*, 2013).

Transgenic expression of hevein, the rubber tree lectin, in *B. juncea* cv. RLM-198 confers protection against *A. brassicae* (Kanrar *et al.*, 2002), β–amino-butyric acid pre treatment of *B. juncea* plants induces *A. brassicae* resistance mediated through an enhanced expression of pathogenesis related protein genes, independent of SA and JA accumulation (Kamble and Bhargava, 2007). The cDNA encoding Pm AMP 1 has been successfully incorporated into the genome of *B. napus*, and it is in planta expression confers greater protection against *A. brassicae*. Pm AMP1 is a cysteine rich antimicrobial peptide isolated from *Pinus monticola* (Verma *et al.*, 2012). Combined expression of a barley class II chitinase and Type I ribosome inactivating protein in transgenic *B. juncea* provides resistance against *A. brassicae* (Chhikara *et al.*, 2012). Transcriptional responses to exposure to the brassicaceous defense metabolites camalexin and allyl-isothiocyanate in *A. brassicicola* have been recorded (Sellam *et al.*, 2007).

Elicitation and accumulation of phytoalexins in crucifers after exposure to Alternaria and their role in disease resistance have been demonstrated (Verma and Saharan, 1994). Calcium sequestration property of *A. brassicae* can be used in enhancing resistance to this pathogen in rapeseed by soil or foliar application of calcium compounds (Tewari, 1991). In-spite of several bottlenecks in the development of resistant cultivars, various methods/techniques including conventional as well as biotechnological approaches are being utilized to incorporate desired traits in cruciferous crops against Alternaria blight.

Biochemistry

During *Alternaria*-crucifers host pathogen interactions number of biochemical changes produce various kinds of primary and secondary metabolites, which influence the host defense system and pathogen virulence. *A. brassicicola* produces compounds like antitumoric depudecin, antibiotic complex brassicicolin and phytotoxic brassicicenes. The production of glucosinolates and phytoalexins has been correlated with host resistance (Verma and Saharan, 1994).

Alternaria species pathogenic to crucifers produce host specific and non-host specific toxins, which facilitate their pathogenic process to become successful pathogen. Prior to colonization, necrotrophs must kill their host cells at a distance by producing both toxins and lytic enzymes often by triggering genetically programmed apoplastic pathways, or by directly causing cell damage resulting in necrosis. *A. brassicae* and *A. brassicicola* pathogenic to crucifers produce number of toxins and metabolites belonging to chemical groups, terpenoides, pyranones, steroids, and nitrogen containing. Effects of toxins on plants at physiological, biochemical and molecular level have been investigated. The role of toxins in the process of infection, their biosynthesis, mode of action, chemical structure, role in host defense, transformation into phytoalexins have been understood (Verma and Saharan, 1994; Bains and Tewari, 1989; Thomma, 2003).

Breeding Approaches for Resistance

Genetics of Alternaria blight resistance in inter and intra-specific crosses of *Brassica juncea* and *Brassica carinata* have been determined for the inheritance of resistance to *Alternaria brassicae*. There is preponderance of additive gene effects compared to dominance gene effects. Therefore, reciprocal recurrent selection or diallel selective matings can be successfully used to develop tolerant/resistant genotypes against Alternaria blight (Saharan and Krishnia, 2001). Inter-mating between resistant plants helps in increasing the level of resistance. The frequency of favourable genes for disease resistance increases in the population thus enhancing the probability of obtaining multiple disease resistance (Saharan and Krishnia, 2001). The selection for resistance against Alternaria blight should be done in inter-mated population rather than in F_2 and F_3 population, since the generations of inter-varietal hybrids would prevent the harmful effects of linkages and linkage disequilibrium, and shuffle the desirable genes in one recombinant.

Genetic Engineering

Non-availability of resistance sources within crossable germplasm of Brassica needs the use of genetic engineering approaches to develop genetic resistance against Alternaria blight. A number of genes for imparting resistance this fungus have been transferred to *B. juncea* via genetic transformation technique. *Osmotin* protein introgressed into Indian mustard delayed the appearance of symptoms of Alternaria blight disease (Taj *et al.*, 2004). Class I basic *glucanase* gene from tomato have been transformed by Mondal *et al.* (2007) into *B. juncea* var. RLM 198 and found that the transgenic plants expressing *glucanase* exhibited restricted number, size and spread of lesions caused by *A. brassicae* under pathogen-challenged conditions and the onset of disease was also delayed as compared to the non-transformed plants. Rustagi *et al.* (2014) reported a significant inhibition in hyphal growth of both *A. brassicae* and *Sclerotinia sclerotiorum* in transgenic Indian mustard plants developed with *msrAI* gene coding for antimicrobial peptides, potential for resistance against a broad spectrum of phytopathogens. Hada *et al.* (2015) developed transgenic *B. juncea* cv. Varuna plants with *thaumatin* gene through *Agrobacterium tumefaciens*-mediated genetic transformation technique. T_1 transgenic lines expressing the *thaumatin* gene showed an enhanced resistance against *Alternaria brassicae* by inhibiting the fungal growth up to 54 per cent as compared non-transformed plants.

Sclerotinia Stem Rot

Sclerotinia sclerotiorum (Lib.) de Bary, a causative pathogen of *Sclerotinia* rot has been discovered and studied for more than 150 years. *Sclerotinia sclerotiorum* has a vast host range which coupled with a lack of host specificity, makes the breeding of resistant varieties a difficult task. Therefore, *Sclerotinia* rot is still one of the most devastating disease in rapeseed mustard crop worldwide and the yield losses estimated as high as 100 per cent depending upon the percentage of plants infected and the stage of growth of the crop at the time of infection (Morral *et al.*, 1976; Purdy, 1979).

In India, *Sclerotinia* rot was first reported by Shaw and Ajrekar (1915). Now it has been a threat to cultivation of oilseed *Brassica* in Rajasthan, Haryana, Madhya Pradesh, Uttar Pradesh, Bihar and Punjab states of India. During past few years, *Sclerotinia* stem rot has come out to be the highly devastating disease causing considerable yield losses to mustard crop. To date, complete resistance to the causative pathogen has not been identified, although few genetic sources with partial resistance to the pathogen are available for breeders. Genetic studies demonstrated that *Sclerotinia* disease is a quantitative trait (Gentzbittel *et al.*, 1998; Zhao and Meng, 2003) and the effective controls for this disease are to avoid the planting on infested soil and to prevent build-up of sclerotia in soils.

Host Range

Sclerotinia sclerotiorum is the most nonspecific, omnivorous and successful plant pathogen. It has a broad host range that makes its control very difficult; hence it restricts the number of non-host crops which can be included in crop rotations. The records of susceptible hosts of *Sclerotinia sclerotiorum* are scattered throughout literature. The most recent host index for *S. sclerotiorum* prepared by Boland and Hall (1994) contains 42 subspecies or varieties, 408 species, 278 genera, and 75 families of plants.

Symptoms

Initially the elongated water-soaked lesions appear at the base of the stem that expand rapidly. The lesions generally become bleached and necrotic and subsequently develop patches of fluffy white mycelium, which is the most obvious sign of mycelial infection (Bolton *et al.*, 2006). When the stem is completely gridled and covered by a cottony mycelial growth, the plant wilts and dries. The affected stem tends to shred and numerous grayish-white to black, spherical sclerotia appear either on the surface or in the pith of the affected stem (Sharma *et al.*, 2015). When the crop is at seed maturation stage, the plants tend to lodge, touching the siliquae to the soil level. Such plants, though remaining free from stem or aerial infection throughout, show rotting of the siliquae with profuse fungal growth, along with sclerotial bodies just above the soil level.

Sources of Resistance

Breeding for resistance against *S. sclerotiorum* is a very challenging because of wide host range of the pathogen. Although, partial resistance has been identified in some *B. napus* and *B. juncea* genotypes from China, Australia and India, breeding to increase the levels of resistance against Sclerotinia disease in *B. napus* and/or *B. juncea* has been ineffective. This is mainly because resistance against *S. sclerotiorum* in existing cultivars and cultivated germplasm appears to be of a complex nature, *i.e.*, it can either be monogenic and/or polygenic depending on the plant species and materials under investigation (Baswana *et al.*, 1991; Zhao *et al.*, 2006). Partial field resistance has been identified in the Chinese variety Zhongyou 821 (Li *et al.*, 1999; Buchwaldt *et al.*, 2003). Nine genotypes *viz.*, Cutton, ZYR-6, PSM-169, PDM-169, Westar, PYM-7, Parkland, Tobin and Candle showed resistance to stem rot in India (Shivpuri *et al.*, 1997). Four genotypes *viz.*, PCR-10, RW-8410, RW- 9401 and RGH-8006 had better resistance against *S. sclerotiorum* as compared to susceptible check (Pathak *et al.*, 2002). The genotypes of *B. juncea* namely, EC 597328 (Montara), EC 597329 (Berry) and EC 597331 (Ringot I) of Chinese origin were also found tolerant (Sharma *et al.*, 2009).

Novel approaches to the introduction of increased *Sclerotinia* resistance into oilseed rape include the development of transgenic plants expressing oxalate oxidase activity (Thompson *et al.*, 1995) capable of degrading oxalic acid, a pathotoxin of the fungus. For evaluation of resistance to rape SR, Sang *et al.* (2013) transferred the *MSI-99m* gene (modified *MSI-99*) into Chinese rape variety Zhongyou 821 using *Agrobacterium*-mediated method. Nine transformed lines carrying a *MSI-99m* expression vector were detected by polymerase chain reaction (PCR), among which seven lines expressed *MSI-99m* gene according to qRT–PCR analysis. Disease resistance analysis consistently showed that the high-level expression of *MSI-99m* increased resistance to *S. sclerotiorum* in transgenic rape lines. It was demonstrated that *MSI-99m* gene might be applied as a resistant gene resource in rape for the improvement of rape varieties. Glucosinolates present in Brassicaceae plants have fungicidal and bactericidal properties which can also give resistance against fungal pathogens. Induced mutagenesis has also proved to be a successful technique for introducing disease resistance genes into crop varieties. Induced mutants exhibiting increased partial (quantitative) disease resistance have been isolated from small populations (Kinane and Jones, 1996).

Lack of effective resistance against Sclerotinia in cultivated species has stimulated the interest of researchers towards exploitation of wild crucifer species to diversify the existing gene pool. Brassicaceae family contains a wide array of different species, only two wild crucifers, *Capsella bursa-pastoris* and *Erucastrum gallicum* have been reported to show high level of resistance against Sclerotinia. Although introgressive hybrids were successfully obtained between *B. rapa and B. napus* species and *Capsella bursa-pastoris* (Chen *et al.*, 2007), it remains to be confirmed if the introgression of resistance against *S. sclerotiorum* from *E. gallicum* into cultivated species has in fact been accomplished. Introgression lines following hybridization of three wild crucifers (*Erucastrum cardaminoides, Diplotaxis tenuisiliqua* and *E. abyssinicum*) with *B. napus* or *B. juncea* have been developed had much higher levels of resistance (Garg *et al.*, 2010).

Screening Techniques

Various screening techniques have been used by various workers for particular disease. However standard methods adopted in All India Coordinated Research Project (rapeseed and mustard) for screening of resistant/tolerant genotypes against white rust, Alternaria blight, Sclerotinia stem rot and downey mildew is detaled below: (Proceedings, 24th AGM of AICRP-RM, ICAR-DRMR Bharatpur).

Method of Artificial Inoculation for White Rust

Test plants (including checks) should be inoculated twice *i.e.* at initiation of flowering and pod formation stage. Inoculum may be prepared by collecting fresh zoosporangia from naturally infected leaves with *Albugo candida*. Petri plates containing zoosporangia suspended in distilled water be kept at 4°C for 2 h to facilitate germination of zoosporangia. To assure germination of the sporangia, the plates may be examined under the low power microscope. Germinating zoosporangia would be emptied and zoospores will be visible in the suspension. Suspension containing zoospores be filtered through double layered muslin cloth and further diluted with distilled water for spraying on leaves of test plants. This may be done with the help of atomizer/small sprayer in the afternoon (after 1500 hrs). Data for disease severity may be recorded 75 DAS/at maximum disease pressure on leaves and 15 days before harvest for staghead.

Method of Artificial Inoculation for Alternaria Blight

Test plants (including checks) should be inoculated twice *i.e.* at initiation of flowering and pod formation in the afternoon (after 1500 hrs) with conidial suspension (10^5 cfu/ml) of pure culture of *Alternaria brassicae* using distilled water. Disease severity should be recorded 90 DAS/at maximum disease pressure on leaves and 15 days before harvest on pods.

Method of Artificial Inoculation for Sclerotinia Rot

Cut 5 mm discs of fungal mycelium along with medium from 7 days old pure culture *Sclerotinia sclerotiorum* grown on thick layer of PDA at 20°C. Place one disc on third internode of plant at flowering stage. Wrap the stem along with the fungal agar disc with a swab of cotton dipped in sterile distilled water. Record observation when the check shows development of long disease lesions (> 3 cm) tentatively 20 days before harvest.

Method of Artificial Inoculation for Downy Mildew

Freshly harvested conidia in distilled water from naturally/artificially infected cotyledonary leaves of susceptible variety should be used for preparation of conidial suspension. Inoculum concentration should be adjusted to 10^4 conidia/ml using distilled water. Prepared conidial suspension of *Hyaloperonospora parasitica* should be inoculated directly to test plants at 2/3 leaf stage in the afternoon (after 1500 hrs). Observations should be recorded after 7-10 days of inoculation.

Observations to be Recorded

1. Date of first appearance of each disease including bacterial rot
2. Data as percent disease severity/percent disease incidence for WR (75 DAS/at maximum disease pressure), AB/PM/BR (90 DAS/at maximum disease pressure) on leaves and pods and number of staghead (15 days before harvest) should be recorded on 10 randomly selected plants from each plot using 0-9 scale. Date of observation and date of sowing should be indicated in data sheet itself.
3. Cotyledonary infection due to downy mildew and pod infection due to Alternaria blight should be recorded separately.

   Staghead formation should be recorded as percent incidence and percent twigs infected. Staghead (per cent twigs affected) = (number of twigs infected/total number of twigs) x 100.
4. Data for all major diseases may be recorded as percent disease severity (AB, WR and PM) on leaves/pods or as percent disease incidence (SR, DM, CR, BR).
5. Date of each observation should be provided in the data sheet.
6. Data should be statistically analysed as per the design using ANOVA after arc sin transformation. Actual and transformed (in parenthesis) values along with mean, CD ($P< 0.05$) and CV (per cent) are to be submitted for report preparation.

Scale (0-9) for rating of entries for reaction to Alternaria blight, white rust and powdery mildew should be used.

0 (Immune for WR) = No lesion

1 (HR) = Non-sporulating pinpoint size or small brown necrotic spots, less than 5 per cent leaf area covered by lesion

3 (R) = Small roundish slightly sporulating larger brown necrotic spots, about 1-2 mm in diameter with a distinct margin or yellow halo, 5-10 per cent leaf area covered by lesions

5 (MR) = Moderately sporulating, non-coalescing larger brown spots, about 2-4 mm in diameter with a distinct margin or yellow halo, 11-25 per cent leaf area covered by the spots

7 (S) = Moderately sporulating, coalescing larger brown spots about 4-5 mm in diameter, 26-50 per cent leaf area covered by the lesions

9 (HS) = Profusely sporulating, rapidly coalescing brown to black spots measuring more than 6mm diameter without margins covering more than 50 per cent leaf area

$$*AVS = \frac{(N\text{-}1 \times 0) + (N\text{-}2 \times 1) + (N\text{-}3 \times 3) + ((N\text{-}4 \times 5) + (N\text{-}5 \times 7) + (N\text{-}6 \times 9)}{\text{Number of leaf samples}}$$

* AVS: Average Severity Score

$$*PDI = \frac{(N\text{-}1 \times 0) + (N\text{-}2 \times 1) + (N\text{-}3 \times 3) + ((N\text{-}4 \times 5) + (N\text{-}5 \times 7) + (N\text{-}6 \times 9)}{\text{No. of leaf samples} \times 9} \times 100$$

* PDI: Percent Disease Intensity

where,

N-1 to N-6 represents frequency of leaves in the respective score

Note

1. The word spots can be read as pustules if the same scale is used for white rust rating
2. In case of white rust, brown spot can be read as creamy white pustule
3. This scale can also be used in management trials
4. For PM, the same rating scale will be followed ignoring the lesion/pustule characteristics

Scale (0-4) for rating of entries for reaction to Sclerotinia rot

Reaction	Rating	Lesion (cm)
Resistant	0	< 3
Moderately resistant	1	3-5
Moderately susceptible	2	5-10
Susceptible	3	10-15
Highly susceptible	4	>15

* Stem diameter and per cent incidence must also be recorded.

References

Adhikari, T.B., Liu, J.Q., Mathur, S., Wu, C.X., and Rimmer, S.R. (2003). Genetic and molecular analyses in crosses of race 2 and race 7 of *Albugo candida*. *Phytopathology*, 93: 959-965.

Bains, P.S. and Tewari, J.P. (1989). Bioassay and mode of action of an *Alternaria brassicae* toxin. *Can. J. Plant Pathol.*, 11: 184-189.

Bansal, V.K., Thiagarajah, M.R., Stringam, G.R. and Tewari, J. P. (1999). Inheritance of partial resistance to race 2 of *Albugo candida* in canola-quality mustard (*Brassica juncea*) and its role in resistance breeding. *Plant Pathol.*, 48: 817–822.

Baswana, K.S., Rastogi, K.B. and Sharma, P.P. (1991). Inheritance of stalk rot resistance in cauliflower (*Brassica oleracea var. Botrytis* L.). *Euphytica,* 57: 93–96.

Bhowmik, T.P. and Munde, P.N. (1987). Identification of resistance in rapeseed and mustard against *Alternaria brassicae* (Berk.) Sacc. and some resistance sources. *Beitrage trop. Landwirtsch. Veterinarmed,* 25: 49-53.

Boland, G.J. and Hall, R. (1994). Index of plant hosts of *Sclerotinia. Can. J. Plant Pathol.,* 16: 93–108.

Bolton, D.M., Thomma, PHJB and Nelson, D.B. (2006). *Sclerotinia sclerotiorum* (Lib.) de Bary: biology and molecular traits of a cosmopolitan pathogen. *Mol. Plant Pathol.,* 7: 1–16.

Buchwaldt, L., Yu, F.Q., Rimmer, S.R. and Hegedus, D.D. (2003). Resistance to *Sclerotinia sclerotiorum* in a Chinese *Brassica napus* cultivar. 8th International Congress of Plant Pathology, Christchurch, New Zealand, p. 289.

Chauhan, Y.S.R. and Kumar, K. (1991) Natural cross pollination in Indian mustard. *Cruciferae Newslett.,* 14/15: 24-25.

Chavan, V., Bhargava, S., Kamble, A. (2013). Temporal modulation of oxidant and antioxidative responses in *Brassica carinata* during β-aminobutyric acid-induced resistance against *Alternaria brassicae. Physiol. Mol. Plant Pathology,* 83: 35–39.

Chen, H.F., Wang, H. and Li, Z.Y. (2007). Production and genetic analysis of partial hybrids in intertribal crosses between *Brassica* species (*B.rapa, B. napus*) and *Capsella bursa-pastoris. Plant Cell Reptr.,* 26: 1791-1800.

Cheung, W.Y., Gugel, R.K. and Landry, B.S. (1998). Identification of RFLP markers linked to the white rust resistance gene (Acr) in mustard (*Brassica juncea* (L.) Czern. and Coss). *Genome,* 41: 626-628

Chhikara, S., Chaudhury, D., Dhankher, O.P. and Jaiwal, P.K. (2012). Combined expression of a barley class II chitinase and type I ribosome inactivating protein in transgenic Brassica juncea provides protection against Alternaria brassicae. *Plant Cell Tiss. Organ Cult.,* 108: 83-89.

Delwiche, P.A. and Williams, P.H. (1974). Resistance to *Albugo candida* race 2 in *Brassica* spp. *Proc. Am. Phytopathol.,* Soc., 1: 66.

Delwiche, P.A. and Williams, P.H. (1981). Thirteen marker genes in *Brassica nigra. J. Hered.,* 72: 289-290.

Fan, Z., Rimmer, S.R. and Stefanson, B.R. (1983). Inheritance of resistance to *Albugo candida* in rape (*Brassica napus* L). *Can. J. Genet. Cytol.,* 25: 420-424.

Ferreira, M.E., Williams, P.H. and Osborn, T.C. (1995). Molecular Mapping of a Locus Controlling Resistance to Albugo candida in *Brassica napus* using molecular markers. *Phytopathology,* 85: 2018-220.

Garg, H., Atri, C., Sandhu, P.S., Kaur, B., Renton, M., Banga, S.K., Singh, H., Singh, C., Barbetti, M.J. and Banga, S.K. (2010). High level of resistance to *Sclerotinia sclerotiorum* in introgression lines derived from hybridization between wild crucifers and the crop *Brassica* species *B. napus* and *B. juncea. Field Crops Res.,* 117: 51-58.

Gentzbittel, L., Mouzeyar, S., Badaoui, S. and Mestries, E. (1998). Cloning of molecular markers for disease resistance in sunflower (*Helianthus annuus* L.) *Theor. Appl. Genet.*, 96: 519-525.

Gupta, K. and Saharan, G.S. (2002). Identification of pathotype of *Albugo candida* with stable characteristic symptoms on Indian mustard. *J. Mycol. Pl. Pathol.*, 32: 46-51.

Gupta, K., Saharan, G.S. and Singh, D. (2001). Sources of resistance in Indian mustard against white rust and *Alternaria* blight. *Cruciferae Newslett.*, 23: 59-60.

Hada, A., Rawat, S., Krishnan, V., Jolly, M., Jeevaraj, T., Manickavasagam, M., Ganapathi, A., Sachdev, A. and Grover, A. (2015). Overexpression of thaumatin gene confers enhanced resistance to *Alternaria brassicae* and tolerance to salinity and drought in transgenic *Brassica juncea* (L.) Czern. *Plant Cell Tiss. Organ Cult.*, DOI 10.1007/s11240-015-0846-8

Hansen, L.H. and Earle, E.D. (1997). Somatic hybrids between *Brassica oleracea* (L.) and *Sinapis alba* (L.) with resistance to *Alternaria brassicae*. *Theor. Appl. Genet.*, 94: 1078-1085.

Hemingway, J.S. (1976). Mustard: Brassica spp. And Sinapis alba (cruciferae). In: N.W. Simmonds (ed.), Evolution of crop plants. Longmans, London, UK, pp. 56-59.

Hill, C.B., Crute, I.R., Sherri, C. and Williams, P.H. (1988). Specificity of *Albugo candida* and Peronospora parasitica pathotypes toward rapid-cycling crucifers. *Cruciferae Newslett.*, 13: 112-113

Kamble, A., and Bhargava, S. (2007). β–Aminobutyric Acid-induced Resistance in Brassica juncea against the Necrotrophic Pathogen *Alternaria brassicae*. *J. Phytopathol.*, 155, 152–158.

Kanrar, S., Venkateswari, J.C., Kirti, P.B., and Chopra, V.L. (2002). Transgenic expression of hevein, the rubber tree lectin, in Indian mustard confers protection against *Alternaria brassicae*. *Plant Science*, 162, 441-448.

Kanrar, S., Venkateswari, J., Kirti, P. and Chopra, V. (2002). Transgenic Indian mustard (*Brassica juncea*) with resistance to the mustard aphid (*Lipaphis erysimi* Kalt.). *Plant Cell Reports*. 20(10): 976-981.

Kaur, P., Sivasithamparam, K. and Barbetti, M.J. (2011). Host range and phylogenetic relationships of *Albugo candida* from cruciferous hosts in Western Australia, with special reference to *Brassica juncea*. *Plant Dis.*, 95: 712-718.

Kinane, S.J. and Jones, P. (1996). Isolation and characterisation of induced wheat mutants exhibiting partial resistance to powdery mildew. *Cereal Rusts and Powdery Mildew Bull.*, 24: 214–217.

Kole, C., Teutonico, R., Mengistu, A., Williams, P.H. and Osborn, T.C. (1996). Molecular mapping of a locus controlling resistance to *Albugo candida* in Brassica rapa. *Phytopathology*, 86: 367-369.

Kolte, S.J. (1985). Diseases of annual edible oilseed crops. Volume II: Rapeseed-mustard and sesame diseases. pp. 135. CRC Press Inc. Boca Raton, Florida.

Krishnia, S.K., Saharan, G.S. and Singh, D. (2000). Genetic variation for multiple disease resistance in the families of interspecific cross of *Brassica juncea* x *B. carinata*. *Cruciferae Newslett.*, 22: 51-53.

Lakra, B.S. and Saharan, G.S. (1988). Morphological and pathological variations in *Albugo candida* associated with *Brassica* species. *Indian J. Mycol. Plant Pathol.*, 18: 149-156.

Lakra, B.S. and Saharan, G.S. (1989). Correlation of leaf and staghead infection intensities of white rust with yield and yield components of mustard. *Indian J. Mycol. Plant Pathol.*, 19: 279-281.

Lakshmikumaran, M., Das, S. and Srivastava, P.S. (2001). Application of Molecular Markers in Brassica Coenospecies: Comparative Mapping and Tagging. In: Nagata, Toshiyuki, Tabata, Satoshi (Eds.) Biotechnology in Agriculture and Forestry, vol 52 Brassicas and Legumes From Genome Structure to Breeding, Springer, Berlin, Heidelberg, New York, pp. 37-68.

Li, Y., Chen, J., Bennett, R., Kiddle, G., Wallsgrove, R., Huang, Y. and He, Y. (1999). Breeding, inheritance, and biochemical studies on *Brassica napus* cv. Zhongyou 821: tolerance to *Sclerotinia sclerotiorum* (stem rot).), Proc 10th International Rapeseed Congress, Canberra, Australia, September 26-29, p. 61.

Liu, J.Q., Parks, P. and Rimmer, S.R. (1996). Development of monogenic lines for resistance to *Albugo candida* from a Canadian *Brassica napus* cultivar. *Phytopathol.*, 86: 1000–1004.

Love, C.G., Robinson, A.J., Lim, G.A.C., Hopkins, C.J., Batley, J., Barker, G., Spangenberg, G.C., Edwards, D. (2005). Brassica ASTRA: an integrated database for Brassica genomic research. *Nucleic Acids Res.*, 33: D656-D659.

Marmath, K.K., Giri, P., Taj, G., Pandey, D. and Kumar, A. (2013). Effect of zeatin on the infection process and expression of MAPK-4 during pathogenesis of *Alternaria brassicae* in non-host and host *Brassica* plants. *African J. Biotech.*, 12 (17): 2164-2174.

Mondal, K.K., Bhattacharya, R.C., Koundal, K.R. and Chatterjee, S.C. (2006). Transgenic Indian mustard (*Brassica juncea*) expressing tomato glucanase leads to arrested growth of *Alternaria brassicae*. *Plant Cell Rep.*, DOI 10.1007/ s00299-006-0241-3.

Mondal, K.K., Bhattacharya, R.C., Koundal, K.R. and Chatterjee, S.C. (2007). Transgenic Indian mustard (Brassica juncea) expressing tomato glucanase leads to arrested growth of Alternaria brassicae. *Plant Cel. Rep.*, 25: 247-252.

Mondal, K.K., Chatterjee, S.C., Viswakarma, N., Bhattacharya, R.C., Grover, A. (2003). Chitinase-mediated inhibitory activity of *Brassica* transgenic on growth of *Alternaria brassicae*. *Current Microbiology*, 47, 171–173.

Morrall, R.A.A., Dueck, J. McKenzie, D.L. and McGee. D.C. (1976). Some aspects of *Sclerotinia sclerotiorum* in Saskatchewan, 1970-75. *Canadian Plant Dis. Surv.*, 56: 56-62.

Mukherjee, A.K., Mohapatra, T., Varshney, A., Sharma, R. and Sharma, R.P. (2001). Molecular mapping of a locus controlling resistance to Albugo candida in Indian mustard. *Plant Breeding*, 120(6): 483-497.

Naumann, T.A., and Wicklow, D.T. (2013). Chitinase modifying proteins from phylogenetically distinct lineages of *Brassica* pathogens. *Physiological and Molecular Plant Pathology*, 82: 1-9.

Oh, I.S., Ae Ran Park, Min Seok Bae, Sun Jae Kwon, Young Soon Kim, Ji Eun Lee, Na Young Kang, Sumin Lee, Hyeonsook Cheong, and Ohkmae K Park. (2005). Secretome analysis reveals an *Arabidopsis* lipase involved in defense against *Alternaria brassicicola*. *The Plant Cell*, 17: 2832–2847.

Panjabi-Massand, P., Yadava, S.K., Sharma, P., Kaur, A., Kumar, A., Arumugam, N., Sodhi, Y.S., Mukhopadhyay, A., Gupta, V., Pradhan, A.K. and Pental, D. (2010). Molecular mapping reveals two independent loci conferring resistance to Albugo candida in the east European germplasm of oilseed mustard *Brassica juncea*. *Theor. Appl. Genet.*, 121: 137-145.

Pathak, A.K., Godika, S., Jain, J.P. and Muralia, S. (2002). Screening of *Brassica* genotypes against stem rot disease of mustard caused by *Sclerotinia sclerotiorum* (Lib.) de Bary. *J. Mycol. Pl. Pathol.*, 32: 111-112.

Petrie, G.A. (1988). Races of *Albugo candida* (white rust and staghead) on cultivated Cruciferae in Saskatchewan. *Can. J. Plant Pathol.*, 10: 142-150.

Pound, G.S. and Williams P.H. (1963). Biological races of *Albugo candida*. *Phytopathol.*, 53: 1146–1149.

Prabhu, K.V., Somers, D.J., Rakow, G. and Gugel, R.K. (1998). Molecular markers linked to white rust resistance in mustard *Brassica juncea*. *Theor. Appl. Genet.*, 97: 865-870.

Purdy, L.H. (1979). *Sclerotinia sclerotiorum*: History, diseases, and symptom pathology, host range, geographic distribution, and impact. *Phytopathol.*, 69: 875-880.

Quiros, C. F., Grellet, F., Sadowski, J., Suzuki, T., Li, G. and Wroblewski, T. (2001). Arabidopsis and Brassica Comparative Genomics: Sequence, Structure and Gene Content in the *ABI1-Rps2-Ck1* Chromosomal Segment and Related Regions. *Genetics*, 157 (3): 1321-1330.

Rai, B. and Singh, A. (1976). Commercial seed production in rapeseed. Ind. Farm 26: 15-17.

Rakow, G. and Woods, D.L. (1987) Outcrossing in rape and mustard under Saskatchewn prairic conditions. *Canadian J. plant Sci.*, 67: 147-151.

Rimmer, S.R. and Buchwaldt, L. (1995). Diseases. In: Kimber D, McGregor DI (eds) Brassica oilseeds production and utilization. CAB International, Oxford, UK, pp. 111-140.

Rustagi, A., Kumar, D., Shekhar, S., Yusuf, M.A., Misra, S. and Sarin, N.B. (2014). Transgenic *Brassica juncea* plants expressing MsrA1, a synthetic cationic antimicrobial peptide, exhibit resistance to fungal phytopathogens. *Mol. Biotechnol.*, 56(6): 535-545.

Sachan, J.N., Kolte, S.J. and Singh, B. (2000). Inheritance of white rust (*Albugo candida* race 2) in *Brassica juncea*. *Indian Phytopathol.*, 53: 206-209.

Saharan, G.S. and Kadian, A.K. (1983). Analysis of components of horizontal resistance in rapeseed and mustard cultivars against *Alternaria brassicae*. *Indian Phytopathol.*, 36: 503-507.

Saharan, G.S. and Krishnia, S.K. (2001). Multiple disease resistance in rapeseed and mustard. In: Role of resistance in intensive agriculture (Eds.), S. Nagarajan and D. P. Singh, Kalyani Publications, New Delhi, pp. 98-108.

Sang, X., Dengwei, J., Liu, Y., Xiao, B., Min, C. and Qing, Y. (2013). Genetic transformation of *Brassica napus* with *MSI-99m* gene increases resistance in transgenic plants to *Sclerotinia sclerotiorum*. *Mol. Pl. Breed.*, 4: 247-253.

Sellam, A., Dongo, A., Guillemette, T., Hudhomme, P., and Simoneau, P. (2007). Transcriptional responses to exposure to the brassicaceous defence metabolites camalexin and allyl-isothiocyanate in the necrotrophic fungus *Alternaria brassicicola*. *Mol. Pl. Pathol.*, 8, 195–208.

Sharma, P., Kumar, A., Meena, P.D., Goyal, P., Salisbury, P., Gurung, A., Fu, T.D., Wang, Y.F., Barbetti, M.J. and Chattopadhyay, C. (2009). Search for resistance to *Sclerotinia sclerotiorum* in exotic and indigenous *Brassica* germplasm. Proc 16th Australian Research Assembly on Brassicas, Ballarat, Australia, September 14-16, pp. 169-173.

Sharma, P., Meena, P.D., Verma, P.R., Saharan, G.S., Mehta, N., Singh, D. and Kumar, A. (2015). *Sclerotinia sclerotiorum* (Lib.) de Bary causing Sclerotinia rot in oilseed Brassicas: A review. *J. Oilseed Brassica*, 6 (Special): 1-44.

Sharma, R., Aggarwal, A.K., Kumar, R., Mohapatra, T. and Sharma, R.P. (2002). Construction of an RAPD linkage map and localization of QTLs for oleic acid level using recombinant inbreds in mustard (*Brassica juncea*). *Genome*, 45(3): 467-472.

Sharma, T.R. and B.M. Singh. (1992). Parameters of resistance to *Alternaria brassicae* in some Brassica species. *Indian Phytopathol.*, 45 (suppl.).

Sharma, T.R. and Singh, B.M. (1992). Transfer of resistance to *Alternaria brassicae* in *Brassica juncea* through interspecific hybridization among *Brassica. J. Genet. Breed.*, 46: 373–378.

Shaw, F.J.P. and Ajrekar, S.L. (1915). The genus Rhizoctonia in India. Mem. Department of Agricultural Indian Bot. Ser. 7: 177-194.

Shivpuri, A., Chhipa, H.P., Gupta, R.B.L. and Sharma, K.N. (1997). Field evaluation of mustard genotypes against white rust, powdery mildew and stem rot. *Annals Arid Zone*, 36: 387-389.

Singh, B.K., Nandan, D., Supriya, A., Ram, B., Kumar, A., Singh, T., Meena, H.S., Kumar, V., Singh, V.V., Rai, P.K. and Singh, D. (2015). Validation of molecular markers for marker-assisted pyramiding of white rust resistance loci in Indian Mustard (*Brassica juncea* L.). *Can. J. Plant Sci.*, 95: 939–945.

Somers, D.J., Rakow, G. and Rimmer, S.R. (2002). *Brassica napus* DNA markers linked to white rust resistance in *Brassica juncea*. *Theor. Appl. Genet.*, 104: 1121-1124.

Spect, C.E. and Diederichsen, A. (2001). Brassica. In: Hanelt. P. (ed) Mansfeld's Encyclopedia of Agricultural and Horticultural Crops. 6 vols. Springer-Verlag, Berlin, Heidelberg, New York vol. 3. pp. 1453-1456.

Taj, G., Kumar, A., Bansal, K.C, Garg, G.K. (2004). Introgression of osmotin gene for creation of resistance against Alternaria blight by perturbation of cell cycle machinery. *Ind. J. Biotechnol.*, 3: 291-298.

Tanhuanpaa, P. (2004). Identification and mapping of resistance gene analogs and a white rust resistance locus in *Brassica rapa* ssp. *oleifera*. *Theor. Appl. Genet.*, 108: 1039–1046.

Tewari, J.P. (1991). Structural and biochemical bases of the blackspot disease of crucifers. *Advances in Structural Biology*, 1: 325-349.

Tewari, J.P. and Conn, K.L. (1993). Reaction of some wild crucifers to *Alternaria brassicae*. Bull OILB CROP 16: 53-58.

Thomma, B. P. H. J. (2003). Pathogen profile- *Alternaria* spp.: from general saprophyte to specific parasite. *Mol. Pl. Pathol.*, 4: 225–236.

Thompson, C., Dunwell, J.M., Johnstone, C.E., Lay, V., Ray, J., Schmitt, M., Watson, H. and Nisbet, G. (1995). Degradation of oxalic acid by transgenic oilseed rape plants expressing oxalate oxidase. *Euphytica*, 85: 169–172.

Tiwari, A.S., Petrie, G.A. and Downey, R.K. (1988). Inheritance of resistance to *Albugo candida* race 2 in mustard [*Brassica juncea* (L.) Czern.]. *Can. J. Plant Sci.*, 68: 297-300.

U, N. (1935). Genome analysis in *Brassica* with special reference to the experimental formation of *Brassica napus* and peculiar mode of fertilization. *Jpn. J. Bot.*, 7: 389-452.

Varshney, A., Mohapatra, T. and Sharma, R.P. (2004). Development and validation of CAPS and AFLP markers for white rust resistance gene in *Brassica juncea*. *Theor. Appl. Genet.*, 109: 153-159.

Verma, P.R., and Saharan, G.S. (1994). Monograph on *Alternaria* Diseases of Crucifers. Saskatoon Research Centre Technical Bulletin 1994–6E, Agriculture and Agri-Food Canada, Saskatoon, SK, Canada.

Verma, P.R., Saharan, G.S., Bartaria, A.M. and Shivpuri, A. (1999). Biological races of *Albugo candiada* on *Brassica juncea* and *B. rapa* var. *toria* in India. *J. Mycol. Pl. Pathol.*, 29: 75-82.

Verma, S.S., Yajima, W.R., Rahman, M.H., Shah, S., Liu J.J., Ekramoddoullah A.K. and Kav, N.N. (2012). A cysteine-rich antimicrobial peptide from *Pinus monticola* (PmAMP1) confers resistance to multiple fungal pathogens in canola (*Brassica napus*). *Plant Molecular Biology*, 79: 61-74.

Verma, V. and Bhowmik, T.P. (1989). Inheritance of resistance to a *Brassica juncea* pathotype of *Albugo candida* in *Brassica napus*. *Can. J. Plant Pathol.*, 11: 443–444.

Vignesh, M., Yadava, D.K., Sujata, V., Mohapatra, T., Jain, N., Yadav, A.K., Malik, D., Yadav, M.S. and and Prabhu, K.V. (2009). Genetics of white rust resistance in [*Brassica juncea* (L.) Czern. and Coss.] and allelic relationship between interspecific sources of resistance. *Indian J. Genet.*, 69(3): 205-208.

Vishwanath, K., and Kolte, S.J. (1997). Variability in *Alternaria brassicae:* Response to host genotypes, toxin production and fungicides. *Indian Phytopathol.*, 50: 373-381.

Warwick, S.I., Francis, A. and Al-Shehbaz, I.A. (2006). Brassicaceae: species checklist and database on CD-Rom. *Pl. Syst. Evol.*, 259: 249-258.

Zhang F.L., Xu, J.B., Yan, H. and Li, M.Y. (1997). A study on inheritance of resistance to black leaf spot in seedlings of Chinese cabbage. *Acta Agri. Boreali Sinica*, 12: 115-119.

Zhao, J. and Meng, J. (2003). Genetic analysis of loci associated with partial resistance to *Sclerotinia sclerotiorum* in rapeseed (*Brassica napus* L.). *Theor. Appl. Genet.*, 106: 759-764.

Zhao, J., Udall, J.A., Quijada, P.A., Grau, C.R., Meng, J. and Osborn, T.C. (2006). Quantitative trait loci for resistance to *Sclerotinia sclerotiorum* and its association with a homeologous non-reciprocal transposition in *Brassica napus* L. *Theor. Appl. Genet.*, 112: 509–516.

10 Biotechnology for Genetic Improvement of Oilseed Crops

D.K. Dwivedi, Archana Devi and Preeti Kumari

Introduction

Adoption of oilseed crops has been growing up significantly due to industry interest in the composition of their seed oils. Oils are used as food/industrial feed (Maheshwari and Kovalchuk, 2014) and as a range of product applications such as surfactants, soap, detergents, lubricants, solvents, paints, inks, chemical feed stocks and cosmetics (Carlsson, 2009). In this chapter we have discussed about the use of biotechnology including tissue culture, molecular marker technology and genetic engineering tools that has allowed the functional study of genes with potential application for breeding in agriculture, focusing on oilseed crop genetic improvement with high precision and less uncertainty and of course, in less time; those scientific efforts where it was sought to upset fatty acids production or biotic tolerance will also be presented.

The term biotechnology is derived from a fusion of biology and technology. It involves use of living organism for the production of something is useful to us. Biotechnology consists of "the controlled use of the biological agents, such as microorganisms or cellular components, for beneficial use. Convention on Biological Diversity (CBD) has defined biotechnology as "any technological application that uses biological systems, living organisms, or derivatives thereof, to make or modify products or processes for specific use" (United Nations, 1992). In fact, biotechnology includes several agricultural as well as food manufacturing tools and techniques. However, when a biotechnology development uses new deoxyribonucleic acid (DNA) techniques, molecular biology, and reproductive technological applications ranging from gene transfer to DNA typing to cloning of plants and animals, it has been considerable modern biotechnology (McCullum, *et al.*, 2003). The potential of modern biotechnology is widely known, as it makes the use of recombinant DNA technology to generate modified microorganisms, plants and animals to make them more suitable for several potential applications: improved crops, production of new antibiotics and hormones, gene therapy, bioremediation, and genetic editing, one of the most recent techniques.

Genetic engineering means the manipulation of genes under highly controllable laboratory conditions. This technology is formally known as recombinant DNA. A central feature is the isolation and selective replication of specific genes. This is called gene cloning. The purpose of this is to obtain multiple copies of a gene itself or its products. Genetic engineering crops based on recombinant DNA technology were first introduced for commercial production in 1990s. This technology uses the identification, isolation and manipulation followed by the introduction of desired gene(s) from one organism (for example, a plant or bacteria) to another, thus giving rise to a transgenic or genetically modified organism. This technique has Genetic Improvement of Oilseed Crops Using Modern Biotechnology been fast replacing plant breeding so as to incorporate characteristics that are impossible to achieve by breeding. Biotechnology has the potential to help overcome many of the short-comings of the species being promoted, especially where exogenous genes are needed because there are characters that are difficult to produce by traditional breeding, or where characters tissue-specific or temporal expression or suppression of endogenous genes would be valuable (Gressel, 2008). For oilseed crops, modern biotechnology should allow the production of plants with specific fatty acids content. In the following paragraphs, main advances in plant genetic improvement using modern biotechnology, focused on oilseed crops, those scientific efforts in rapeseed (*Brassica napus*), peanut (*Arachis hypogaea*), soybean (*Glycine max*), sunflower (*Helianthus annuus*), linseed (*Linum usitatissimum* L.) and cotton (*Gossypium* spp.), where it was sought to upset fatty acids production and biotic tolerance will be presented.

Rapeseed

Rapeseed (*Brassica napus*) is considered one of the most important oil sources for edible or industrial uses, being the research to get better oil quality is important to improve rapeseed as a high-quality vegetable oil. Canola oil contains multiple fatty acids, such as palmitic acid, stearic acid, oleic acid, linoleic acid, α-linolenic acid, arachidic acid, erucic acid (Wang *et al.*, 2017). Due to its elevated dietary value, Canola oil is included in human diets where it has been shown to diminish plasma cholesterol levels in contrast with diets containing higher levels of saturated fatty acids. It has established that use of canola oil also control biological purpose that influence various other biomarkers of disease risk (Lin *et al.*, 2013).

A assembly of researchers developed transgenic canola seeds with notably increasing of oil content (Tan *et al.*, 2011). Seed-specific over expression of BnLEC1 and BnL1L genes (from canola), placed under the control of the truncated canola storage protein 2S-1 promoter, which is also known as the *napA* promoter, at an appropriate level significantly boost the seed oil content of the transgenic oilseed plant without noticeable negative effects on other core agronomic traits. In the same way to improve canola oil production, Qi *et al.* (2012) isolated the RNA-binding motifs No2 (RRM2) of the flowering control locus A (FCA) protein (FCA-RRM2) from variety Nannongyou of Canola, and then, it was introduced in cotyledon nodes using *Agrobacterium rhizogenes*, placed under a 35S-35S promoter (a variant of the cauliflower mosaic virus 35S promoter with higher transcriptional activity) to drive transgenic expression, into pBin438 vector with kanamycin resistance gene (for bacterial selection) and the hygromycin phosphotransferase gene (for plant transformation selection). Canola FCA-RRM2 increases plant size, organ size, cell size, plant productivity and oil content. These results provide a practical approach for the genetic improvement of this plant. Aimed at not good perception of erucic acid (cis-13-docosenoic acid) in the canola oil triglycerides, because of presumptive effects on growth retardation and pathogenic changes to internal organs when fed at high concentrations to laboratory animals, a research was made to decrease erucic acid level in Canola plants. Shi *et al.* (2015) reported the development of canola transgenic with change in fatty acids compositions, using *B. napus* cultivar "CY2" as the transgenic recipient of BnFAE1, a fragment involved in the synthesis of very long-chain fatty acids. These authors placed a BnFAE1 fragment driven by *napin A* promoters and then, they co-cultured hypocotyls with *Agrobacterium tumefaciens* EHA105 to introduce the genetic construct in canola cells. Due to CY2 that has high erucic acid (about 40 per cent) and low oleic acid (about 20 per cent) content, the researchers made seed-specific knockdown of BnFAE1, significantly changing the fatty acid composition. They demonstrated that the RNAi construct of BnFAE1 could effectively interfere with mRNA levels of BnFAE1 gene in F_1 hybrid seeds derived from crosses between BnFAE1-Ri lines and high erucic acid cultivars. At the end of canola transgenic lines with a dramatically decreased erucic acid (less than 3 per cent) were developed.

Disease Resistance through Microspore Mutagenesis

The non-availability of several desirable genes or alleles within the *B. juncea* gene pool and limitations associated with the conventional breeding demand biotechnological interference for induction of novel genetic variation. Among these the most notable ones include the transgenic technology (Murphy 1999) and mutation breeding (Maluszynski *et al.*, 2000). Several scientists have working on the use of tissue specific expression of transgenes to change endogenous biochemical profiles or to add novel biosynthetic pathways for qualitative traits; however, manipulating quantitative traits is as yet not feasible with this approach. During recent years, mutation research for genetic manipulation has been fairly successful in generation of novel genetic variation in *brassicas* (Potts *et al.*, 2001). The use of seed mutagenesis is limited due to the chimeric nature of the mutant plants, adverse linkages and undesirable pleitropic effects, thus rendering this technique unpredictable and many times unreliable (Maluszynski *et al.*, 1995). These bottlenecks of the seed mutagenesis technique can be overcome by the use of doubled haploids (DH), which are produced from doubling the chromosome number of haploid plants (Kotts, 1998). The doubled haploid plants, were selected with good agro-morphological traits; high number of primary and secondary branches, high number of pods on the main shoots, and improved seeds/siliqua and seed size. The seeds were evaluated for biochemical profile and a high variability was observed for different fatty acids with palmitic acid ranging from 3.22-16.0 per cent, oleic acid from 18.4 - 44.0 per cent, linoleic acid from 18.0- 37.0 per cent, linolenic acid from 4.0-16.0 per cent and erucic acid ranging from < 2.0 to 40.0 per cent. Mutant plants have been identified with low disease score; white rust disease score ranged from 0.6 to 2.6 and for *Alternaria* blight disease index (DI) ranged from 0.03 to 1.0 under field conditions and 1.3 to 3.3 under *in vitro*, assessed by detached leaf method.

Disease Resistance through Embryo Rescue

The genetic base for the fungal diseases white rust and *Alternaria* blight are limited, and none of the cultivars of *B. juncea* are resistant or immune to fungal diseases (Kolte 2002; Yadav and Kumar 2004). Several *in vitro* techniques such as somaclonal variation, somatic hybridization and transgenics have been utilized to transfer resistance to *Alternaria* blight and white rust from secondary/tertiary gene pools/wide relatives. However, most of these techniques are either restricted to species other than *B. juncea* or the desired level of resistance has not been achieved. The most successful and widely used approach to realize incompatible hybrids is wide hybridization intervened with *in vitro* embryo rescue technique (ovary/ovule culture or sequential embryo culture) allowing transfer of desired genes from related species (Shivanna 1996). Resistance/high tolerance to the most devastating fungal diseases, white rust (*Albugo candida*) and Alternaria blight (*Alternaria brassicae*), has been transferred from *B. carinata* to *B. juncea* through inter-specific hybridization aided by ovule culture. The hybrids were characterized through detailed morphological traits; leaf (shape, size, texture, colour, tip) and floral morphology (colour of petals, arrangement of flowers, anther shape and structure), they resembled more to the male parent in all morphological traits (Gupta *et al.*, 2007). This was substantiated through molecular studies. The molecular markers ISSRs have been routinely utilized for genetic diversity, phylogenetic studies, and somaclonal variants in brassicas (Sarla *et al.*, 2001). The ISSRs were used for the first time for hybrid characterization. The study clarified that hybrids resembled more to the male donor (*B. carinata*) with a genetic similarity value of more than 60 per cent in comparison to the female parent thus indicating a strong influence of male donor (Gupta *et al.*, 2004). The elite genotypes, resembling *B. juncea* phenotype derived from advanced backcross progenies BC_3F_2 and BC_2F_3 of crosses from *B. juncea* genotype [RESJ 830/RESJ 837/TERI (OE) M21-1]/*B. carinata* Var. Kiran have been selected for low erucic/high oleic acid and good tolerance to fungal diseases white rust and *Alternaria* blight (DI < 2). These genotypes can be used as valuable sources for developing Indian mustard with improved oil quality and reduced dependency on harmful chemical pesticides for oilseed sustainability in an environment friendly manner.

Peanuts

Peanut (*Arachis hypogaea*) is grown worldwide as an oilseed crop. In many countries, peanut seeds do an important contribution to the people diet because they are a good source of proteins and lipids for human nutrition. Grosso *et al.* (1997) determined that peanut seeds are rich in oil (about 50 per cent of seed composition) and oleic acid (18: 1), linoleic acid (18: 2), palmitic acid (16: 0), behenic acid (22: 0), eicosenoic acid (20: 1), stearic acid (18: 0), arachidic (20: 0) and lignoceric acid (24: 0). However, the fatty acid composition of peanut oil varies depending on the seed maturity, genotype, growth location and climatic conditions (Carrín and Carelli, 2010). Research about transgenic peanut crops has been undertaken for the development of fungi resistant. This crop is susceptible to many types of pathogens including those caused by fungi. Chenault *et al.* (2002) reported the development of transgenic peanuts which were introduced two hydrolase genes, a glucanase from alfalfa (*Medicago sativa* L.), and a chitinase from rice (*Oryza sativa* L.) into somatic embryos using gene gun. Although the study focused on seedlings characterization (found up to 37 per cent of hydrolase activity in transgenic lines), it is assumed transgenic lines obtained could be promising due to high transgene expression what would exhibit some level of resistance to a broad range of fungal pathogens. Following with the same modified peanut lines, Chenault *et al.* (2003) developed an assay under greenhouse conditions where these lines were tested for resistance to *Sclerotinia minor* by inoculation with a mycelial plug. There were lines up to 84 per cent of resistance to the pathogen. On the other hand, the peanut lines considered more resistance kept going in race and were tested for *S. minor* resistance under field conditions (Chenault *et al.*, 2005). In this study three transgenic lines showed a significant resistance to the pathogen compared with the wild-type cultivar. Finally, Jonnala *et al.* (2005) determined the oil composition of the best three transgenic lines obtained in the previous report. Chenault *et al.* (2005) reported similar oil content of all transgenic peanut lines to that wild-type line, indicating that genetic modification did not cause substantial unintentional changes in peanut chemical composition. In the same way, Ng *et al.* (2008) examined chemical characteristics, and volatile components of those three GM peanut lines using gas chromatograph/ mass spectrometer (GC/MS) equipped with an olfactory detector. Result showed minimal variations in nutritional composition between GM peanuts and wild type, indicating that genetic modifications did not cause significant change in peanut.

Soybean

Soybean, *Glycine max* L. Merr., is a major crop that produces the best vegetable oil and protein for use in food and beverage production worldwide. Among legume species, soybean has the highest protein content (around 40 per cent), while other species have a protein content between 20 and 30 per cent. On the other hand, cereals have a protein content ranging from 8 to 15 per cent. Other interesting point for oleochemistry is that soybean also contains about 20 per cent oil. Soybean oil is a complex mixture of five fatty acids: palmitic, stearic, oleic, linoleic, and linolenic acids.

Nowadays, soybean oil is currently found in food products such as margarine, salad dressings and cooking oils, and industrial products such as plastics and biodiesel fuel. Lecithin, a natural emulsifier and lubricant extracted from soybean oil, is used in applications from pharmaceuticals to protective coatings (American Soybean Association, 2017). Due to its importance as a crop, genetic transformation techniques have been used extensively to improve the crops valuable traits. Herbicide-tolerant (Roundup Ready) Soybean (*Glycine max* L. Merrill) resistant to glyphosate (N-phosphonomethylglycine) was the first transgenic variety introduced for commercial production by Padgette, *et al.* (1995). In the contrary way to seek increase in fatty acids production, the goal was to give a competitive advantage to soybean favoring desirable plants and inhibiting undesirable plants by the application of glyphosate, the active ingredient of the non-selective herbicide Roundup. A glyphosate-tolerant soybean line was obtained through expression of the bacterial 5-enolpyruvylshikimate-3-phosphate synthase (EPSP synthase, EPSPS) enzyme from *Agrobacterium* sp. strain CP4, under the cauliflower mosaic virus 35S promoter (P-E35S), with the *Petunia hybrida* EPSPS chloroplast transit peptide (CTP) and a portion of the 32 nontranslated region of the nopaline synthase gene (NOS 32) terminator. This soybean line was highly tolerant to glyphosate, showing no visual injury after application of up to 1.68 kg acid equivalent ha^{-1} of glyphosate under field conditions.

In terms of genetic transformation methods, many reports describing soybean transformation by particle bombardment using meristems as the target tissue have been published. The Biolistics particle delivery system for soybean transformation was evaluated in two different regeneration systems: from shoot tips obtained from immature zygotic embryos of the cultivar Williams 82, and the second was somatic embryogenesis from a long-term proliferative suspension culture of the cultivar Fayette (Sato, *et al.*, 1995). A method for high-frequency recovery of transgenic soybean by combining resistance to the herbicide imazapyr as a selectable marker, multiple shoot induction from embryonic axes of mature seeds and biolistics techniques was made by Rech, *et al.* (2008). A targeting method to insert genes with biolistics to predefined soybean genome sites using the yeast FLP-FRT recombination system was made by Li *et al.* (2009). A double-barreled gene *gus* device was used to improve transformation efficiency and then study soybean resistance (R) gene-mediated responses to effectors, induction and suppression of cell death by a wide variety of pathogen and host molecules (Kale and Tyler, 2011).

The abscisic acid (ABA)-independent dehydration responsive element binding (DREB) gene family from *Arabidopsis thaliana* was inserted into soybean plants using biolistics to improve tolerance to abiotic stresses (Engels, 2013). In the same way, Leite, (2014) introduced an activated form of abscisic acid-responsive element binding protein (AREB1) into soybean plants to improve water deficit stress. Finally, the biotechnological potential of plastid genetic engineering was used to develop a reproducible method to generate plastid transformants in soybean. To sum up, transformation vectors were delivered to embryogenic cultures by the particle gun method and selection performed using the *aadA* antibiotic resistance gene, getting early homoplasmy and avoiding further selection cycles (Dubald, 2014).

Genetic engineering approaches have been applied to enrich the content of soybean oil for a particular fatty acid or class of fatty acids. Transgenic soybean seeds were developed by down-regulating the expression of *FAD2* genes that encode the enzyme that converts the monounsaturated oleic acid to the polyunsaturated linoleic acid. Those transgenic soybean seeds had oleic acid content of approximately

80 per cent of the total oil, whereas conventional soybean oil contains oleic acid at levels of 25 per cent of the total oil. With the same aim, Haun *et al.* (2014) reported the creation of high oleic acid soybean varieties using targeted mutagenesis with transcription activator-like effectors nucleases (TALENs) to bind and cleave specific DNA sequence targets in the *FAD2-1A* and *FAD2-1B* genes with high efficiency. Mutant soybean plants produced nearly four times more oleic acid than the wild-type parents (80 per cent vs. 20 per cent, respectively). Furthermore, because they use a technique considered "genetic editing," the soybean lines lacked foreign DNA in their genomes and are thus not transgenic. Rather, they only have small deletions of coding sequence in the FAD2-1 gene targets. On the other hand, regarding to biotic factors, Kim *et al.* (2016) developed transgenic soybean to improve resistance against SMV. HC-Pro coding sequences were introduced within a RNAi inducing hairpin construct and *Agrobacterium*-mediated transformation system. Then, their response to viral infection was analyzed. The inhibition of HC-Pro expression enhanced viral resistance after viral infection, when compared to the resistance of virus-susceptible non-transgenic plants. RNAi induced by the hairpin construct of the SMV HC-Pro sequence effectively confers viral resistance. Among others, these results have proven the usefulness of RNAi-mediated resistance for crop improvement. Synthetic *Bacillus thuringiensis cry* genes were used to develop transgenic agronomic loses caused by insects from Lepidoptera order such as *Anticarsia gemmatalis, Pseudoplusia includes* and *Helicoverpa zea.*

Sunflower

Sunflower (*Helianthus annuus* L.) is one of the most important oilseed crops cultivated on a global level. Its seeds have always been ground and pounded into flour for making bread, cracked and eaten as snacks, mixed with vegetables, and extracted for oil. The seeds are also a source of purple dye and have medicinal uses (Skoric, 2008). Sunflower seeds are composed by 20 per cent protein and 50 per cent fat. In this crop oil, up to 90 per cent of its fatty acids are unsaturated, namely oleic (C18: 1, 16–19 per cent) and linoleic (C18: 2, 68–72 per cent) acids. The remaining 10 per cent of its fatty acids are palmitic (C16: 0, 6 per cent), stearic (C18: 0, 5 per cent), and minor quantities of myristic (C14: 0), myristoleic (C14: 1), palmitoleic (C16: 1), arachidic (C20: 0), behenic (C22: 0) (Davey and Jan, 2010). Several scientific efforts have been made to develop genetic improvement methods in sunflower, using modern biotechnology. Perhaps one of the earliest works in sunflower was developed by Moyne *et al.* (1989), which introduced plasmid into isolated sunflower protoplast. Another effort was made by Knittel *et al.* (1994) who used microprojectile bombardment of half-shoot apices followed by co-culture with *Agrobacterium tumefaciens*, to obtain transgenic shoots. However, Grayburn, and Vick (1995) modified a step in which shaking of explants with glass beads replaced the microprojectile bombardment stage used by Knittel *et al.* (1994). In an attempt to reduce or eliminate the *in vitro* regeneration component of a sunflower-transformation protocol, infected 2-day-old seedlings, each with one cotyledon detached, with *A. Tumefaciens* strain LBA4404 carrying a specific plasmid. On the other hand, to overcome the generation of chimeric plants, Muuler *et al.* (2001) used zygotic embryos, the latter being 4–6 mm in size and cut transversely below the cotyledons, and then, explants were cultured in the dark for 1 day before being bombarded with gold particles and co-cultured with *A. tumefaciens*. Weber *et al.*, 2003 reported an alternative procedure to wound cells of target sunflower explants that involved treatment of the explants with the cell wall-degrading enzymes Cellulase Onozuka R-10 (0.1 per cent w/v) and pectinase Boerozyme M5 (0.05 per cent w/v). After that, but before *Agrobacterium* inoculation, a sonication (50 MHz, 2, 4, 6 s) step of explants showed that transient expression of *gus* or *gfp* transgenes was increased. One of the most important aspects of any transformation protocol is an efficient selection of transgenic plants. On the way to develop a procedure to minimize the number of transgenic escapes, Radonic *et al.* (2008) germinated sunflower seeds for 24 h on half-strength MS-based medium, before cutting the seeds to give two half embryos, each with one cotyledon. Once that, cotyledon explants were inoculated with *A. tumefaciens* carrying a vector with the *nptII* and the *gus* genes. Leaving aside developments made around transformation methods, and focusing on advances toward genetic improvement with some functional characteristics, some efforts to improve oil production in sunflower have been made recently. Dagustu *et al.* (2008) introduced the *Erwinia uredovora* phytoene desaturase (*crtl*) and hydroxymethylglutaryl-CoA (*Hmgr-CoA*) genes into sunflower, which have potential to increase oil quality. On the other hand, Radonic *et al.* (2008) developed transgenic sunflower plants resistant to *Verticillium dahlia* and *Sclerotinia sclerotiorum* introducing antifungal genes, including *gln2* (a glucanase) from *Nicotiana tabacum*, a chitinase (*ch5B*) from *Phaseolus vulgaris*, an osmotin gene (*ap24*) from *N. tabacum*, and a gene coding for a ribosome inhibitor protein (rip). In the same way, Neskorodov *et al.* (2010) developed transgenic sunflower resistant to the herbicide phosphinothricin, herbicide resistance also being exploited to select the transgenic plants. Some research interests have been around decreasing levels of palmitic and stearic acid of sunflower, due to their contribution on increasing the plasma cholesterol level in humans, associated with heart disease. Skoric *et al.* (2008) induced mutations via seed treatment with γ-rays, X-rays, and mutagenic chemicals such as ethyl methane sulfonate (EMS) and dimethyl sulfate (DMS) to generate sunflower genotypes with high levels of C 18: 2, C 18: 1, C 18: 0, C 16: 1 and C 16: 0.

Linseed

Linseed (*Linum usitatissimum* L.) is also known as Flax, which is widespread in the temperate zone, is used for the production of oil and fiber. The contents of unsaturated fatty acids in linseed seed can reach 45 per cent; it is also rich in lignin, glue, vitamins A and B, *etc.*, which makes its supplementations very valuable for human health. Linseed fibre possesses air permeability, hygroscopicity, and antibacterial characteristics, which makes flax textile products exceedingly popular in the international market (Qiao and Long, 2013). The main reason for shortage of flax accessions, we have not developed cultivars with high yield and resistance to diseases. Now our modern varieties with narrow genetic back ground may be susceptible to certain flax major diseases, for example, powdery mildew, no any variety are completely resistant to it. Hybridization was the main flax breeding method, but in the recent years, along with the development of biotechnology, advanced molecular techniques have been utilized, resulting in a quickly increased breeding level. Here, we introduce the achievements in field of linseed research

Biotechnology in Linseed Improvement

Gene Transfer System

A linseed gene transfer system was established by *Agrobacterium tumefaciens* – mediated method. According to the specific procedures, flax hypocotyl is used as an ex-plant and Murashige and Skoog (MS) medium as the optimal medium for flax trans-genesis, with the addition of 50 mg/L kanamycin for selection (Song, 2009 and Ji *et al.*, 2005). Callus selected through induction medium, regeneration culture medium, and rooting medium. After transferring the resistant genes and the gene responsible for the synthesis of cellulose synthase into the explants, GMO plants containing the target gene have been acquired (Wu, 2014 and Wen *et al.*, 2011).

Haploid Utilization

(1) Distant Hybridization

hybridization between wild flax as the female parent (*Limum grandiflorum* L.), and cultivated linseed as the male parent 30 per cent crosses produced inflate bolls containing green embryos without endosperm. Young embryos were peeled off at the 8–20 days after crossing and were transferred onto MS culture medium (Cheng and Guan, 2015). Through this a number of resistance lines have been obtained, shortening the breeding time.

(2) Polyembryonic Seed Utilization

Haploid breeding is the main method in flax breeding programs. China was the first country to perform experiments on flax haploidization *via* anther culture. Each polyembryonic seed can produce at least two seedlings, one of which is commonly haploid, whereas the other is diploid. The aim of study was to double the haploid one. A series of high-quality polyembryonic lines were obtained with good, with a polyembryonic rate of more than 10 per cent.

(3) Anther and Microspore Culture

F_1 plants were used as experimental material. When microspores were at the late uninucleate stage, the buds were disinfected. Then, the culture media for callus induction and differentiation were selected. Further, rooting was induced, and plants were subsequently transplanted.

Exogenous DNA Induction

Exogenous DNA induction work was to obtain DNA of properconcentration, purity, and length. Then, pollen tunnel technology was utilized, and DNA was injected into the ovary, to realize certain gene transfer.

Molecular Research on Linseed Diseases

Powdery mildew is a minor disease of linseed, but many varieties are susceptible to this disease. A cross between resistance and susceptible was made.

Then, the parents, F_1, and F_2 were sown in the field. Pathogenic bacterium was multiplied in the greenhouse during the fast growing period of flax plants. Further, the seedlings in the field were inoculated with conidia. Disease occurrence was investigated, the plants in F_1 generation expressed HR, which indicated that the trait of resistance of line 9801-1 was completely dominant and was transferred via nucleus inheritance. In F_2 generation, the ratio of resistant to susceptible plants was nearly 3: 1. Therefore, after summarizing the performance in F_1, scientists drew the conclusion that powdery mildew resistance should be a trait of dominant single gene inheritance.

Molecular Markers for Linseed Wilt Resistance Gene

Linseed wilt disease is one of the main flax diseases worldwide, caused by *Fusarium oxysporum* linen (Boba *et al.*, 2014). The research work on molecular markers related to genes responsible for resistance to high wilt disease, not only contribute to the development of the work on wilt resistance marker-assisted selection and improve the efficiency of linseed disease resistance breeding, but also lay the foundation for the isolation and cloning of wilt resistance genes. The flax wilt disease resistance genes FuJ7(t) was analyzed through AFLP markers by Bo *et al.* (2006), who crossed 'Jin Ya 7' (highly resistant) variety with 'Jin Ya 1' (highly susceptible) variety to wilt. It was found that 'Jin Ya 7' was controlled by two dominant genes for wilt resistance; thus, it was transferred via nucleus inheritance (Bo *et al.*, 2003).

AFLP analysis with 48 primers was conducted on the resistance and susceptibility gene pool in F_2 population and their parents ('Jin Ya 7' and 'Jin Ya 1'). There were three stable differences in the totally amplified identifiable bands. These three specific bands and the target gene linkage relationships were analyzed using F_2 segregation population. The specific bands AG/CAG and FuJ7(t) wilt resistance gene were found to be closely linked, and the genetic distance between them was 5.2 cM. The AG/CAG segments were recovered, cloned, sequenced, and successfully transformed into a SCAR marker. At present, the specific SCAR marker genes FuJ7(t) are used for molecular detection and marker-assisted selection breeding.

Cotton

Cotton (*Gossypium* spp.) is one of most important fiber crops at world level. According to Liu *et al.* (2002) cottonseed oil represents approximately 16 per cent of the seed weight, and perhaps it is the most valuable product derived from cottonseed. Likewise, cottonseed oil is composed by 26 per cent of saturated palmitic acid (C16: 0), 15 per cent of mono-unsaturated oleic acid (C18: 1), and 58 per cent of polyunsaturated linoleic acid (C18: 2) [80]. In fact, this complementary product (cottonseed oil) has some advantages over soybean oil and rapeseed oil, like good quality and price, so that it is used in foods or as a raw material for biodiesel production (Wang *et al.*, 2017). However, some reports warn that cottonseed oil content around oversaturated, polyunsaturated, and monounsaturated fatty acids is unbalanced (Wang *et al.*, 2017 and Chapman *et al.*, 2001). Chapman *et al.* (2001) reported the development of transgenic cotton plants with increased seed oleic acid levels. Using an *Agrobacterium*-mediated system transformation, these authors introduced a binary vector previously designed to suppress expression of the endogenous cottonseed enzyme fatty acid desaturase 2 (Fad2) by subcloning a mutant allele from a rapeseed fad2 gene. It is known that FAD2 enzyme, in the endoplasmic reticulum of plant cells, catalyzes conversion of oleic acid to linoleic acid so that, decreasing this enzyme activity would be an increase of oleic acid content in cottonseed oil. At the end of the research, increased seed oleic acid content ranged from 21 to 30 per cent (by weight) of total fatty acid content in primary transformants and 47 per cent of oleic acid content in their progeny, which represent an increasing of three times comparing with standard cottonseed oil. Due to consumption of the saturated fatty acid, overall cholesterol levels increases, more specifically low-density lipoprotein (LDL) which is considered "bad cholesterol," and it is well known worldwide that its consumption increases risk of cardiovascular disease (Baum *et al.*, 2012). a group of researchers started a study to improve the quality of cottonseed oil. Liu *et al.* (2017) used RNAi technology to regulate fatty acid metabolism of cottonseed inhibiting GhFAD2-1 and GhFATB gene expression levels, simultaneously. These genes encoding the microsomal oleate desaturase and palmitoyl-acyl carrier protein thioesterase, respectively, play significant roles in regulating the proportions of saturated and polyunsaturated fatty acids in cottonseed lipids. Using this technology, they decreased palmitic acid and linoleic acid content and increased oleic acid content, but unfortunately, they got an adverse effect on seed germination and seed vigor. In spite of achieving an adequate balance in the content of fatty acids, thinking in human consumption of cottonseeds oil, it is necessary to explore others effective regulating strategies to improve the quality of cottonseed oil.

On the other hand, recently, Wang *et al.* (2017) reported a genome-wide analysis in several *Gossypium* species and possible ancestral diploids. In that study, scientist analyzed a total of 40 Lysophosphatidic acid acyltransferase (LPAAT) genes and found that this gene is involved in increasing oil composition and content which was demonstrated in some experiments in transgenic yeast. This report shows an important way for further studies due to LPAAT genes that are involved in natural cottonseed oil content and variation which should open a possible strategy in development of genetically modified cotton crops with improvement of seed oil content and composition.

Quality Improvement through Genetic Engineering

The potential use of genetic engineering to modify plant seed oil composition has been recognized for a number of years. The oilseed crops have the potential to produce high quality edible oils as well as speciality oils having commercial applications. For instance jojoba is a rich source of wax esters; coconut is rich in capric, lauric and myristic fatty acid, palm has a high proportion of palmitic, oleic and stearic acid, whereas linseed is a rich source of linoleic acid. These fatty acids are used in a wide range of products ranging from the production of soaps, detergents, cosmetics, surfactants, lubricants, plastics, varnishes and pharmaceuticals. Due to the no domestication of most of the potential sources and their restricted availability, at present the fatty acids for industrial applications are mostly derived from petrochemicals. However, in the near future, with the biased use of global reserves of fossil derived hydrocarbons alternative sources of industrial fatty acids from the environment friendly oil crops are sought after. This can be achieved either by altering the existing fatty acids profile or by adding new genes for synthesis of novel fatty acids. In most of the oil bearing crops, the biosynthetic pathway of fatty acid synthesis is similar and their differential accumulation in the seed is genetically controlled depending upon the species. During the seed development process, photosynthetically fixed carbon is imported into the seed in the form of sucrose, and is converted into the storage products with the help of enzymes present in the seed. The seed contains all the enzymes that are required for the conversion of sucrose into any of the storage products. However, it is the rate of sucrose uptake by the various biosynthetic pathways that lead to the differential accumulation of a particular storage product in the seed. Thus the genetic manipulation of any of the biosynthetic pathway can lead to a specific ratio of seed storage product, according to the end use of the seed. This can be done either by modifying the length of the existing hydrocarbons in fatty acid chain (modifying the chain elongation enzymes) or by changing the position of double bonds (modifying desaturase enzymes). The seed specific or tissue specific genetic modifications may be used for creating changes in endogenous fatty acid biosynthesis pathway or addition of new biosynthetic pathways. The use of seed specific antisense technology has allowed for the selective modulation of key enzyme activities in the developing seed, while keeping the rest of the genetic background of the plant absolutely constant. Co-suppression based on post transcriptional gene silencing of endogenous desaturase gene has shown promising results in developing high oleic acid genotypes of rape seed mustard. The recently derived RNAi approach has also shown a great potential for endogenous desaturase silencing. Using this concept total silencing of the $\Delta 12$ desaturase gene in *B. napus* has been acheived, resulting in the production of genotypes accumulating 89 per cent oleic acid in the seed oil. The rapeseed oil normally contains low levels of lauric acid (C12) and stearic acid (C18) at a concentration of 1–2 per cent and 0.1–0.2 per cent, respectively. High lauric rapeseed can be used as a substitute in detergent markets, leading to displacement of conventional lauric oils derived from coconut or palm kernel, whereas high stearic rapeseed is a useful substitute in margarine markets and replaces conventional hydrogenated rapeseed oil. The two most notable achievements in oil modification to-date are the 40 per cent stearic and 40 per cent lauric rapeseed varieties (laurical) first produced and entered in field trials by Calgene in 1993–94. Thus Laurical was the first genetically manipulated rapeseed variety given permission for commercial cultivation in 1995 in US. The $\Delta 9$ stearoyl ACP desaturase gene which normally converts stearic to oleic acid was partially inactivated in rapeseed using antisense technology, resulting in the accumulation of a seed oil containing up to 40 per cent stearic acid. This high stearic variety contains an antisense copy of a *Brassica* stearate desaturase gene which inhibits the function of the normal rapeseed stearic desaturase gene, resulting in an accumulation of stearic acid, rather than their saturation to oleate. The resulting high stearic oil has many advantages over the normal rapeseed oil for the production of certain solid fats, such as margarines. With the advent of transgenic technology, the genes coding for enzymes that synthesize industrially important fatty acids can be transferred from non-traditional crops into more important oil crops. The canola oil having low erucic acid has food applications in margarine, salad and salad dressings while the high erucic rapeseed has industrial application. The canola quality rapeseed has also been genetically modified for containing high levels of β-carotene. This high carotenoid canola oil may prove very beneficial to combat the vitamin A deficiency in developing world. Various species of Brassicaceae have been transformed with mutated Sn-2 acyltransferase gene from yeast and have been reported to show increase in seed oil content, seed weight and erucic acid content. Lauric oils are mainly used in soaps and detergents although their use in confectionary fats and milk formulas is also being investigated. Lauric acid which is present at insignificant levels in rapeseed is found at high levels in the seed oil of the California Bay plant, *Umbelluria californica*, due to the presence in the latter species of a lauryo-ACP thioesterase. This gene has been cloned from the Bay plant and inserted into rapeseed causing premature chain-termination, resulting in a novel variety with seed oil containing almost 25 per cent lauric acid. Following this an Sn-2 acyl transferase gene (LPAAT) from coconut has been introduced in lauric rapeseed to increase the accumulation of lauric acid in the seed triacyl glycerol molecules. Similar to the development of lauric acid producing rapeseed, several novel genes coding for altered fatty acid synthesis have been used for altering seed fatty acid profile. Some worthy examples are Caprilic and Capric acid (from *Cuphea* spp.), myristic acid (from *Myristica fragrans*), Crepenylic acid (from *Crepis alpira*), Richinolic acid (from Castor), Vernolic acid (from *Crepis palaestina*) and petroselenic acid (from *Coriandrum sativum*). Thus, in future, plant derived oils may be an important source of industrial oil derived chemical or oleo chemicals.

References

Baum, S.J., Kris-Etherton, P.M., Willett, W.C., *et al.* (2012). Fatty acids in cardiovascular health and disease: A comprehensive update. *Journal of Clinical Lipidology.* 6: 216-234.

Bo, T., Yang, J., Ren, Y. and Chen, J. (2006). Evaluation of flax germplasm for resistance to wilt. *Chinese journal of oil crop sciences,* 28: 470- 475.

Bo, T., Ye, H., Li, X. and Zhu, L. H. (2003). Identification of molecular markers linked to the wilt resistance gene FuJ7 (t) in flax. *Scientia Agricultura Sinica* 36: 287-291.

Boba, A., Kulma, A. and Kostyn, K. (2014). The influence of carotenoid biosynthesis modification on the *Fusarium culmorum* and *Fusarium oxysporum* resistance in flax [J]. *Physiological and Molecular Plant Pathology,* 76: 39-47.

Carlsson, A.S. (2009). Plant oils as feedstock alternatives to petroleum—A short survey of potential oil crop platforms. *Biochimie,* **91**: 665-670.

Carrín, M.E. and Carelli, A.A. (2010). Peanut oil: Compositional data. *European Journal of Lipid Science and Technology.* **112**: 697-707.

Chapman, K.D., Austin-Brown, S. and Sparace, S.A., *et al.* (2001). Transgenic cotton plants with increased seed oleic acid content. *Journal of the American Oil Chemists' Society.* **78**: 941-947.

Chenault, K.D., Burns, J.A. and Melouk, H. A. *et al.* (2002). Hydrolase activity in transgenic peanut. *Peanut Science.* **29**: 89-95.

Chenault, K.D., Melouk, H.A. and Payton, M.E. (2005). Field reaction to Sclerotinia blight among transgenic peanut lines containing antifungal genes. *Crop Science.* **45**: 511-515.

Chenault, K.D., Payton, M.E. and Melouk, H.A. (2003). Greenhouse testing of transgenic peanut for resistance to *Sclerotinia minor. Peanut Science.* **30**: 116-120.

Cheng, L. and Guan, F. Z. (2015). Interspecific hybridization and immature embryo rescure between *Linum usitatissimum* and *Linum grandiflorum. Plant fiber sciences in China,* 1: 1-4.

Dagustu, N., Fraser, P. and Enfssi, E., *et al.* (2008). Screening for high callus induction and agrobacterium-mediated transformation of sunflower (*Helianthus annuus* L.). *Biotechnology and Biotechnological Equipment.* **22**: 933-937.

Davey, M.R. and Jan, M. (2010). Sunflower (*Helianthus annuus* L.): Genetic improvement using conventional and in vitro technologies. *Journal of Crop Improvement.* **24**: 349-391.

Dubald, M., Tissot, G. and Pelissier, B. (2014). Plastid transformation in soybean. *Methods in Molecular Biology.* **1132**: 345-354.

Engels, C., Fuganti-Pagliarini, R. and Marin, S.R.R., *et al.* (2013). Introduction of the rd29A: AtDREB2A ca gene into soybean (*Glycine max* L. Merril) and its molecular characterization in leaves and roots during dehydration. *Genetics and Molecular Biology.* **36**: 556-565.

Gressel, J. (2008). Transgenics are imperative for biofuel crops. *Plant Science.* **174**: 246-263 American Soybean Association (2017). SoyStats: A reference guide to important soybean Facts and Figures. http: //soystats.com/wp-content/uploads/17ASA-006_Soy-Stats-_1F-web.pdf.

Grosso, N.R., Julio, A.Z. and Lamarque, A.L., *et al.* (1997). Proximate, fatty acid and sterol compositions of aboriginal peanut (*Arachis hypogaea* L) seeds from Bolivia. 1997. [Epub ahead of print]. DOI: 10.1002/(SICI)1097-0010(199703)73: 3<349: : AID-JSFA736>3.0.CO;2-E.

Gupta, K., Prem, D. and Negi, M.S. *et al.* (2004). ISSRs: An efficient tool to screen interspecific F₁ hybrids in *Brassicas.* In Proceedings: 4th International Crop Science Congress, 26th Sept - 1st Oct. 2004, Brisbane, Australia.

Gupta, K., Prem, D. and Nashaat, N.I. *et al.* (2007). Phenotypic variations in plant progenies of interspecific crosses involving Brassica juncea/B. carinata, In: Proceedings, 12th International Rapeseed Congress, Wuhan, China; Sustainable Development in Cruciferous Oilseed Crops Production, (ed.) Tingdong Fu and Chunyun Guan, Science Press USA Inc.1: 239-242.

Haun, W., Coffman, A. and Clasen, B.M., *et al.* (2014). Improved soybean oil quality by targeted mutagenesis of the fatty acid desaturase 2 gene family. *Plant Biotechnology Journal.* **12**: 934-940.

Jonnala, R.S., Dunford, N.T. and Chenault, K. (2005). Nutritional composition of genetically modified peanut varieties. *Journal of Food Science*. **70**: S254-S256.

Kale, S.D. and Tyler, B.M. (2011). Assaying effector function in planta using double-barreled particle bombardment. *Methods in Molecular Biology*. **712**: 153-172.

Kim, H.J., Kim, M.J. and Pak, J.H., *et al.* (2016). RNAi-mediated Soybean mosaic virus (SMV) resistance of a Korean Soybean cultivar. *Plant Biotechnology Report*. **10**: 257-267.

Knittel, N., Gruber, V. and Hahne, G., *et al.* (1994). Transformation of sunflower (*Helianthus annuus* L.): A reliable protocol. *Plant Cell Reports*. **14**: 81-86.

Kolte, S.J. (2002). Diseases and their management in oilseed crops- New paradigm. In: Oilseeds and oils- research and development needs. (Eds) M. Rai, H. Singh and D.M. Hegde. Indian Society of oilseeds Research, Hyderabad. pp. 244- 253.

Kotts, L.S. (1998). Application of doubled haploid technology in breeding of oilseed *Brassica napus*. *Agbiotech news and information*. 10: 69-73.

Leite, J.P., Barbosa, E.G.G. and Marin, S.R.R., *et al.* (2014). Over-expression of the activated form of the AtAREB1 gene (AtAREB1ΔQT) improves soybean responses to water deficit. *Genetics and Molecular Research*. **13**: 6272-6286.

Li, Z., Xing, A. and Moon, B.P., *et al.* (2009). Site-specific integration of transgenes in soybean *via* recombinase-mediated DNA cassette exchange. Plant Physiology. 151. [Epub ahead of print]. DOI: 10.1104/pp.109.137612.

Lin, L., Allemekinders, H. and Dansby, A., *et al.* (2013). Evidence of health benefits of canola oil. *Nutrition Reviews*. **71**: 370-385.

Liu, F, Zhao, Y.P. and Zhu, H.G., *et al.* (2017). Simultaneous silencing of GhFAD2-1 and GhFATB enhances the quality of cottonseed oil with high oleic acid. [Epub ahead of print]. 2017. DOI: 10.1016/j.jplph.2017.06.001.

Liu, Q., Singh, S.P. and Green, A.G. (2002). High-stearic and high-oleic cottonseed oils produced by hairpin RNA-mediated post-transcriptional gene silencing. *Plant Physiology*. **129**: 1732-1743.

Maheshwari, P. and Kovalchuk, I. (2014). Genetic engineering of oilseed crops. *Biocatalysis and Agricultural Biotechnology*. **3**: 31-37.

Maluszynski, M. B.S., Ahloowalia and Sigurbjornsson, B. (1995). Application of *in vivo* and *in vitro* mutagenesis technique for crop improvement. *Euphytica*. 85 (1-3): 303-315.

McCullum, C., Benbrook, C. and Knowles, L., *et al.* (2003). Application of modern biotechnology to food and agriculture: Food systems perspective. *Journal of Nutrition Education and Behavior*. **35**: 319-332.

Moyne, A. L., Tagu, D. and Thor, V. *et al.* (1989). Transformed calli obtained by direct gene transfer into sunflower protoplasts. *Plant Cell Reports*. 97-100.

Murphy, D.J. (1999). Production of novel oils in plants. *Curr. Opinion Biotech*. 10: 175-180.

Muuler, A., Iser, M. and Hess, D. (2001). Stable transformation of sunflower (*Helianthus annuus* L.) using a non-meristematic regeneration protocol and green fluorescent protein as a vital marker. *Transgenic Research.* **10**: 435-444.

Neskorodov, Y.B., Rakitin, A.L. and Kamionskaya, A.M., *et al.* (2010). Developing phosphinothricinresistant transgenic sunflower (*Helianthus annuus* L.) plants. *Plant Cell, Tissue and Organ Culture.* **100**: 65-71.

Ng, E.C., Dunford, N.T. and Chenault, K. (2008). Chemical characteristics and volatile profile of genetically modified peanut cultivars. *Journal of Bioscience and Bioengineering,* **106**: 350-356.

Padgette, S.R., Kolacz, K.H. and Delannay, X. *et al.* (1995). Development, identification, and characterization of a glyphosate-tolerant soybean line. *Crop Science.* **35**: 1451.

Qi, W., Zhang, F., Sun, F., *et al.* (2012). Over-expression of a conserved RNA-binding motif (RRM) domain (csRRM2) improves components of *Brassica napus* yield by regulating cell size. *Plant Breeding.* **131**: 614-619.

Qiao, R. Q. and Long, S. H. (2013). Research progress of molecular biology on flax 5: 265-269.

Radonic, L.M., Zimmermann, J.M. and Zavallo, D., *et al.* (2008). Introduction of antifungal genes in sunflower via Agrobacterium. *Electronic Journal of Biotechnology.* **11**: 8-9.

Rech, E.L., Vianna, G.R. and Aragão, F.J.L. (2008). High-efficiency transformation by biolistics of soybean, common bean and cotton transgenic plants. *Nature Protocols.* **3**: 410-418.

Sarla, N., Reddy, M.P. and Siddiq, E.A. (2001). Application of Inter Simple Sequence Repeat (ISSR) polymorphism in assessing diversity in crop plants. *Ind. J. Plant Genet. Res.* 14: 246- 247.

Sato, S., Newell, C. and Kolacz, K., *et al.* (1995). Stable transformation via particle bombardment in two different soybean regeneration systems. *Plant Cell Reports.* **12**: 408-413.

Shi, J., Lang, C. and Wu, X., *et al.* (2015). RNAi knockdown of fatty acid elongase1 alters fatty acid composition in *Brassica napus. Biochemical and Biophysical Research Communications.* **466**: 518-522.

Shivanna, K. R. (1996). Incompatibility and wide hybridisation. In: V. L. Chopra and S. Prakash (eds). Oilseed and vegetable Brassicas: Indian perspective 77-102 Oxford and IBHPublishing Co. Pvt. Ltd., New Delhi.

Skoric, D., Jocic, S. and Sakac, Z., *et al.* (2008). Genetic possibilities for altering sunflower oil quality to obtain novel oils. *Canadian Journal of Physiology and Pharmacology.* **86**: 215-221.

Song, S. (2009). Application of the genetic engineering in breeding for flax disease resistance. *Chinese Agricultural Science Bulletin.* 25: 227-229.

Tan, H., Yang, X. and Zhang, F., *et al*. (2011). Enhanced seed oil production in canola by conditional expression of *Brassica napus* LEAFY COTYLEDON1 and LEC1-LIKE in developing seeds. *Plant Physiology*. **156**: 1577-1588.

United Nations (1992). Convention on biological diversity. The Secretariat of the Convention on Biological Diversity. Rio de Janeiro, p.30.

Wang, N., Ma, J. and Pei, W. *et al*. (2017). A genome-wide analysis of the lysophosphatidate acyltransferase (LPAAT) gene family in cotton: organization, expression, sequence variation, and association with seed oil content and fiber quality. *BMC Genomics*. **18**: 218.

Wang, Z., Qiao, Y. and Zhang, J., *et al*. (2017). Genome wide identification of microRNAs involved in fatty acid and lipid metabolism of *Brassica napus* by small RNA and degradome sequencing. *Gene*. **619**: 61-70.

Weber, S., Friedt, W. and Landes, N., *et al*. (2003). Improved Agrobacterium-mediated transformation of sunflower (*Helianthus annuus* L.): Assessment of macerating enzymes and sonication. *Plant Cell Reports*. **21**: 475-482.

Wen, X. Shao, T., Li, X. J., Zhan, K. and Weina, A. S. *et al*. (2011). Progress on genetic transformation of foreign gene in flax. *Biotechnology bulletin* 12: 14-20.

Wu, J. (2014). Cloning the key gene of cellulose synthase (LusiCesA1) in flax. *Crops* 6: 35-38.

Yadav, R. and Kumar, A. (2004). Genetic diversity for white rust resistance in Indian mustard varieties and introgressions of resistance from other species. In: Proceedings of 11th International Rapeseed Congress, 6- 10th July, Copenhagen, Denmark.

11 Biological Management of Oilseed Brassica Diseases

P.D. Meena, Ibandalin Mawlong, H.S. Meena and Ashish Sheera

Introduction

An outlook on our agriculture due to climate change has poise concern to every scientist on the rise of not only abiotic stress but also biotic stress on plant system. The evolving pathogens along with already present pathogens have burden the plants to combat defense for each stress.

Rapeseed-mustard crops are confronted by various major diseases like Sclerotinia rot, white rust, Alternaria blight, downy mildew, powdery mildew and many others. Severe occurrence of diseases deteriorate the quantity as well as quality of seed and oil content drastically in different oilseed brassica crops. The yield reduction in oilseed brassica crops due to biotic stresses is about 19.9 per cent, out of which diseases contribute severe yield reduction at various plant growth stages which can be hardly overcome by conventional methods of pest management. Many of the well known synthetic pesticides have lost its efficacy due to development of resistance by the pathogen. The uncheck application of pesticides have lead an awareness among concern consumers with respect to environment sustainability, food safety and quality.

Biological control is an environmentally sound and effective means of reducing or mitigating pathogens and their effects through the use of natural enemies. Treatment of plants with various agents, including cell wall fragments, plant extracts and synthetic chemicals, can induce resistance to subsequent pathogen attack both locally and systemically (Walters and Fountaine, 2009). However, inconsistent effects under field conditions are one of the main problems in translation of biocontrol studies into practical approaches. To overcome this hurdle, ecological knowledge about biocontrol agents and their interactions with abiotic and biotic factors is necessary (Koberl *et al.*, 2011). Oilseed crop management strategies have been prioritised on increasing yield, crop protection, human health, environmental hazards and social aspects (Meena *et al.*, 2003b). Since, the development of disease resistant varieties is far from desired level, the alternative left with the plant pathologists is to develop effective biocontrol methods. This chapter reviews the current status of bio-control of diseases of oilseed brassica and the foreseeable future.

Sclerotinia Rot

Sclerotinia sclerotiorum a soil borne plant pathogen having extremely wide host range more than 400 plant species (Boland and Hall, 1974) it has caused tremendous damage to plants such as sclerotenia stem rot on oilseed brassica, while white mold on bean, head rot on sunflower and white rot on carrot. Sclerotenia rot, not only lead to reduction in yield but also oil quality in oilseed brassica.

Management of *S. sclerotiorum* is a major challenge faced by plant pathologists. There have been many reports on conventional methods such as the use of chemical fungicides like Carbendazim, a benzimidazole which have proved effective for a short time scale but its efficacy decline with increasing development of resistant isolates (Zhang *et al.*, 2003). Crop rotation for three to four years has been found effective control measure but because the longevity of this soil borne pathogen may not be sufficient. Conventional breeding for screening out resistant oilseed brassica lines have become less effective with the restricted gene pool.

In light of the present day scenario particularly pertaining to environmental, human health and development of resistance to funcicides, biological control has become an attractive control measure for plant disease management. Biocontrol is the reduction of the amount of inoculum or disease-producing activity of a pathogen accomplished by or through one or more organisms other than man (Cook and Baker, 1983). Concerns associated with the use of chemical pesticides have made the biological control of *Sclerotinia* diseases an interesting area of research amongst plant pathologists (Bardin and Huang, 2001).

Fungal Antagonists

In general, the use of biocontrol agent is restricted to controlled environment as they need stable environment to thrive successfully in the infection court (Whipps, 1994). Several biocontrol agents have been reported. The mycoparasitic fungi parasitizing sclerotia include *Coniothyrium minitans*, *Trichoderma* spp., *Gliocladium* spp. *Sporidesmium sclerotivorum*, *Talaromyces flavus*, *Epicoccum purpurescens*, *Streptomyces* sp., *Fusarium*, *Hormodendrum*, *Mucor*, *Penicillium*, *Aspergillus*, *Stachybotrys*, and *Verticillium* (Makkonen and Pohjakallio, 1960; Adams and Ayers, 1979; Singh and Kaur, 2001; Martinson and del Rio, 2001; Baharlouei *et al.*, 2011). The aim of these biocontrol agents is to reduce the sclerotial load in soil build up by the soil–borne pathogens. Both fungi and bacteria near the soil surface play an important role in degrading the sclerotial bodies. Diurnal fluctuation of soil temperature, moisture and relative humidity cause cracks on the sclerotial rinds, resulting not only in leakage of the cell constituents, but also providing avenues of infection by the antagonistic microbes dwelling in the soil. Soil application of *C. minitans* to different host crops reduced carpogenic germination and viability of sclerotia under wide range of temperature, relative humidity and soil moisture (Sandys-Winsch *et al.*, 1993; McLaren *et al.*, 1996; Budge *et al.*, 1995; Hedke and Tiedemann, 1998). Although, soil incorporation of *C. minitans* in microplots of oilseed rape reduced soil inoculum, it neither led to disease control nor yield improvement (McQuilken *et al.*, 1995). Wettable Granule of *C. minitans* containing 1×10^9 viable conidia per gram colonized and decayed sclerotia within 3 months of its application, and was effective against *S. sclerotiorum* infecting vegetables, ornamentals, oilseed rape and beans (Luth, 2001). Hannusch and Bolland, (1996) reported that change in air temperature by 4 °C, or relative humidity by 5 per cent, adversely affected the *S. sclerotiorum*- suppressing ability of the fungi isolated from anthroplanes of bean and rapeseed. Singh and Kaur (2001) reported mycoparisitism of *T. harzianum* (Th38), *T. viride*, and *E. purpurescens* against *S. sclerotiorum*; *T. harzianum* controls *S. sclerotiorum* by hyphal mycoparasitism rather than by sclerotial parasitism. The isolates of *T. harzianum* 3, *T. harzianum* 4, *T. virens* and *T. viride* were most potent in reducing the linear growth of mycelium and apothecial production of *S. sclerotiorum in-vitro*; *T. harzianum* was most effective in reducing the lesion length and disease intensity when applied simultaneously, or seven days prior to the pathogen under the screen house conditions; antagonist applied at the rate of 15 g/kg soil as wheat bran inoculum was superior in reducing SR of rapeseed-mustard (Mehta *et al.*, 2012). Srinivasan *et al.* (2001) reported that even the cultural filtrate of *T. viride* and *T. harzianum* significantly reduced the growth of *S. sclerotiorum in vitro* conditions. Pathak *et al.* (2001) also reported reduced SR incidence with soil application of *T. viride, T. harzianum* and *G. virens* combined with bavistin seed treatment. Soil-borne strains of *Gliocladium roseum, T. harzianum* and *Aspergillus* sp. also showed inhibitory effect against the pathogen (Rodriguez and Godeas, 2001). Application of Kalisena SL (*Aspergillus niger*) formulation in *S. sclerotiorum*-infested plots saved the cauliflower seedlings, and soil application of *A. niger* AN27 parasitized and killed the existing sclerotial population (Sen, 2000). *Streptomyces* sp isolate 422 significantly reduced SR incidence (Baharlouei *et al.*, 2011) by increasing the level of hydrolytic enzymes including chitinase and β-1, 3 glucanase in canola plants (Fernando *et al.*, 2007). Soil incorporation of sclerotial parasite *S. sclerotiorum* was also found effective up to 5 years in controlling Sclerotinia rot (Martinson and Del Rio, 2001).

Considering the epidemiology of the disease, any biological control which only reduces primary source of soil inoculum by reducing sclerotial germination, apothecial development and discharge of ascospores will be of very limited importance for the control of SR in oilseed rape and mustard. This is simply because there is still enough sclerotial inoculum left in the soil to infect rapeseed-mustard plants at flowering stage. The potential mycoparasite, applied aerially, must also be effective in inhibiting germination of ascospores on petals. For successful control of SR in oilseed rape and mustard the potential mycoparasite must therefore, be apllied both aerially and in the soil. In this regard, results of several studies by Meena *et al.* (2009, 2011, 2013, 2014) showing significant effectiveness of *T. harzianumi* applied as soil inoculants, seed treatment and foliar spray, singly or in combination, in controlling SR in mustard is a step in the right direction. Soil inoculation of

T. harzianum isolate GR and soil application of FYM- infected with *T. harzianum* isolate SI-02 reduced SR incidence by 69 per cent and 60.8 per cent, respectively (Meena *et al.*, 2009). Similarly, seed treatment combined with foliar spray of *T. harzianum*, and seed treatment with *T. harzianum* and foliar spray with garlic (*Allium sativum*) bulb extract, not only significantly reduced SR incidence, but also gave higher cost: benefit ratio (Meena *et al.*, 2011, 2013, 2014). The disease suppression was due to the effective saprophytic colonization of petals.

Bacterial Antagonists

Antagonists like plant growth promoting rhizobacteria are exploited for the management of both foliar and soil borne pathogens of various economically important crop plants. Several bacterial antagonists including *Bacillus*, *Pseudomonas* and *Agrobacterium* species are commercialized, for their effective role in disease management. But, research on the use of bacterial antagonists for the management of white mold fungus still remains to be explored fully. Strains of *Bacillus* spp., frequently isolated from the sclerotia of *S. sclerotiorum* from North Dakota in the USA, not only reduced germination of infected sclerotia, but also adversely affected integrity and colour of medulla (Wu, 1988); fifty three per cent of sclerotial bodies of *S. sclerotiorum* recovered from the soils of North Dakota were infected by *Bacillus* species. *Bacillus* strains isolated from canola and wheat plants also showed antifungal activity to *S. sclerotiorum in vitro* (Zhang and Fernando, 2003). Apart from pre-colonization, several *Bacillus* spp. also produced the antibiotic Zwittermicin-A (Zhang and Fernando, 2004), which increased degradation and reduced germination of sclerotia of *S. sclerotiorum* (Nelson *et al.*, 2001). *Erwinia herbicola* and *B. polymyxa* inhibited the growth of *S. sclerotiorum in vitro* (Godoy *et al.*, 1990).

Antagonistic *Pseudomonas* spp. (DF41) and *P. chlororaphis* (PA23) inhibited the germination of ascospores of *S. sclerotiorum* (Savchuk and Fernando, 2004). Application of DF41 and PA23 on to petals increased bacterial population after 24 h, but population decreased between 96 and 120 h after application; significant differences in disease severity were correlated to timing of ascospore applications in the control treatments. Results from these studies indicate that PA23 and DF-41 are effective biocontrol agents against *S. sclerotiorum* (Savchuk and Fernando, 2004). The bacterial strains/species *viz.*, *B. subtillis*, *P. fluorescens* 132, *P. maltophila* and *P. fluorescens* M have been reported to be most effective by inhibiting mycelial growth and sclerotial formation against *S. sclerotiorum in vitro* conditions. Per cent mycelial growth inhibition of the pathogen was negatively correlated with the number of sclerotia formation. Regression equation developed revealed Y = -1.9882x + 65.418, with $R^2 = 0.819$ (Mehta and Hieu, 2014). The *P. chlororaphis* (PA23), *B. amyloliquefaciens* (BS6), and *Pantoea agglomerans* exert multiple mode of action and lead to the suppression of carpogenic germination and mycelial growth through the production of volatile and non-volatile antimicrobial antibiotics. Moreover, bacteria PA23 and BS6 trigger/induce resistance via the production of defense related gene products. *P. agglomerans* degrades oxalic acid through the production of oxalate oxidase. The above-mentioned promising strains would pave the way for the management of *S. sclerotiorum* in both agricultural and horticultural crops (Fernando *et al.*, 2004). *P. chlororaphis* strain PA-23 controlled ascospore germination, and stem rot of canola in both greenhouse and field studies (Fernando *et al.*, 2007). Further, antibiotics extracted from PA23 caused inhibition of sclerotial and spore germination, hyphal lysis, vacuolation, and protoplast leakage in a number of plant pathogens, including *S. sclerotiorum* (Zhang *et al.*, 2004). *P. agglomerans* isolated from leaves and flowers of canola produces oxalate oxidase and degrades oxalic acid produced by *S. sclerotiorum*, the pathogencity factor required for the successful establishment of the host-parasite relationship. Pre-colonization of infection court including blossoms and leaves by *P. agglomerans* would probably be very effective in preventing the infection process.

White Rust

White rust characterised by the formation of white to cream coloured zoosporangial pustules on cotyledons, leaves, stems, and inflorescences. Staghead galls are formed as the result of inflorescence infection (Verma and Petrie, 1980) is caused by *Albugo candida* which widely seen in cruciferous vegetables.

The information on the management of white rust (WR) of cultivated brassica crops have been generated on the aspects of biological control. The bacterium, *Pseudomonas syringae*, caused a soft rot disease preferentially on the green developing stagheads, of field-grown Canola (*B. rapa*) in Alberta, Canada, and on *B. juncea* in India, leading to abortion of the developing oospores. This natural biological control of *A. candida* may regulate its overwintering population in the field, and may have implications towards developing planned biological control strategies for this pathogen (Tewari *et al.*, 2000). Bulb extract of *Allium sativum* and bioagent *Trichoderma viride* individually as seed treatment and in combination with their respective sprays were at par with mancozeb for both white rust severities on leaves as also for stagheads per plot (Meena *et al.*, 2003a). Meena *et al.* (2005) found leaf extract of *Azadirachta indica* very effective in controlling-WR under field conditions. Seed treatment with *T. harzianum* followed by foliar spray of *P. fluorescens* also gave significantly lower disease index (20.0 and 29.8 per cent) and was highly effective as compared to control (25.7 and 68.3 per cent) during the year 2005-06 and 2006-07, respectively. Garlic bulb extract as seed treatment with successive foliar spray was also equally effective (Bhatt *et al.*, 2009). Crude extracts of *Rhus coriaria, Anagallis arvensis* and *Mesphilus germanica* inhibits zoospore release from sporangia of *A. candida* at 50 ppm (Omranpour *et al.*, 2011).

Alternaria Blight

Alternaria blight diseases caused by *Alternaria brassicae* (Berk) Sacc. is one of the main causing agent with average yield reduction by 32 to 57 per cent in oilseed brassica (Meena *et al.*, 2010). The severity of this disease threatens not just yield loss but seed size and oil content in rapeseed mustard (Prasad *et al.*, 2003), plant growth and development and oil filling processes. Alternaria affects most cruciferous crops and the disease severity is greatly influenced by wet area, relatively high rainfall (Humpherson-Jhones and Phelps, 1989) and date of sowing adds to the disease severity (Meena *et al.*, 2002). No single method or approach is feasible, viable, stable, effective, and economical in dealing with any Brassica-Alternaria system. Therefore, it is necessary to integrate all the methods of plant disease control which are available to effectively manage Alternaria diseases of brassicaceous plants. A lot of work on the biocontrol activity of fungal and bacterial agents in controlling the phytopathogens has been well documented (Whipps 2001; Ozbay and Newman, 2004).

Seed dressing with a formulation of *Streptomyces* species provided 80–90 per cent control of seed-borne *A. brassisicola* even after storing treated seeds in dry place for 35–40 days (Tahvonen and Avikainen, 1987). Foliar application of diffusate of *Streptomyces rochei* also reduced Alternaria blight intensity in *B. rapa* (Sharma and Gupta, 1978a, b). In other oilseed crops like linseed seed-dressing with *Bacillus* spp. and *T. harzianum* effectively increased survival of healthy plants grown from *A. linicola* infected linseed seeds with weekly sprays with *T. viride* isolate is seen to be at par to the fungicide iprodione (rovral) in controlling *A. linicola* infestation on linseed capsules (Mercer *et al.*, 1992, 1993). Alternaria blight control was as effective with fungicide mancozeb as with spraying *T. viride* at critical growth stages in Indian mustard (Meena *et al.*, 2004). Foliar application of *Auvobasidium pullulens* and *Epicoccum nigrum*, 14 h before pathogen attack, reduced the infection by *A. brassisicola* (Pace and Campbell, 1974); *B. amyloliquefaciens* applied as foliar spray also effectively reduced Alternaria blight in *B. napus* (Danielsson *et al.*, 2006). Several antagonists of *A. ricini, viz., Mucor varians, T. viride, Rhizopus nigricans, Pseudomonas fluovescans, Aspergillus niger* and *A. sativus* (Silva *et al.*, 1998), of *A. helianthi, viz., B. mycoides, P. fluovescens, P. putida, P. cepocid* and *B. polymyxa* (Prasad and Kulshrestha, 1999), have been reported and, if tested, some may prove effective in controlling Alternaria blight in *B. juncea*. *Allium sativum* bulb extract was found useful in reducing Alternaria blight disease in Indian mustard (Meena *et al.*, 2004).

Plant extracts of roots, leaves, stems, inflorescence and fruits of several species have shown fungicidal activities. Fungitoxicants such as essential oil from the roots of radish inhibits (1: 2500) *A. brassicae* (Nehrash, 1961). The deproteinized leaf extracts of *Acacia nilotica, Enicostema hyssopifolium, Mimosa hamata*, and *Vitis vinifera* have shown fungistatic activity against *A. brassicicola* (Umalkar *et al.*, 1977). The extracts prepared from the leaves of *Lawsonia alba*, roots of *Datura stramonium*, and inflorescence of *Mentha piperita* have fungitoxic activity against *A. brassicae* isolated from cauliflower leaves. Extracts of the ferns *Adiantum caudatum, Diplazium esculentum*, and *Pteris vittata* reduces growth, and germination of *A. brassicicola* (Yasmeen and Saxena, 1990; Ahmed and Agnihotri, 1977).

In vitro conditions, most of the plant extracts restrict mycelial growth, sporulation, and conidial germination. Some of them have shown efficacy under field conditions by restricting the infection points, spot size, and disease intensity (Khurana *et al.*, 2005; Patni *et al.*, 2005; Muto *et al.*, 2006; Ho *et al.*, 2006, 2007; Patni and Kolte, 2006; Guleria and Kumar, 2009; Guleria *et al.*, 2010; Baka, 2010; Meena and Sharma, 2012; Singh *et al.*, 2013).

Eucalyptus spray gives significantly lesser number of spots per leaf (2.05), minimum size of spots (1.28mm), minimum sporulation intensity (1.22 x 10^5) and minimum disease index followed by *Carotropis, Ocimum* and *Polyanthai* extracts @ 5 per cent (Patni and Kolte, 2006). *Alternaria* blight of mustard can be managed by spraying aqueous extract of *Azadirachta indica, Allium sativum* and *Zingiber officinale*

@ 5, 10, 15 and 20 per cent along with higher yield. However, spraying of 15 per cent *Azadirachta indica* extract gives highest yield with best cost benefit ratio (Mahapatra and Das, 2013). While, according to Meena *et al.* (2004, 2008) application of bulb extract of *Allium sativum* at 45 and 75 days after sowing gave highest seed yield at Sewar, India. The *Allium sativum* extract is most effective in controlling seed borne fungi of mustard also (Latif *et al.*, 2006).

Antagonists for Biocontrol

Saprophytic phylloplane fungi such as *Aureobasidium pullulans*, and *Epicoccum nigrum* are pathogenic to *A. brassicicola* (Pace and Campbell, 1974). The *Verticillium*-state of *Nectria inventa* Pethybridge, a destructive mycoparasite, is one of the dominant phylloplane fungi of rapeseed (Tsuneda and Skoropad, 1978a). Amongst the leaf surface mycoflora, the most antagonistic fungi are *E. purpurascens, A. pullulans* and *Cladosporium cladosporioides* in case of *A. brassicae*. The most significant effects observed is when spores of the leaf surface fungi and metabolites of *Acremonium roseogriseum, Aspergillus terreus* and *C. cladosporioides* are sprayed prior to inoculation of pathogen on leaves which is reported to inhibit growth of *A. brassicae* (Rai and Singh, 1980).

Pre-treatment application with spore suspension of *Streptomyces rochei* or its diffusate results in a marked reduction of leaf blight intensity caused by *A. brassicae* and *A. brassicicola* on brown sarson, *B. rapa* var. *dichotoma* (Sharma and Gupta, 1978a, b, 1979). With the decline in the population of *S. rochei* there is a rise in the population of *Alternaria* from December onwards indicating the possibility of antagonism. Several factors, such as climatic conditions, age of the host plant, maturity of the leaves, and variety of the host have profound influence on the deposition of conidia of both the pathogens, and *S. rochei* (Sharma and Gupta, 1980). Jayant and Sinha (1981) reported that *S. hygroscopicus* is strongly antagonistic to *A. brassicae* and *A. brassicicola*. When the culture filtrate is sprayed over the sarson plants, a week before or a week after spraying the spore suspension of *A. brassicae* and *A. brassicicola*, germination, and disease development is reduced. Whenever the spore population of *Streptomyces arabicus* is higher on leaves of Yellow Sarson and Taramira, the population of conidia of *A. brassicae* and *A. brassicicola* declines. The population of antagonistic *S. arabicus* is higher in the young leaves as compared to old ones (Sharma *et al.*, 1984). The antifungal substance in the diffusate of the antagonist is thermolabile (Sharma *et al.*, 1985a). A pigmented and xylose-utilizing strain of *S. bobili* is found to be active against, *A. brassicae, A. brassicicola,* and *A. raphani* (Sharma and Sinha, 1989).

An isolate of *Streptomyces* spp. obtained from light-coloured, Finnish Sphagnum horticultural peat has proved effective biological control agent against plant pathogens (Tahvonen, 1982a, b). Treating cauliflower seeds with *Trichoderma viride* and *Streptomyces* spp. isolates inhibits or reduces damping-off caused by *A. brassicicola* (Tahvonen, 1982b, 1988). Seed dressing with Mycostop, a powdery formulation prepared from spores and mycelium of *Streptomyces*, is 80-90 per cent successful in controlling damping-off from seeds artificially infected with *A. brassicicola*. The dressing remains effective on seeds stored under dry conditions for 5-6 weeks, but subsequently decreases slowly. *Streptomyces* dressing controls in a manner comparable to chemical dressing with Thiram, preventing damping-off caused by *Alternaria* fungi in seedlings, which are grown from commercial seed lots of different origin (Tahvonen, 1985; Tahvonen and Avikainin, 1987). White *et al.* (1990) reported that Mycostop, a biofungicide based on a selected *S. griseoviridis* isolate from Finnish Sphagnun peat, introduced either by seed treatment or soil treatment, controls *A. brassicicola* of cauliflower and cabbage.

Brassicaceous seeds treated with either *Gliocladium virens*-19, *Trichoderma harzianum*-22, *T. harzianum*-50, *Penicillium corylophilum*-36, and *P. oxalicum*-76 have a significantly higher emergence of healthy seedlings. The hyphae of antagonistic fungi are able to adhere to conidia or coil around or penetrate germ tubes or hyphae of *A. brassicicola*. Conidia of *A. brassicicola* shrivel, and plasmolyze in the presence of antagonistic fungi (Wu and Lu, 1984b). For biocontrol of seed-borne *A. raphani*, and *A. brassicicola* of radish antagonists like *Chaetomium globosum*, *T. harzianum*, *T. koningii*, and *Fusarium* spp. treated plants show consistently increased numbers of healthy seedlings (Vannacci and Harman, 1987). Wu and Lu (1984b) found *Trichoderma*, *Gliocladium*, and *Penicillium* spp. parasitizing *A. brassicicola*. Spraying of *T. viride* conidial suspension causes reduction in Alternaria blight of mustard on leaves (76 per cent) and pods (68 per cent). The bioagents survives on phylloplane upto 30 days at 80-90 per cent RH and 20-35°C temperature (Reshu and Khan, 2012). In Poland, Madej (1986) reported *Gonatobotrys simplex* as a hyper-parasite of *A. brassicae*.

Mechanism of Biocontrol

Antagonistic mechanism of bio-control has been studied in few selected host-parasite systems. The parasitism of *A. brassicae* by the *Verticillium-state* of *Nectria inventa* occurs either by penetration or contact without penetration. Upon contact of Parasitic hyphae induce abnormal responses in host cells. A reaction consisting largely of an electron-dense transparent matrix, and dispersed tubule-like electron-dense material develops between the cell wall, and the invaginated plasma membrane. The tubule-like elements subsequently aggregate to form electron-dense deposits below the cell wall. The affected cell forms a septal plug, accumulates membranes, and finally degenerates. Hyphae of *N. inventa* penetrate the conidial cells of *A. brassicae* mainly by a process, which appears primarily enzymic in nature. The cytoplasm of the penetrated cell becomes progressively less dense, and the cell eventually appears empty (Tsuneda *et al.*, 1976; Tsuneda, 1977; Tsuneda and Skoropad, 1980). Penetration of *A. brassicae* hyphae causes separation of the cells, penetration of conidia occurs most frequently at the septum or at the basal pore in juvenile conidia (Tsuneda and Skoropad, 1977a, b). Conidia of *N. inventa* require at least 24 h to initiate germination and 4 days to parasitize *A. brassicae* on intact leaves (Tsuneda and Skoropad, 1978b). Conidia of *A. brassicae* leak various amino acids, and sugars when they are exposed to an alternate dry wet condition, the longer the drying period, the larger the amount of leakage. Among the amino acids, glutamine is exuded in the largest amount followed by aspartic acid, and glutamic acid. The sugar fraction consists mainly of glucose and fructose. These leaked nutrients stimulate germination, and growth of *N. inverta*.

As a result, germination of *Alternaria* conidia is drastically reduced to less than 5 per cent, and they are eventually destroyed by *N. inventa* (Tsuneda and Skoropad, 1978c). The treatment of oilseed rape seeds with results in significant protection against all fungal pathogens tested. The aftermath effect of seeds treated with Bacillus innoculum (*Bacillus* strains UCMB-5113 and UCMB-5036) showing higher survival rate with no visible signs of infection such as chlorosis or yellowing are observed on plants that did germinate after bacterial treatment. The germination efficiency is the same as the seeds harvested from different plants including control (Danielsson *et al.*, 2006). *Bacillus subtilis* strain UK-9, an isolate from reclaimed soils, was studied for its biological control activity against Alternaria leaf spot disease of mustard. In dual culture, production of antifungal metabolites by the bacteria causes morphological alterations of vegetative cells, and spores, disruption, and lysis of their cell wall. The antagonists reduce spore germination on leaves, and disease incidence of the pathogen in plant trials as well as it also demonstrated plant growth promoting ability. Many antagonistic species produce antibiotic substances, which have the ability to produce enzymes causing the lysis of cell wall components of the pathogenic fungus. This helps the antagonist to penetrate the host hypha, and grow on it as a hyper-parasite (Tapio and Pohto-Lahdenpera, 1989).

Biological Control v/s Biochemical Change

A comparison studies on Indian mustard seeds treated separately with three bioagents *viz.*, *Trichoderma harzianum*, *Pseudomonas fluorescens*, and *Bacillus subtilis*, grown in experimental fields, followed by spraying on 30 and 60 days after sowing. . The use of biocontrol agents shows enhanced growth in comparison to control, decreased disease index on leaves as well as the pods of Indian mustard. These biocontrol agents enhances the content of dry matter, total phenol, ortho-dihydroxyphenol, starch, total soluble sugars, reducing sugars, total lipids, and different membrane lipids in the leaves. The total protein content decreases after treatment with biocontrol agents at 30 and 60 days after sowing. It suggests that treatment with biocontrol agents initiates a number of biochemical changes, which can be considered to be a part of plant defense response (Sharma *et al.*, 2010a). The proportionate increase of various lipid fractions *i.e.* phospholipids, glycolipids, and sterol content in seed with a corresponding decrease in total glycerides. And further proportionate increase of individual fatty acids like 18: 3, 20: 1 and 22: 1 while that of 18: 1 and 18: 2 fatty acids decreased in seeds with application of biocontrol agents indicating their involvement in signal transduction for defence against stress. There are both qualitative, and quantitative differences in the banding patterns of storage protein albumin and globulin after application of biocontrol agents. The biochemical alterations in the host induced by treatment with biocontrol agents can be associated with defense mechanisms and enhanced growth of the plant (Sharma *et al.*, 2010b).

Downy Mildew

Garlic juice or aqueous extracts of garlic was reported to be toxic to *H. parasitica* which causes downy mildew of radish (Ark and Thompsom, 1959). Bacteria were observed on the mycelium, conidiophores and conidia of *H. parasitica* on *Lepidium graminifolium* (Nicolas and Aggery, 1940). This was associated with a reduction in conidial germination. Bedlan (1987) obtained good control of *H. parasitica* in radish with an isolate of *Trichoderma*. Paraffin oil based *Trichoderma* formulations reduced downy mildew, white rust and Alternaria blight of mustard under organic farming (Saxena and Tewaria, 2017).

Powdery Mildew

Bacillus velezensis YP2 inhibited the mycelial growth of several plant pathogens including *Cercespora* spp., *Septoria* sp., *Phoma* sp., *Botrytis cinerea* and *Sclerotinia scleotiorum* occurring in leafy vegetables. The culture broth of *B. velezensis* YP2 was very effective to control the powdery mildew of leaf mustard caused by *Erysiphe cruciferarm* (Lee *et al.*, 2016).

Conclusion

Increase in public concern about the environment has increased the need to develop and implement effective biocontrol agents for crop protection. Several potential biocontrol agents used for plant disease management behave as opportunistic human pathogens. Oilseed Brassica management strategies are prioritized on increasing productivity and production, crop protection, doubling farmer's income, human health, environmental hazards and social aspects. This sequential rather than integrated approach can contribute to imbalance problems related to sustainability in oilseed production. Accordingly, there is a need to explore each factor to set priorities and in case of a negative trend in sustainability, necessary adjustments can be made in disease management strategy. It is necessary to understand disease epidemic in variable environmental conditions. The integrated disease management strategy including cultural, biological and host resistance should be refined, retested and revalidated under changing environmental conditions. These technologies if effectively utilized and industrialized can lead to early substitution of use of chemical fungicides for an eco-friendly and sustainable disease management programme in oilseed crops. All these can strengthen our commitment, confidence and ability to generate more of success stories for making on-farm bio management of oil crop disease a reality.

References

Adams, P. B. and Ayers, W. A. (1979). Ecology of *Sclerotinia* species. *Phytopathology*, 69: 896-899.

Ahmed, S. R., and Agnihotri, J. P. (1977). Antifungal activity of some plant extracts. *Indian Journal Mycology Plant Pathology*, 7: 180-181.

Ark, P. V., and Thompson, J. P. (1959) Control of certain diseases of plants with antibiotics from garlic (*Allium sativum* L.). *Plant Disease Reporter*, 43: 276-282.

Baharlouei, A., Sharifi-Sirchi, G. R. and Shahidi Bonjar, G. H. (2011). Biological control of *Sclerotinia sclerotiorum* (oilseed rape isolate) by an effective antagonist *Streptomyces*. *African Journal of Biotechnology*, 10: 5785-5794.

Baka, Z. A. M. (2010). Antifungal activity of six Saudi medicinal plant extracts against five phyopathogenic fungi. *Archives of Phytopathology and Plant Protection*, 43, 736-743.

Bardin, S. D. and Huang, H. C. (2001). Research on biology and control of *Sclerotinia* diseases in Canada. *Canadian Journal of Plant Pathology*, **23**: 88–98.

Bedlan, G. (1987) The most important diseases of radishes. Pflanzenschutz (Wien) 3: 9-10.

Bhatt Rashmi, Awasthi, R. P. and Tewari, A. K. (2009). Management of downy mildew and white rust diseases of mustard. *Pantnagar Journal of Research*, **7**(1): 54-59.

Boland, G. J. and Hall, R. (1994). Index of plant hosts of *Sclerotinia*. *Canadian Journal of Plant Pathology*, **16**: 93–108.

Budge, S. P., McQuilken, M. P., Fenlon, J. S. and Whipps, J. M. (1995). Use of *Coniothyrium minitans* and *Gliocladium virens* for biological control of *Sclerotinia sclerotiorum* in glass-house. *Biological Control* **5**: 513-522.

Cook, R. J. and Baker, K. F. (1983). Approaches to biological control. In: The natural and practice of biological control of plant pathogens. St.Paul, Minnesota, USA, American Phytopathological Society. pp. 84-131.

Danielsson, J., Reva, O., and Meijer, J. (2006). Protection of Oilseed Rape (*Brassica napus*) toward fungal pathogens by strains of plant-associated *Bacillus amyloliquefaciens*. *Microbial Ecology*, **54**: 134–140.

Fernando, W. G. D., Nakkeeran, S. and Yilan, Z. (2004). Eco-friendly methods in combating *Sclerotinia sclerotiorum* (Lib.) de Bary. *Recent Research and Development in Environment Biology*, **1**: 329-347.

Fernando, W. G. D., Nakkeeran, S., Zhang, Y. and Savchuk, S. (2007). Biological control of *Sclerotinia sclerotiorum* (Lib.) de Bary by *Pseudomonas* and *Bacillus* species on canola petals. *Crop Protection*, **26**: 100-107.

Godoy, G., Steadman, J. R., Dickman, M. B. and Dam, R. (1990). Use of mutants to demonstrate the role of oxalic acid in pathogenicity of *Sclerotinia sclerotiorum* on *Phaseolus vulgaris*. *Physiology of Molecular Plant Pathology*, **37**: 179–191.

Guleria, S., and Kumar, A. (2009). Antifungal activity of *Agave americana* leaf extract against *Alternaria brassicae*, causal agent of Alternaria blight of Indian mustard (*Brassica juncea*). *Archives of Phytopathology and Plant Protection*, **42**: 370-375.

Guleria, S., Kumar, A., and Tiku, A. K. (2010). Toxicity of *Solanum xanthocarpum* fruit extract against *Alternaria brassicae*, causal agent of *Alternaria* blight of Indian mustard (*Brassica juncea*). *Archives of Phytopathology and Plant Protection*, **43**: 283- 289.

Hannusch, D. J. and Bolland, G. J. (1996). Influence of air temperature and relative humidity on biological control of white mold of bean (*Sclerotinia sclerotiorum*). *Phytopathology* **86**: 156-162.

Hedke, K. and Tiedemann, A. V. (1998). Environmental influences on the decomposition of sclerotia of *Sclerotinia sclerotiorum* by *Coniothyrium minitans*. *Mitteilungen aus der Biologischen Bundesanstalt* **357**: 352.

Ho, W. C., Su, H. J., Li, J. W., and Ko, W. H. (2006). Effect of extracts of Chinese medicinal herbs on spore germination of *Alternaria brassicicola*, and nature of an inhibitory substance from gallnuts of Chinese sumac (*Rhus chinensis*). *Canadian Journal of Plant Pathology*, **28**: 519-525.

Ho, W. C., Wu, T. Y., Su, H. J., and Ko, W. H. (2007). Effect of Oriental medicinal plant extracts on spore germination of *Alternaria brassicicola* and nature of inhibitory substances from speedweed. *Plant Disease*, **91**: 1621-1624.

Humpherson-Jones, F. M. and Phelps, K. (1989). Climatic factors influencing spore production in *Alternaria brassicae* and *Alternaria brassicicola*. *Annals of Applied Biology*, **114**: 449-458.

Jayant, M., and Sinha, S. K. (1981). Control of leaf spot disease of Brown Sarson caused by *Alternaria brassicae* and *Alternaria brassicicola* by the antibiotic substance produced by a strain of *Streptomyces hygroscopicus*. 3rd International Symposium of Plant Pathology, New Delhi, India, 21 (Abstr.)

Khurana, A. K., Mehta, N., and Sangwan, M. S. (2005). Variability in the sensitivity of *Alternaria brassicae* isolates to plant extracts. *Journal of Mycology and Plant Pathology*, **35**: 76-77.

Koberl, M., Ramadan, E. M., Roßmann, B., Staver, C., Furnkranz, M., Lukesch, B., Grube, M., and Berg, G. (2011). Using ecological knowledge and molecular tools to develop effective and safe biocontrol strategies. In: Pesticides in the Modern World/Book 5.

Latif, A. M., AbuKoser, S. M., Khan, A. I., Habibur, M., Rahman, M., and Anwar, H. M. (2006). Efficacy of some plant extracts in controlling seed borne fungi infection of mustard. Bangladesh. *Journal of Microbiology*, **23**: 168-170.

Lee, S. Y., Weon, H. Y., Jeong, J. K., and Han, J. H. (2016). Biocontrol of Leaf Mustard Powdery Mildew Caused by *Erysiphe cruciferarm* using *Bacillus velezensis* YP2. *The Korean Journal of Pesticide Science*, **20**: 369-374.

Luth, P. (2001). The biological fungicide Contans WG - a preparation on the basis of the fungus *Coniothyrium minitans*. Proc XI International Sclerotinia Workshop, (Young CS and Hughes K. J. D., eds), Central Science Laboratory, York, UK, July 8–12, p.127.

Madej, T. (1986). Incidence of diseases affecting cruciferous plants in Poland. *Wissenschaftliche Zeitschrift der Wilhelm-Pieck-Universitat Rostock, Naturwissenschaftliche Reihe*, **35**: 50-51.

Mahapatra, S., and Das, S. (2013). Evaluation of fungicides and botanicals against Alternaria leaf blight of mustard. *Indian Journal of Plant Protection*, **41**: 61-65.

Makkonen, R. and Pohjakallio, O. (1960). On the parasites attacking the sclerotia of some fungi pathogenic to higher plants and on the resistance of those sclerotia to their parasites. *Acta Agril Scandinavica* **10**: 105-126.

Martinson, C. A. and del Rio, E. L. (2001). Prolonged control of *Sclerotinia sclerotiorum* with *Sporidesmium sclerotivorum*. Proc XI International Sclerotinia Workshop, (Young CS and Hughes K. J. D., eds), Central Science Laboratory, York, UK, July 8–12, p.133.

McLaren, D. L., Huang, H. C. and Rimmer, S. R. (1996). Control of apothecial production of *Sclerotinia sclerotiorum* by *Coniothyrium minitans* and *Talaromyces flavus*. *Plant Disease*, **80:** 1373-1378.

McQuilken, M. P., Mitchell, S. J., Budge, S. P., Whipps, J. M., Fenlon, J. S. and Archer, S. A. (1995). Effect of *Coniothyrium minitans* on sclerotial survival and apothecial production of *Sclerotinia sclerotiorum* in field-grown oilseed rape. *Plant Pathology*, **44:** 883-896.

Meena, P. D., Awasthi, R. P., Godika, S., Gupta, J. C., Kumar, A., Sandhu, P. S., Sharma, P., Rai, P. K., Singh, Y. P., Rathi, A. S., Prasad, R., Rai, D. and Kolte, S. J. (2011). Eco-friendly approaches managing major diseases of Indian mustard. *World Applied Science Journal*, **12:** 1192-1195.

Meena, P. D., Chattopadhyay, C., Meena, P. S., Goyal, Poonam and Kumar, V. R. (2014). Shelf life and efficacy of talc-based bio-formulations of *Trichoderma harzianum* isolates in management of *Sclerotinia* rot of Indian mustard (*Brassica juncea*). *Annals of Plant Protection Sciences* 22: 127-135.

Meena, P. D., Gour, R. B., Gupta, J. C., Singh, H. K., Awasthi, R. P., Netam, R. S., Godika, S., Sandhu, P. S., Prasad, R., Rathi, A. S., Rai, D., Thomas, L., Patel, G. A. and Chattopadhyay, C. (2013). Non-chemical agents provide tenable, eco-friendly alternatives for the management of the major diseases devastating Indian mustard (*Brassica juncea*) in India. *Crop Protectection*, **53:** 169-174.

Meena, P. D., Kumar, A., Chattopadhyay, C. and Sharma, P. (2009). Eco-friendly management of Sclerotinia rot in Indian mustard (*Brassica juncea*). Proc. 16th Australian Research Assembly on Brassicas, Ballarat, Australia. September 14-16, pp. 202-204.

Meena, P. D., and Sharma, P. (2012). Antifungal activity of plant extracts against *Alternaria brassicae* causing blight of *Brassica* spp. *Annals of Plant Protection Science*, **20:** 256-257.

Meena, P. D., Bineeta Sen, Sanjeev Kumar, Sahana Majumdar and Chattopadhyay C. (2003b). Biological control of oilseed crop diseases - current status and prospects. In: *Oilseed Crop Diseases*, Eds. Saharan, G.S.; Naresh Mehta and M.S. Sangwan, Indus Co. Publ., New Delhi, pp. 519-529.

Meena, P. D., C. Chattopadhyay, F. Singh, B. Singh and A. Gupta. (2002). Yield loss in Indian mustard due to White rust and effect of some cultural practices on Alternaria blight and White rust severity. *Brassica* 4: 18-24.

Meena, P. D., Meena, R. L., and Chattopadhyay, C. (2008). Eco-friendly options for management of Alternaria blight in Indian mustard (*Brassica juncea*). *Indian Phytopathology*, **62:** 65-69.

Meena, P. D., Meena, R. L., Chattopadhyay, C., and Kumar, A. (2004). Identification of critical stage for disease development and biocontrol of Alternaria blight of Indian mustard (*Brassica juncea*). *Journal of Phytopathology*, **152:** 204-209.

Meena, R. L., Meena, P. D. and Chattopadhyay, C. (2003a). Potential for biocontrol of white rust of Indian mustard (*Brassica juncea*). *Journal of Mycology and Plant Pathology*, **33:** 493.

Meena, R. L., Meena, P. D. and Chattopadhyay, C. (2005). Management and Factors Affecting White Rust Development in Indian mustard (*Brassica juncea* L.). *Resistant Pest Management Newsletter*, **13**: 33-36.

Mehta, Naresh and Hieu, N. T. (2014). Evaluation of Bacterial strains and species against *Sclerotinia sclerotiorum* responsible for white stem rot of mustard. *PAU Journal of Research*, **51**: 93-95.

Mehta, Naresh, Hieu, N. T. and Sangwan, M. S. (2012). Efficacy of various antagonistic isolates and species of *Trichoderma* against *Sclerotinia sclerotiorum* causing white stem rot of mustard. *Journal of Mycology and Plant Pathology* **42**: 244-250.

Mercer, P. C., Ruddok, A. and Mc Gimpsey, H. C. (1992). Evaluation of iprodione and Trichoderma viride against Alternaria linicola. *Tests of Agro chemicals and cultivars* **13**: 20-21.

Mercer, P. C., Ruddok, A., Nee, E. and Papadopolous, S. (1993). Biological control of Alternaria diseases of linseed and oilseed rape. *Bulletin OILB SROP* **16**: 89-99.

Muto, M., Mulabagal, V., Huang, H. C., Takahashi, H., Tsay, H. S., and Huang, J. W. (2006). Toxicity of black nightshade (*Solanum nigrum*) extracts on *Alternaria brassicicola*, causal agent of black leaf spot of Chinese cabbage (*Brassica pekinensis*). *Journal of Phytopathology*, **154**: 45-50.

Nehrash, A. K. (1961). The antimicrobial properties of cultivated radish. Report I. The anti-microbial activity of extracts and essential oil from cultivated and wild radish. *Journal of Microbiology, Kiev*, **23**: 32-37.

Nelson, B. D., Christianson, T. and Mc Clean, P. (2001). Effects of bacteria on sclerotia of *Sclerotinia sclerotiorum*. Proc XI International Sclerotinia Workshop, (Young CS and Hughes K. J. D., eds), Central Science Laboratory, York, UK, July 8–12, p. 39.

Nicolas, G., and Aggery, B. (1940) On some *Peronospora* species parasitized by bacteria. *Review of Mycology*, NS **5**: 14-19.

Omranpour, M., Abbasi, S. and Bahraminejad, S. (2011). Evaluation of the inhibitory effect of some plant crude extracts against *Albugo candida*, the causal agent of white rust. *World Academy of Science, Engineering and Technology*, **58**: 362-364.

Ozbay, N. and Newman Steven, E. (2004). Biological control with *Trichoderma* spp. with emphasis on *T. harzianum*. *Pakistan Journal of Biological Sciences*, **7**: 478-484.

Pace, M. A., and Campbell, R. (1974). The effect of saprophytes on infection of leaves of *Brassica* spp. by *Alternaria brassicicola*. *Transactions of British Mycological Society*, **63**: 193-196.

Pathak, A. K., Godika, S., Jain, J. P. and Muralia, Suresh. (2001). Effect of antagonistic fungi and seed dressing fungicides on the incidence of stem rot of mustard. *Journal of Mycology and Plant Pathology* **31**: 327-329.

Patni, C. S., and Kolte, S. J. (2006). Effect of some botanicals in management of Alternaria blight of rapeseed-mustard. *Annual Plant Protection Science*, **14**: 151-156.

Patni, C. S., Kolte, S. J., and Awasthi, R. P. (2005). Efficacy of botanicals against Alternaria blight (*Alternaria brassicae*) of mustard. *Indian Phytopathology*, **58**: 426-430.

Prasad, R. D. and Kulshreshta, D. D. (1999). Bacterial antagonists of *Alternaria helianthi* of sunflower. *Journal of Mycology and Plant pathology*, **29**: 127-128.

Prasad, R., Saxena Deepa and Chandra S. (2003). Yield losses by Alternaria blight in promising genotypes of Indian mustard. *Indian Phytopathology*, **56**: 151-152.

Rai, B., and Singh, D. B. (1980). Antagonistic activity of some leaf surface microfungi against *Alternaria brassicae* and *Drechslera graminea*. *Transactions of British Mycological Society*, **75**: 363-369.

Reshu, and Khan, M. M. (2012). Role of different microbial-origin bioactive antifungal compounds against *Alternaria* spp. causing leaf blight of mustard. *Plant Pathology Journal*, 11, 1-9.

Rodriguez, M. A. and Godeas, A. M. (2001). Comparative study of fungal antagonist *Sclerotinia sclerotiorum*. Proc XI International Sclerotinia Workshop, (Young CS and Hughes KJD, eds), Central Science Laboratory, York, UK, July 8–12, pp. 125-126.

Sandys-Winsch, D. C., Whipps, J. M., Gerlagh, M. and Kruse, M. (1993). World distribution of the sclerotial mycoparasite *Coniothyrium minitans*. *Mycological Research*, **97**: 1175-1178.

Savchuk, S. and Fernando, W. G. D. (2004). Effect of timing of application and population dynamics on the degree of biological control of *Sclerotinia sclerotiorum* by bacterial antagonists. *FEMS Microbiology and Ecology*, **49**: 379-388.

Saxena, D., and Tewari, A. K. (2017) *Trichoderma* formulations for the management of mustard diseases under organic forming. *Proceeding of 3rd National Brassica Conference*-2017, ICAR, IARI, New Delhi, pp.249-250.

Sen, B. (2000). Biological control: A success story. *Indian Phytopathology*, **53**: 243-249.

Sharma, A. K., Gupta, J. S., and Maheshwari, R. K. (1984). The relationship of *Streptomyces arabicus* to *Alternaria brassicae* (Berk.) Sacc. and *Alternaria brassicicola* (Schew.) Wiltshire on the leaf surface of yellow sarson and taramira. *Geobios*, **3**: 83-84.

Sharma, A. K., Gupta, J. S., and Singh, S. P. (1985). Effect of temperature on the antifungal activity of *Streptomyces arabicus* against *Alternaria brassicae* (Berk.) Sacc. and *Alternaria brassicicola* (Schew.) Wiltshire. *Geobios*, **12**: 168-169.

Sharma, N., Rahman, M. H., Liang, Y., and Kav, N. N. V. (2010b). Cytokinin inhibits the growth of *Leptosphaeria maculans* and *Alternaria brassicae*. *Canadian Journal of Plant Pathology*, **32**: 306-314.

Sharma, R., and Sinha, S. K. (1989). A pigmented, xylose-utilizing strain of *Streptomyces bobili*. *Current Science*, **58**: 1405-1406.

Sharma, S. K., and Gupta, J. S. (1978a). Biological control of leaf blight disease of Brown Sarson caused by *Alternaria brassicae* and *A. brassicicola*. *Indian Phytopathology*, **31**: 448-449.

Sharma, S. K., and Gupta, J. S. (1978b). Effect of Brown Sarson leaf leachates on the germination of the conidia of *Alternaria brassicae* and *Alternaria brassicicola*. *Proceedings of the Indian National Science Academy B*, **44**: 57-58.

Sharma, S. K., and Gupta, J. S. (1979). Role of surface microorganisms of brown sarson in relation to *Alternaria brassicae* and *Alternaria brassicicola*. *Agra University Journal of Research Science*, **28**: 109-111.

Sharma, S. K., and Gupta, J. S. (1980). *Streptomyces rochei* in relation to *Alternaria brassicae* and *A. brassicicola* on the surface of brown sarson. *Journal of Indian Botanical Society*, **59**: 161-163.

Sharma, S., Singh, J., Munshi, G. D., and Munshi, S. K. (2010a). Biochemical changes associated with application of biocontrol agents on Indian mustard leaves from plants infected with *Alternaria Blight*. *Archives of Phytopathology and Plant Protection*, **43**: 315-323.

Silva, F. A. G., Piexoto, C. N., Assis, S. M. P., Mariano, R. L. R. and Padovan, I. P. (1998) Potential of fluorescent Pseudomonas spp. for biological control of Alternaria ricini on castor bean. *Brazilian Archives of Biology and Technology*, **41**: 91-102.

Singh, R. S. and Kaur, J. (2001). Comparative antagonistic activity of *Trichoderma harzianum*, *T. viride* and *Epicoccum purpurescens* against *Sclerotinia sclerotiorum* causing white rot of brinjal. Proc XI International Sclerotinia Workshop, (Young CS and Hughes KJD, eds), Central Science Laboratory, York, UK, July 8–12, p. 141.

Singh, S., Godara, S. L., and Gangopadhyay, S. (2013). Studies on antifungal properties of plant extracts on mustard blight caused by *Alternaria brassicae*. *Indian Phytopathology*, **66**: 172-176.

Srinivasan, A., Kang, I. S., Singh, R. S. and Kaur, J. (2001). Evaluation of selected *Trichoderma* isolates against *Sclerotinia sclerotiorum* causing white rot of *Brassica napus L*. Proc XI International Sclerotinia Workshop, (Young CS and Hughes KJD, eds), Central Science Laboratory, York, UK, July 8–12, pp. 143-144.

Tahvonen, R. (1982a). Preliminary experiments into the use of Streptomyces spp. isolated from peat in the biological control of soil and seed-borne disease in peat culture. *Journal of Science Agricultural Society, Finland*, **54**: 357-369.

Tahvonen, R. (1982b). The suppressiveness of Finnish light coloured sphagnum peat. *Journal of Science Agricultural Society, Finland*, **54**: 345-356.

Tahvonen, R. (1985). Mycostop-a biological formulation for control of fungal diseases. *Vaxtskyddsnotiser*, **49**: 86-90.

Tahvonen, R. T. (1988). Microbial control of plant disease with *Streptomyces* spp. *Bulletin OEPP*, **18**: 55-59.

Tahvonen, R., and Avikainen, H. (1987). The biological control of seed-borne *Alternaria brassicicola* of Cruciferous plants with a powdery preparation of *Streptomyces* sp. *Journal of Agricultural Science, Finland, Helsinki*, **59**: 199-208.

Tapio, E., and Pohto-Lahdenpera, A. (1989). Interaction between antagonists and pathogenic fungi. *Vaxtskyddsnotiser*, **53**: 12-18.

Tewari, J. P., Tewari, I. and Chatterjee, S. C. (2000). Inhibition of oospore development in *A. candida* by *Pseudomonas syringae*. *Bulletin Oilb/Srop*, 23 (6): 51-53.

Tsuneda, A. (1977). Mycoparasitism of *Alternaria brassicae* by *Nectria inventa*. Ph.D. Thesis, Univ. of Alberta, Edmonton, Canada.

Tsuneda, A., and Skoropad, W. P. (1977a). Formation of micro sclerotia and chlamydospores from conidia of *Alternaria brassicae*. *Canadian Journal of Botany*, **55**: 1276-1281.

Tsuneda, A., and Skoropad, W. P. (1977b). The *Alternaria brassicae-Nectria inventa* host- parasite interfaces. *Canadian Journal of Botany*, 55: 448-454.

Tsuneda, A., and Skoropad, W. P. (1978a). Behavior of *Alternaria brassicae* and its mycoparasite *Nectria inventa* on intact and exicised leaves of rapeseed. *Canadian Journal of Botany*, **56**: 1333-1340.

Tsuneda, A., and Skoropad, W. P. (1978b). Nutrient leakage from dried and rewetted conidia of *Alternaria brassicae* and its effect on the mycoparasite *Nectria inventa*. *Canadian Journal of Botany*, **56**: 1341-1345.

Tsuneda, A., and Skoropad, W. P. (1978c). Phylloplane fungal flora of rapeseed. *Transaction of British Mycological Society*, 70: 329-334.

Tsuneda, A., and Skoropad, W. P. (1980). Interactions between *Nectria inventa*, a destructive mycoparasite, and fourteen fungi associated with rapeseed. *Transaction of British Mycological Society*, 74: 501-507.

Tsuneda, A., Skoropad, W. P., and Tewari, J. P. (1976). Mode of parasitism of *Alternaria brassicae* by *Nectria inventa*. *Phytopathology*, **66**: 1056-1064.

Umalkar, G. V., Mukadam, D. S., and Nehemiah, K. M. A. (1977). Fungistatic properties of some deproteinized leaf extracts. *Science and Culture*, **43**: 437-439.

Vannacci, G., and Harman, G. E. (1987). Biocontrol of seed-borne *Alternaria raphani* and *A. brassicicola*. *Canadian Journal of Microbiology*, **33**: 850-856.

Verma P. R., and Petrie G. A. (1980) Effect of seed infestation and flower bud inoculation on systemic infection of turnip rape by *Albugo candida*. *Canadian Journal of Plant Science*, **60**: 267–271

Walters, D. R. and Fountaine, J. M. (2009). Practical application of induced resistance to plant diseases: an appraisal of effectiveness under field conditions. *Journal of Agricultural Science*, **147**: 523–535.

Whipps, J. M. (1994). Advances in biological control in protected crops. Brighton Crop Protection Conference, *Pests and Diseases* 2: 1259-1264.

Whipps John, M. (2001). Microbial interactions and biocontrol in the rhizosphere. *Journal of Experimental Botany*, **52**: 487–511.

White, J. G., Linfield, C. A., Lahdenpera, M. L., and Uoti, J. (1990). Mycostop a novel biofungicide based on *Streptomyces griseoviridis*. Proceedings of the British Crop Protection Conference, Pests and Diseases, Brighton, 1: 221-226.

Wu, H. (1988). Effects of bacteria on germination and degradation of sclerotia of *Sclerotinia sclerotiorum* (Lib) de Bary. M.Sc., thesis, North Dakota State University, Fargo, ND. 312p.

Wu, W. S., and Lu, J. H. (1984). Seed treatment with antagonists and chemicals to control *Alternaria brassicicola*. *Seed Science and Technology*, **12**: 851-862.

Yasmeen, and Saxena, S. K. (1990). Effect of fern extracts on growth and germination of fungi. *Current Science*, **15**: 798-799.

Zhang, X. L., Sun, X. M. and Zhang, G. F. (2003). Preliminary report on the monitoring of the resistance of *Sclerotinia libertinia* to carbendazim and its internal management. *Chinese Journal of Pesticide Science and Administration*, **24**: 18-22.

Zhang, Y. and Fernando, W. G. D. (2003). Biological control of *Sclerotinia sclerotiorum* infection in canola by *Bacillus* sp. *Phytopathology*, **93**: 94.

Zhang, Y. and Fernando, W. G. D. (2004). Presence of biosynthetic genes for phenazine-1-carboxylic acid and 2, 4-diacetylpholoroglucinol and pyrrolnitrin in *Pseudomonas cholroraphis* strain PA-23. *Canadian Journal of Plant Pathology*, **26**: 430-431.

Zhang, Y., Fernando, W. G. D., Kavitha, K., Nakkeeran, S. and Ramarathnam, R. (2004). Combination of mechanisms in *Pseudomonas chlororaphis* strain PA-23 results in control of multiple pathogens. *Phytopathology*, **94**: 115.

12 Major Diseases of Niger and their Management

H.K. Singh, Mahesh Singh, R.B. Singh,
J.K. Yadav, K.N. Maurya and M.K. Maurya

Niger (*Guizotia abyssinica* Cass) belongs to the family Compositae, is one of the important minor oilseed crops of India. It is also known by various names such as *Ramtil, Kalatil, Khurasani, Jagni* etc. In various regional Indian languages and Noog in Ethiopia. It is mainly cultivated in tribal pockets of Gujarat, M.P., Orissa, Maharashtra, Bihar, Karnataka and Andhara Pradesh in India. Niger is a crop of dry areas grown mostly by tribal. India is the prime producer of niger in the world, which was grown over an area of 3.8 lakh ha with a production of 1.2 lakh tonnes and productivity of 310 kg/ha. In Odisha, it is grown in an area of 0.86 lakh ha with a production of 0.36 lakh tonnes and productivity of 420 kg/ha (Anonymous, 2012). The Niger seed contains 33.3 per cent protein, 34.2-39 per cent total carbohydrates and 13.5 per cent fiber. This crop is found infested by number of diseases and pests, which causes great damage to the crop. The crop is affected by number of fungal diseases. The important diseases of Niger are Alternaria blight (*Alternaria porii* and *A. alternata*), leaf spot (*Cercospora guizoticola*), seedling blight (*Alternaria tenuis*), seed rot (*Rhizotonia bataticola*), rust (*Puccinia guizotiae*), powdery mildew (*Sphaetheca* sp.), root rot (*Macrophomina phaseolina*) and cuscuta as *Phanerogamic* parasite (Rajpurohit, 2004, Bradley and Del Rio, 2007 and Rajpurohit and Dubal, 2009).

Alternaria Blight

It is caused by *Alternaria porri* (Ell.). *Alternaria alternata* has also been reported from India. *Alternaria* blight of niger is more serious in Ethiopia as compared to India.

Symptoms

The disease symptoms appear as concentric rings on the leaves, which turns brown with grey centre later on. As the disease advances, the spots become oval or circular and finally become irregular in shape. Latter the infected leaves dry and fall off. In case of floral buds if infected, they become black and fail to open. Further it spreads to other plant parts and results into premature drying of the plant (Yirgu, 1964[a]).

Causal Pathogen

Alternaria porii and *A. alternata.*

The pathogen is seed borne and survives in plant debris. The fungus is disseminated within and among fields by splashing water and wind, and overwinters in and on infested crop debris (Nagaraja and Krishnappa, 2016).

Management

Increased ploughing between seasons reduced the disease by reducing the amount of soilborne inoculums. Increased spacing between plants also reduced disease development. Use disease free clean seed for sowing. Destruction of crop residues by burning. Two sprays with Zineb or Dithane M-45 at the rate of 0.3 per cent at disease initiation can effectively manage the disease (Saharan, 2005). Spraying of Mancozeb @ 0.2 per cent at 15 days interval was also found effective (Hegde, 2005). Use of resistant varieties like JNC-6, IGP-76, Deomali, GA-11, ONS-8, PCU-197 (Hegde, 2005, Rajpurohit *et al.*, 2005, Sandipan *et al.*, 2014).

Sclerotium Wilt or Collar Rot

The disease was first reported from Dharwad, Karnataka (Siddarmaiah *et al.*, 1979).

Symptoms

The initial symptoms are partial or complete wilting of the stem or branches that are in contact with the infected soil. White mycelial growths with brown colour sclerotia are visible on the infected plant parts. The leaves turn brown and show wilting but remain attached to the plant. The tissues of collar region become soft and depressed. In severe condition root and stem rot appear (Kolte, 1985). The diseased plants turn yellow and dry. As the plant number is reduced, the disease causes yield loss.

Causal Pathogen

Sclerotium rolfsii Sacc.

The perfect stage of the fungus is *Corticium rolfsii*. The disease is soil borne. The pathogen has a wide host range. *S. rolfsii* can colonize either living plant tissues or plant debris. Deeply buried sclerotia survive a year or less while those near soil surface remain viable for many years. White fungus grows on the diseased part and forms mustard seed like sclerotia. Defoliated leaves can also serve as a bridge to facilitate plant to plant spread. The pathogen spreads through infected soil, wind, splashed rain and sclerotia.

Management

☆ Cultural practices such as deep burial of organic matter, plant debris before sowing is particularly useful in reducing the sclerotium wilt.

☆ Deep ploughing and cultivation of groundnut in flat or slightly raised beds is helpful.

☆ Crop rotation with wheat, corn and soya bean may minimize the incidence of stem rot.

☆ Soil application of castor cake or neem cake @ 10q/ha.

☆ Seed treatment with Carbendizim/Thiram/Capton @ 2-3 g/kg seed is effective.

☆ Seed treatment with 4g *Trichoderma viride* formulation followed by application of 2.5 kg *Trichoderma viride* formulation mixed with 100kg farm yard manure before sowing is recommended.

Ozonium Wilt

The disease was reported around Varanasi (India) where it appeared in great severity (Saharan, 1989; Rao and Pavgi, 1975). Heavy losses are incurred as the diseased plants dry.

Symptoms

Necrotic lesions develop on stem of well grown adult plants near at the soil level, which enlarge and progress upwards. A whitish fungal mycelium grows on these necrotic areas under high humidity. The apical portion, including branches, leaves; inflorescence head *etc.* become soft and pulpy and start rotting. The diseased stem breaks. The plant may collapse, breaking the stem.

Causal Pathogen

Ozonium texanum Neel and Wester var. *parasiticum* Thirum.

The mycelium grows on dead plants if enough moisture is present. Root rot is also commonly observed. The fungus forms rhizomorphs and sclerotia. The sclerotia are oval, grey to dark brown. The disease perpetuates through soil borne sclerotia and mycelium.

Management

Seed treatment with Thiram or Emisan-6 at the rate of 2 g/kg seed before sowing is quite effective to manage the disease (Saharan *et al.*, 2005).

Damping Off or Root Rot

The fungus attacks stem of the seedling at ground level, makes water soaked soft and incapable of supporting the seedling which ultimately falls and dies.

Symptoms

The pathogen attacks the seedlings as well as all stage of plant growth. On older seedling elongated brownish black lesions appear which increase in length and width girdling the stem and causing root rot which are black and brittle. Infection occurs in cold, wet weather and moderate moisture level. If the affected plants are pulled out and examined, the entire root system may be found rotting. The lateral root and thinner root get completely rotted. Fungal sclerotia appear as minute black dots on the surface of woody tissues and on the rotting bark.

Causal Pathogen

Rhizoctonia solani

The fungus can survive in the soil for many years in the form of sclerotia. Sclerotia of Rhizoctonia have thick outer layers to allow for survival, and they function as the overwintering structure for the pathogen. In some rare cases (such as the teleomorph) the pathogen may also take the form of mycelia that reside in the soil, as well. The process of penetration of a host can be accomplished in a number of ways. Entry can occur through direct penetration of the plant cuticle/epidermis or by means of natural openings in the plant. Hyphae come in contact with the plant and attach to the plant by which through growth they begin to produce an appressorium which penetrates the plant cell and allows for the pathogen to obtain nutrients from the plant cell. Sclerotia/mycelium overwinters in plant debris, soil, or host plants.

Management

Good Agricultural Practices such as balanced fertilization, timely irrigation and pest management encourage good crop growth which may help in reducing the disease. The seed should be treated with Thiram or Captan @ 3.0 g/kg seed (Sharma, 1982). The disease can also be reduced by drenching the plants with Captan 50 WP @ 0.25 per cent. Crop rotation should be followed.

Macrophomina Root Rot and Blight

Symptoms

Typical root rot, stem rot, charcoal rot, leaf blight symptoms are produced. The symptoms appear as water soaked necrotic lesion that girdles the stem just above the ground level and wilting follows. The recurrence of the disease takes place through seed and soil borne inoculum, later spread through workers, tools and insects like *Melanoagromyza* stem borer. The pathogen causes severe seedling mortality resulting in patchy crop stand and thus reduces the yield.

Causal Pathogen

Macrophomina phaseolina (Maubl) Ashby

The imperfect stage of the fungus is *Rhizoctonia bataticola*. It is seed and soil borne. Symptoms appear on the root as black dots. The pathogen has wide host range. The pathogen is a facultative saprophyte and a soil dweller. Infected soil, plant debris and pods serve as sources of inoculums. The sclerotia are disseminated via plant debris, soil, infected pods *etc.*

Management

- ☆ Seed treatment with spores and mycelial fragments of *Trichoderma viride* has been shown to prevent invasion by *M. phaseolina*.
- ☆ Seed treatment with Carbendazim 2g/kg seed or Captan 3g/kg seed or Thiram @ 4g/kg seed is most useful.
- ☆ Good Agricultural Practices (GAP) such as balanced fertilization, timely irrigation and pest management encourage good crop growth which may help in reducing the disease.
- ☆ Spot drench with Carbendazim at 0.5 g/lit effective in managing the disease.
- ☆ PCNB (Brassicol 75 per cent WP) @0.5 per cent can also be applied or in the form of soil dust 25 kg/ha in two split applications, 12.5 kg/ha before sowing and the rest 12.5 kg/ha at 15 days after first application.

Cercospora Leaf Spot

This disease is prevalent in all the niger growing areas and caused yield reduction. The disease is severe under warm and humid weather. The fungus is active throughout the year where niger is grown in *rabi* and *kharif* seasons.

Symptoms

Symptoms appear as small straw coloured to brownish spots on both the leaf surfaces. Later the spots increase in number and size and cover the entire lamina, spots may coalesce to cause defoliation. Elongated dark brown spots are produced on the stem. The capsules are also affected.

Causal Pathogen

Cercospora guizoticola Govindu and Thirmulachar

The pathogen is seed and soil borne. The primary infection occurs through seed borne inoculum as well as the inoculum present on diseased plant left over in the soil. The disease dissiminates is through conidia formed on infected plant parts. Fungus survives through plant debris left after the harvest and also on the seed. The pathogen remains active on other collateral hosts also.

Management

Cultural practices which promote vigorous growth of the plants help in minimising infection. Crop residues should be removed as they may harbour infection. Use of resistant varieties like PCU-197, 5-4, CHERU No-1, IGP-234, AISI-2, NC-63586, PCU-179, JN-6, JN-9, JN-10, JN-13, JN-14, JN-21, JN-93, JN-94, JN-99, JN-113, JN-116, JN-130 and BMD-69 exhibited the grade of 1.0 against the Cercospora infection (Sandipan *et al.*, 2014[a]). Seed treatment with thiram @ 0.3 per cent is effective to manage the seed borne infection. Application of Carbendazim @ 0.03 per cent or Benomyl 0.05 per cent, spray at 10-12 days interval after appearance of the disease were found effective in managing the disease.

Powdery Mildew

It is major problem in niger. The disease occurs in *rabi* as well as *kharif* season but it is more severe in *rabi*. The yield losses are due to early defoliation as a result of the disease.

Symptoms

Infection by the fungus is favored by high humidity but not by free water. All the aerial parts develop symptoms. Small cottony spot develops on the infected leaves which gradually spread on the whole lamina (Vyas *et al.*, 1981). Some times on stem a purple rings also develops. The diseased leaves turn yellow followed by defoliation in severe cases. The seed formed on diseased plants are small and shrivelled.

Causal Pathogen

Sphaerotheca spp.

Powdery mildew fungi are obligate, biotrophic parasites of the phylum Ascomycota of Kingdom Fungi. The powdery mildew is caused by *Sphaerotheca* sp. in Ethiopia (Chavan, 1961), whereas, it is known to be caused by *Oidium* sp. (*Erysiphe cichoracearum* DC) in India as reported by Vasudeva (1954). The pathogen survives through collateral hosts. During the growing season, hyphae are produced on both upper and lower leaf surfaces. Infections can also occur on stems, flowers, or fruit. Specialized absorption cells, termed haustoria, extend into the plant epidermal cells to obtain nutrition.

Management

The disease can be managed by burning the infected plant parts after the harvesting of the crop. The disease can also be effectively managed by spraying with sulfex at the rate of 0.3 per cent, Carbendazim + Mancozeb (0.2 per cent), Hexaconazole (0.1 per cent) as the disease starts appearing. Another spray can be done after 10-15 days intervals depending upon the disease intensity (Sandipan *et al.*, 2014[a]).

Rust

Rust was first reported from Ethiopia (Yirgu, 1964). The disease is more severe in Ethiopia as compared to India.

Symptoms

Infections begin when a spore lands on the plant surface, germinates, and invades its host. Infection is limited to plant parts such as leaves, petioles, tender shoots, stem, fruits, *etc.* Brown pustules up to 7 mm in diameter appear on the leaves. The lesions consist of densely aggregated brown telia that measure 0.16 to 0.37 mm in diameter. The uredosori are formed on the lower leaf surface and the corresponding upper surface becomes chlorotic. Later teliospores are formed (Kolte, 1985).

Causal Pathogen

Puccinia guizotiae Cumm.

Rust fungi are obligate plant pathogens that only infect living plants. The teliospore measures 37-49 x16-20 um in size (Kolte, 1985).

Management

Generally many rusts are host specific. Therefore, crop rotation can break the disease cycle. Use of rust resistant/tolerant variety. The disease can be managed by spraying crop with Mancozeb @ 0.25 per cent as the disease starts appearing.

Bacterial Leaf Spot

This disease was first reported from Ethiopia (Yirgu, 1964[a])

Symptoms

Small brownish spots are formed on leaves which are surrounded by yellow halos. Most of the spots are observed on the leaf margin. The lesions increase in size, coalesce and give appearance of blight. All the leaves get destroyed. Under severe cases, scabby brown needle pricks lesion formed on the stem (Yirgu, 1964). The bacterium *Xanthomonas campestris* pv. *guizotiae* var. *indicus* also produces similar type of the symptoms on the plants but the bacterium requires low temperature for growth and infection (Moniz *et al.*, 1968).

Causal Pathogen

Xanthomonas campestris pv. *guizotiae* (Yirgou) Dye

Bacterial leaf spot is caused by two pathogens *viz.*, *Xanthomonas campestris* pv. *guizotiae* which is severe in Ethiopia whereas in India it is incited by *X. campestris* pv. *guizotiae* var. *indicus* in India (Moniz *et al.*, 1968).

Management

The disease can be managed by spraying the crop with Blitox-50 at the rate of 0.2 per cent (Saharan *et al.*, 2005).

Cuscuta

Cuscuta, a phanerogamic plant parasite has been observed to attack niger badly.

Symptoms

The plants remain stunted, become pale yellow and bear a very small number of flowers and fruits due to the association of *Cuscuta*.

Causal Pathogen

Cuscuta hyalina

It is a complete stem parasite depend for shelter and food totally on the host. The food is obtained through haustoria's from the host. Usually the Cuscuta seed gets mixed with Niger seed and such mixtures are planted.

Management

The Cuscuta seeds can be removed by sieving before sowing. Removal of Cuscuta infected niger plant at the early crop growth. Pre sowing application of fluchloralin @ 1 kg a.i./ha or pre emergence soil application of pendimethelin @ 1-1.25 kg a.i./ha (Hegde, 2005). Sieving by separation of Cuscuta seeds by 1 mm sieve + 10 per cent brine solution seed treatment (Anonymous, 2009).

References

Anonymous (2009). Annual Progress Report Sesame and Niger 2008-2009. All India Coordinated Research Project on Sesame and Niger, J.N. Agricultural University Campus, Jabalpur (M.P.). pp. 249-251.

Anonymous (2012). Frontline Demonstrations on Oilseeds. Annual 2011-12. Directorate of Oilseed Research, Hyderabad. pp. 148.

Bradley, C.A. and Del Rio, L.E. (2007). First report of *Sclerotia sclerotiorum* on niger (*Guizotia abyssinica Cass*). *Plant Disease*, 91: 1077-1082.

Chavan, V. M. (1961). Niger and Safflower. Indian Central Oilseeds Committee, Hyderabad.

Hegde, D.M. (2005). IPM in oilseed crops, Directorate of Oilseeds Research (ICAR) Rajendranagar, Hyderabad. pp. 24.

Kolte, S. J. (1985). Niger seed diseases In: Diseases of Annual Edible Oilseed Crops. Vol. III. CRC Press, Inc. 139p.

Moniz, L., Syed, G. M. and Rai, H. H. (1968). *Xanthomonas quizotiae* var *indicus* causing leaf spot disease of *Guizotia abyssinica* Cass. *Indian J. Microbiol.* 8: 263.

Nagaraja, O and Krishnappa M. (2016). Detection seed-borne nature Alternaria porri and its location in niger *Guizotia abyssinica* L (F). Cass. *International J. Scientific Research and Engineering Studies*, 3 (7): 3-7.

Rajpurohit T.S. (2004). *Ramtil ke rog avam unki roktham*. Narmada Krishi Parivar, 2004. 16 (1): 3.

Rajpurohit, T.S. and Shraddha, D. (2009). *Ramtil ki fasal ko rogon se bacheyen. Modern Kheti,* 7 (13): 17-19.

Rajpurohit, T.S., Sushma N. and Khare, M.N. (2005). Current status of diseases of sesame and niger and their management Paper presented in National seminar on Strategies for enhancing production and export of sesame and niger April 7-8,2005 Abstract page 44-45.

Rao, R. and Pavgi, M.S. (1975). Two sclerotial diseases of crops in Varanasi. *Indian Phytopath.* **28**: 401-404.

Saharan, G.S. (1989). The present status of niger and linseed pathology research work in India. Proc. Three Meetings, Oilseeds held at Pantnagar and Hyderabad, India. IDRC/CRDI/CIID, 192-202.

Saharan, G.S. Mehta N., and Sangwan, M.S. (2005). Diseases of Oil-seeds- Indus publishing company New Delhi., pp. 475-479.

Sandipan P. B., Jagtap, P. K. and Patel M. C. (2014ᵃ). Sources of resistance in screening of elite material in niger (*Guizotia abyssinica* cass) genotypes against *Alternaria* and *Cercospora* leaf spot diseases under natural condition. *Asian J. Science and Techol.* 5 (8): 491-496.

Sandipan P.B., Jagtap, P.K. and Patel M.C. (2014). Relevance of various fungicides for the control of powdery mildew leaf spot disease of niger (Guizotia abyssinica Cass) under South Gujarat Region. *International J. Scientific Res. Eng. Studies.*, **1** (3): 8-10.

Sharma, S. M. (1982). Improved technology for sesamum and niger. *Indian Fmg.* **32**: 72-77.

Siddarmaiah, A.L., Kulkarni, S. and Basavarajaiah, A.B. (1979). Occurrence of a new collar rot disease of Niger (*Guizotia abyssinica* Cass.) in India. *Curr. Sci.* **18**: 174.

Vasudeva, R. S. (1954). The fungi of India. Indian Council of Agricultural Research, (ICAR), New Delhi, India.

Vyas, S.C., Prasad, K.V.V. and Khare, M.N. (1981). Diseases of Sesame and Niger and Their Control. Bull Directorate of Research, J.N.K.V.V., Jabalpur pp. 16.

Yirgu, D. (1964). Some diseases of *Guizotia abyssinica* in Ethiopia. *Plant Disease Reporter* **48**: 672.

Yirgu, D. (1964ᵃ). *Xanthomonas guizotia* sp. nov. on *Guizotia abyssinica. Phytopathology* **54**: 1490-1491.

Index

Figure 1.1: Bacterial Wilt Symptoms on Groundnut (Photo courtesy jnkvv. ac.in). p. 3

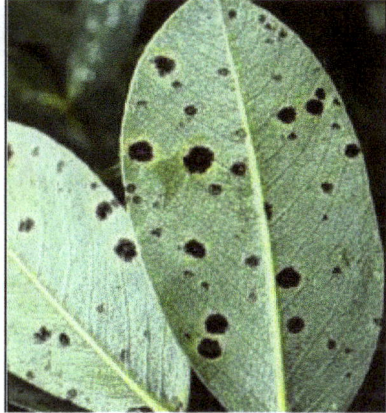

Figure 1.2: Anthracnose Symptoms and Conidia of Fungus (Photo courtesy icar. tripura.center). p. 5

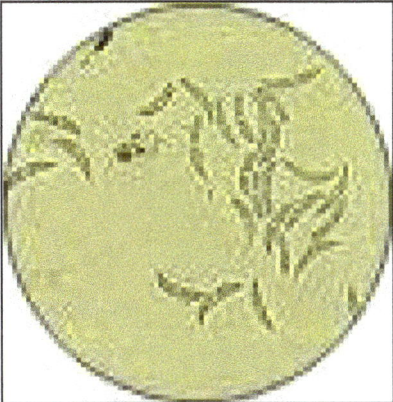

Figure 1.2: Anthracnose Symptoms and Conidia of Fungus (Photo courtesy icar. tripura.center). p. 5

Figure 1.3: Crown Rot of Groundnut Plants (Photo courtesy plantwise. org). p. 6

Oospores

Figure 1.4: Damping Off of Groundnut (Photo courtesy infonet.biovision.org). p. 7

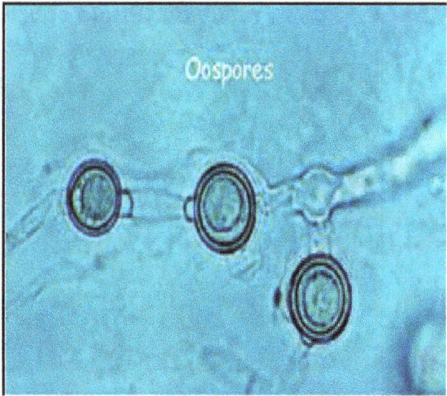

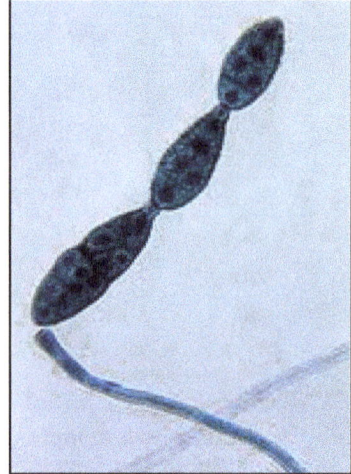

Figure 1.5: Leaf Spot on Groundnut and Conidia of *Alternaria* spp. (Photo courtesy jnkvv.nic.in) p. 9

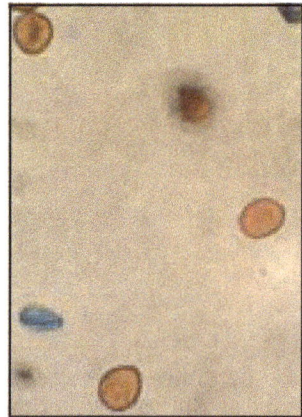

Figure 1.6: Rust Disease of Groundnut and Urediospores of *Puccinia arachidis* (Photo courtesy icar.tripura.center). p. 11

Figure 1.7: Stem Rot of Groundnut (Photo courtesy oca.testhead.blogspot. com). p. 13

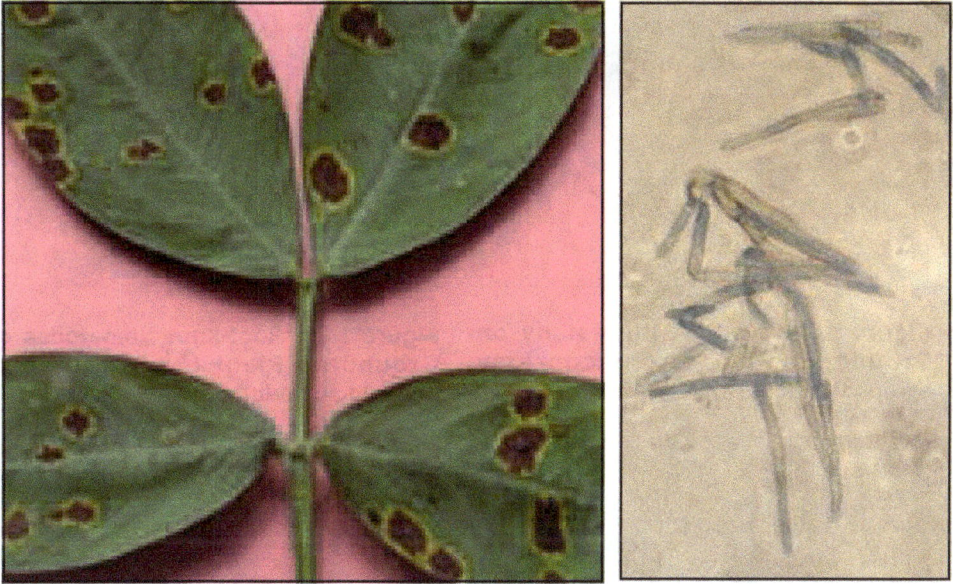

Figure 1.8: Early Leaf Spot Disease of Groundnut and Conidia of *Cercospora arachidicola* (Photo courtesy icar.tripura.center). p. 14

Figure 1.9: Late Leaf Spot Disease of Groundnut and Conidia of *Phaeoisariopsis personata* (Photo courtesy icar.tripura.center). p. 15

Figure 1.10: Kalahasthi Malady on Groundnut Pod (Photo courtesy agropedia.iitk.ac.in). p. 17

Figure 1.11: Root-knot Nematodes on Groundnut Plant (Photo courtesy peanut.nscu.edu). p. 18

Figure 1.11: Root-knot Nematodes on Groundnut Plant (Photo courtesy peanut.nscu.edu). p. 18

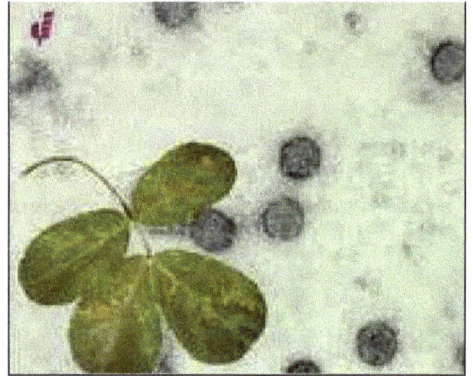

Figure 1.12: Peanut Bud Necrosis Virus on Groundnut Leaf (Photo courtesy printasia.com). p. 18

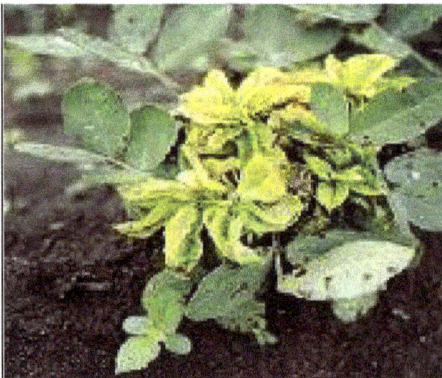

Figure 1.13: Groundnut Rosette Disease Symptoms on Leaf (Photo courtesy icrisat.agropedia.in). p. 20

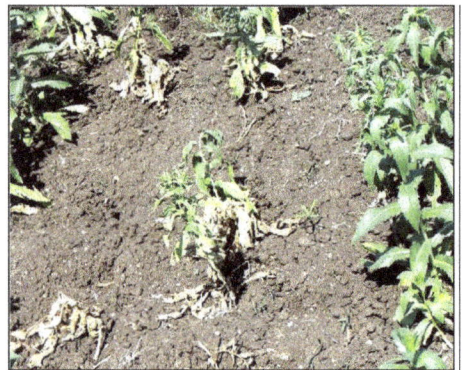

Figure 4.1: Wilt Affected Safflower Plants. p. 51

Figure 4.2: Partial Wilt. p. 51

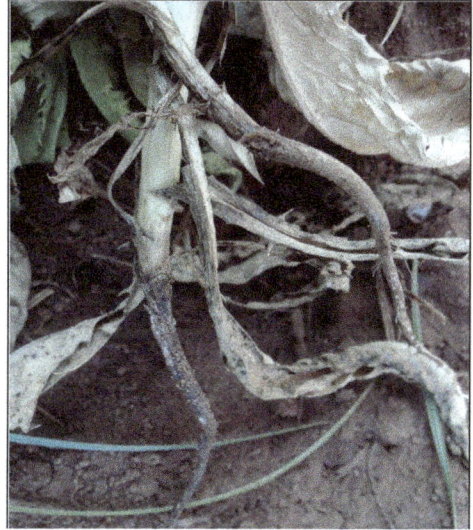

Figure 4.3: Root Rot. p. 54

Figure 4.4: Alternaria Leaf Spot. p. 56

Figure 4.5: Concentric Pattern in Mature Lesion. p. 56

Figure 4.6: Cercospora Leaf Spot. p. 62

Figure 6.1: Collar Rot. p. 96

Figure 6.2: Soybean Rust. p. 102

Figure 6.3: Charcoal Rot. p. 103

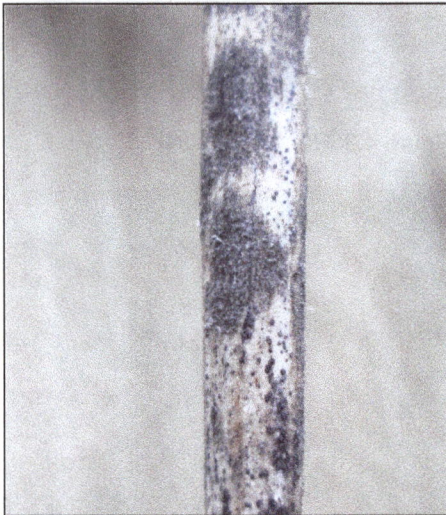

Figure 6.4: Anthracnose Stem Blight. p. 106

Figure 6.5: Bacterial Pustule. p. 110

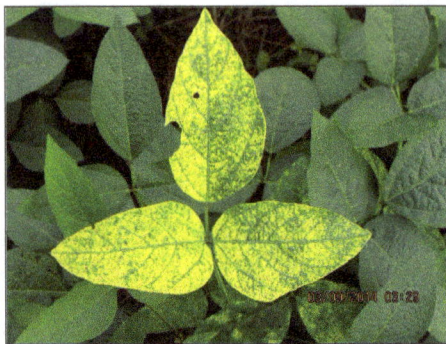

Figure 6.6: Soybean Yellow Mosaic. p. 112

Figure 7.1: Downy Mildew of Sunflower. p. 118

Figure 7.2: Rust. p. 121

Figure 7.3: Outer and Inner Stem Symptoms of Charcoal Rot. p. 124

Figure 7.4: Powdery Mildew. p. 125

Figure 7.5: *Sclerotium* Wilt or Sclerotinia Rot. p. 127

Figure 7.6: Sunflower Mosaic Disease. p. 130

Figure 7.7: Tobacco Streak Virus. p. 131

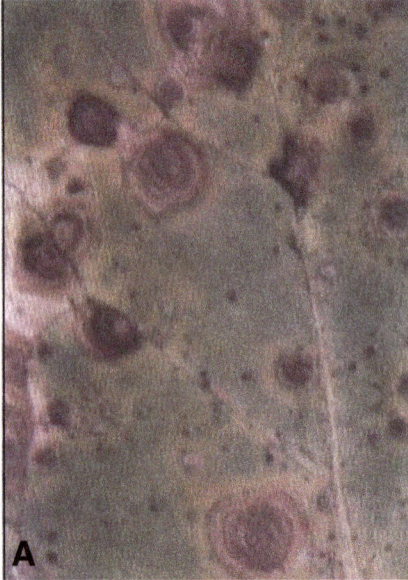 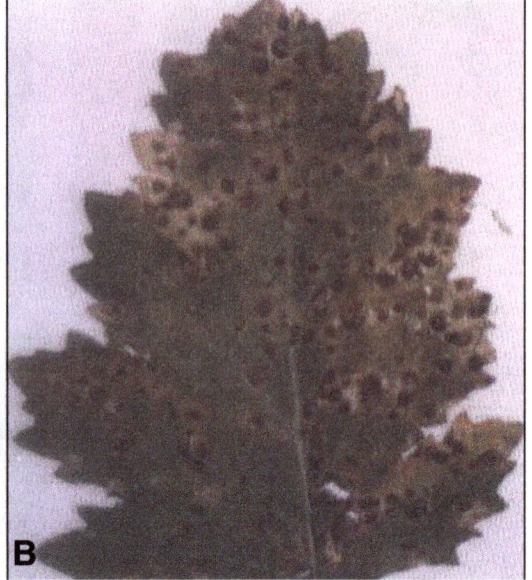

A: Concentric rings on the spots. p. 139

B: Alternaria spots on leaf. p. 139

 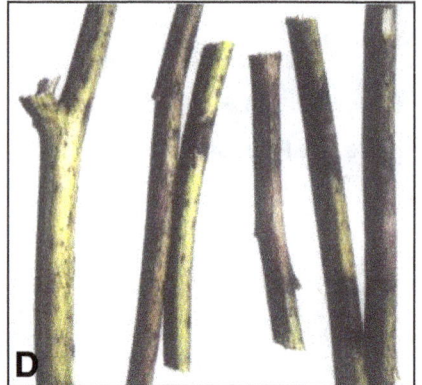

C: Alternaria spots on siliquae. p. 139

D: Alternaria infection on stem. p. 139

   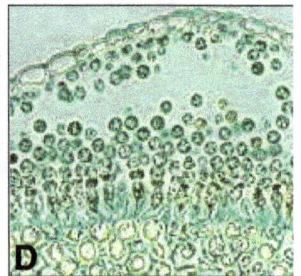

Figures 8.5A-D: A: White Rust Affected Leaf (Lower Surface); B: White Rust Affected Leaf (Upper Surface); C: White Pustules on Swollen Floral Parts (Staghead); D: *Albugo candida* Producing Chains of Unicellular Sexual Sporangia (Courtesy: Mehta, 2014). p. 151

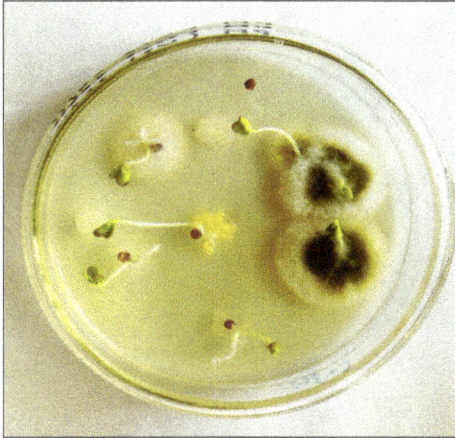

Figure 8.3: *Alternaria* Perpetuates through Mustard Seeds. p. 142

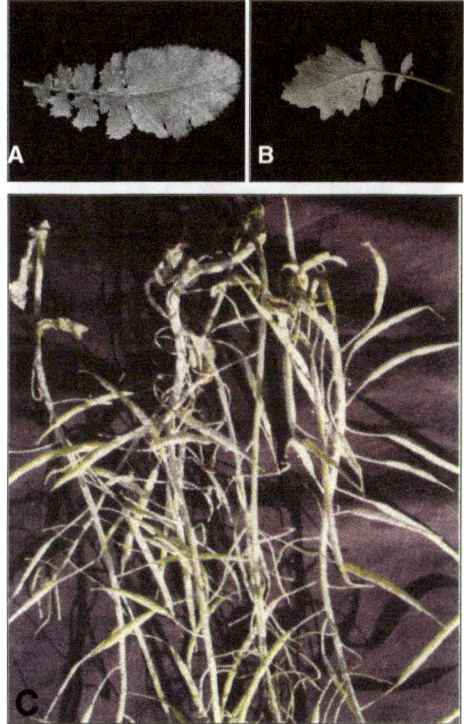

Figures 8.7A-C: A: Powdery Mildew Symptom on Upper Surface, B: Lower Leaf Surface and C: Powdery Infection on Siliquae. p. 163

Figure 8.10A-C: A: Downy Mildew Infection Suppreses the Siliquae Formation; B: Mixed Infection of Downy Mildew and White Rust; C: Heavy Mixed Infection on Siliquae (Courtesy: Shukla *et al.*, 2003). p. 169 (Courtesy: Shukla *et al.*, 2003). p. 169

Figure 8.14: *Sclerotinia sclerotiorum.*
p. 181

Figure 8.12: Symptom of Club Root Affected Root of the Plant. p. 177

Figures 8.13A-B: A: Sclerotinia Stem Rot Affected Plants Sowing Fungal Growth on Girth; B: Sclerotinia Stem Rot Affected Fungal Growth on Stem; (Courtesy: Shukla *et al.,* 2003). p. 180

Figures 8.13C-D: C: Sclerotinia Stem Rot Affected Plant having Sclerotia in the Pith; D: Sclerotia in the Pith in Large View (Courtesy: Shukla *et al.,* 2003). p. 180

Figure 8.16: Typical Symptoms of Bacterial Rot Affected Leaf. p. 188

Figure 8.16: Typical Symptoms of Bacterial Rot Affected Leaf. p. 188

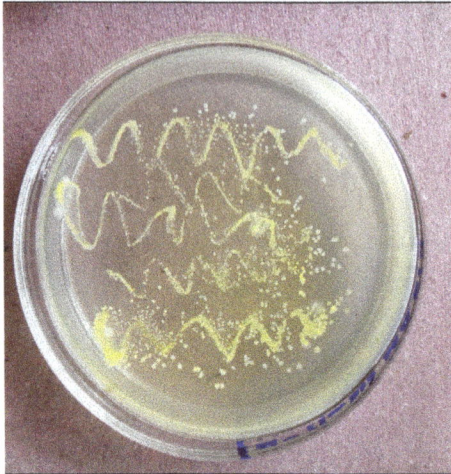

Figure 8.17: Pure Culture of *Xanthomonas compestris* pv. *compestris*. p. 189

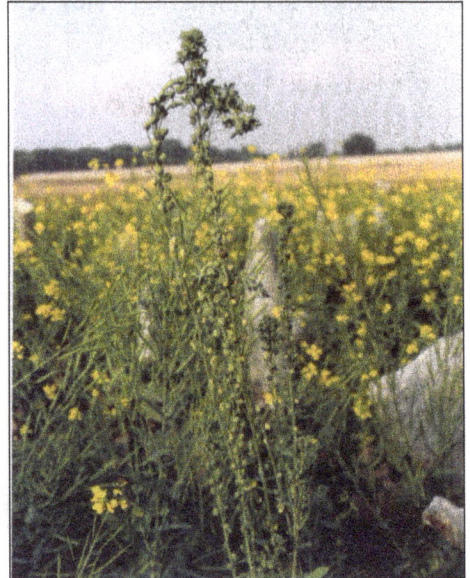

Figure 8.18: Phyllody Afftected Plants in the Field (Courtesy: Shukla *et al.*, 2003). p. 192

www.ingramcontent.com/pod-product-compliance
Lightning Source LLC
Chambersburg PA
CBHW050511190326
41458CB00005B/1496